Ions and Ion Pairs
in Organic Reactions

Contributors

EGBERT DE BOER, *Department of Physical Chemistry, University of Nijmegen, Nijmegen, The Netherlands*

WALTER F. EDGELL, *Department of Chemistry, Purdue University, Lafayette, Indiana*

PAUL KEBARLE, *Chemistry Department, University of Alberta, Edmonton, Canada*

L. DENNIS MCKEEVER, *Physical Research Laboratory, The Dow Chemical Company, Midland, Michigan*

J. HOWARD SHARP, *Department of Chemistry, The University, Leicester, England*

JOHANNES SMID, *Department of Chemistry, State University College of Forestry, Syracuse, New York*

JAN L. SOMMERDIJK, *Department of Physical Chemistry, University of Nijmegen, Nijmegen, The Netherlands*

MARTYN C. R. SYMONS, *Department of Chemistry, The University, Leicester, England*

MICHAEL SZWARC, *Polymer Research Center, State University College of Forestry, Syracuse, New York*

Ions and Ion Pairs in Organic Reactions

VOLUME 1

MICHAEL SZWARC, F.R.S., Editor

Polymer Research Center
State University College of Forestry
Syracuse, New York

WILEY-INTERSCIENCE, a division of John Wiley & Sons, Inc.

New York · London · Sydney · Toronto

Preface

The importance of ionic reactions in organic processes has been recognized for a long time, and we are well acquainted with species such as carbonium ions or carbanions. Although the dominant role of these ions in numerous chemical transformations is fully appreciated, the problem of their structure, as well as that of other ionic species, seemed to be oversimplified, at least until 1950. In the past, and perhaps even today, some chemists have not realized the fundamental difference between free ions and ion pairs, and if the distinction was made, the chemical behavior of both was tacitly assumed to be closely similar. This is erroneous. A large body of experimental data amassed during the last 10 or 15 years clearly demonstrates how drastically different the chemistry of free ions may be from that of ion pairs. For example, the ion pairs of quinoline radical anions rapidly and quantitatively associate forming diamagnetic, covalently bonded dimers, whereas their free ions are stable; the free polystyryl carbanions propagate polymerization 10,000 times faster than some of their ion pairs; the stereochemical course of many reactions becomes controlled when a free ion is converted into an ion pair.

The problem of ions and ion pairs is even more complex than might have been initially anticipated. The behavior of free ions is affected significantly by the nature of the ion's solvation shell, and in any discussion of ion pairs we should take cognizance of the variety of forms they are capable of acquiring. Moreover, further aggregation of ions and ion pairs leads to larger clusters having their own characteristic physical and chemical individualities. Hence a deep understanding of the nature of solvation shells and the structure of ion pairs and their aggregates seems imperative for comprehending their chemistry.

It is desirable, therefore, to be acquainted with the physical methods applied in studies of these problems, especially because so much of the knowledge acquired through them is fascinating and revealing. With this in mind, we divided this monograph into two volumes: the first deals with the physical techniques fruitful in unravelling the problems of structure, energetics, and dynamics of ions and ion pairs; the second volume is devoted to the role of ions and ion pairs in chemical reactions such as proton transfer, electron transfer, or ionic polymerization.

In the first chapter of this volume, I have attempted to clarify the meaning of the concepts of ions, ion pairs, and their aggregates. The ramifications as well as the limitations of these concepts are outlined, their status is examined in thermodynamic and statistical terms, and the effect of the methods applied in the investigations of ions and ion pairs upon the results of observations is reviewed.

The significance of solvents and solvation shells can be fully appreciated by comparing the bare gaseous ions with those present in solutions where they ceaselessly interact with solvent molecules. A chemist accustomed to solution reactions will be surprised to learn, while reading the second chapter, "Ions and Ion-Solvent Molecule Interactions in the Gas Phase" by Paul Kebarle, that in gaseous reactions methanol is a much stronger acid and a better solvating agent than water. The ingenious and original techniques of studies of the stepwise formation of solvation shells, developed by Kebarle, are outlined in the same chapter and they reveal the changes occurring in the solvent clusters when the solvating molecules become crowded around the ions. In a liquid phase only the crowded solvation shells exist; therefore the conventional studies of ionic solutions acquaint us merely with the properties of these fully developed shells.

Johannes Smid describes in the third chapter, "Spectroscopic Studies of Ion-Pair Equilibria" the achievements of visible and UV spectroscopy permitting the differentiation between the various types of ion pair. Most of this elegant work is his own and it is indeed encouraging to realize how much can be learned about the intricate relations between various ion pairs from simple and straightforward experiments.

In the fourth chapter, "Infrared and Raman Studies of Ions and Ion Pairs," Walter Edgell unravels his new and quite unexpected approach to the problem of the interaction of ions and ion pairs with their surroundings. This work makes possible the study of the degree of participation of the repulsive forces in the solvation processes, while the ordinary thermochemical investigations reveal basically the contribution of the powerful attractive forces.

The application of magnetic resonance led to the development of superior and most penetrating techniques of studies of ions and ion pairs. The depth to which they fathom the pertinent problems is indeed amazing. The magnetic techniques not only offer direct and unambiguous methods of detecting ion pairing but also allow us to determine the structure of ion pairs, recognize the different degree of solvent participation in the pairing, provide detailed information about the location of one ion in respect to the other, furnish evidence about the modes of their motion, their frequency, etc. The rich and highly diversified experimental material accumulated by the persistent efforts of many workers who applied the ESR techniques to studies of ion pairs is comprehensively and critically reviewed by Howard Sharp and Martyn Symons in Chapter 5, "Electron Spin Resonance Studies of Ion Pairs." Its lucidly written text includes a wealth of data systematically and clearly presented in many tables and graphs. Moreover, the significance of these data is explained in simple language and the basic physical principles required for arriving at the conclusions are elucidated. Thus the reader easily finds out what could be learned from an ESR experiment and how this method could be utilized in solving his problem.

Then Chapter 6, "Nuclear Magnetic Resonance Studies of Carbon-Lithium Bonding in Organolithium Compounds," by Dennis McKeever deals with an intriguing problem. On increasing the interaction between the ions of a pair their bonding may be varied from purely electrostatic to basically covalent. Organolithium compounds, unique in many respects, furnish a fertile field for studies of a whole spectrum of such bonds. These compounds aggregate also into a variety of clusters and the results of NMR studies concerned with proton, ^{13}C and ^{7}Li resonance shed much light on the character of C—Li bonds and on the structure of the aggregates as well as on the dynamics of their interaction.

Chapter 7, "Nuclear Magnetic Resonance Studies of Alkali Ion Pairs" by Egbert de Boer and Jan Sommerdijk, although specialized, is fascinating. The authors show how NMR studies of paramagnetic species supplement and amplify their ESR investigations. Many fine details of structure and dynamics of ion pairs are detected by this approach.

Finally, in Chapter 8, "Electron Spin and Nuclear Magnetic Resonance Studies of Ion Pairs—Quantitative Approach," the authors of Chapter 7 recast in a rigorous and mathematical language the material presented in a qualitative fashion by Sharp and Symons in Chapter 5. Although the chemist who is not mathematically minded might be lost in the numerous equations, he surely should be able to get the gist of the message: how the use of quantum mechanics solves the quantitative problems of structure, shape, and energetics of ion pairs and accounts for their reactions. For a chemist versed in the subject this chapter may serve as a gold mine of important information about the techniques and approaches recommended in theoretical studies of ion pairs and undoubtedly will inspire him to further work in this field.

In conclusion, the material presented in this volume not only demonstrates the existence of ion pairs but endows them with spirit of life and motion.

MICHAEL SZWARC

Syracuse, New York
June 1971

Contents

Ions and Ion Pairs
in Organic Reactions

1

Concept of Ion Pairs

MICHAEL SZWARC

Department of Chemistry, State University College of Forestry at Syracuse University, Syracuse, New York

1. IONS AND THEIR SOLVATION SHELLS

Inasmuch as our understanding of the concept of ions is profound, a lengthy discussion may seem superfluous. But, to outline briefly, ions are formed when atoms, molecular fragments, or even intact molecules acquire electric charge—positive for cations and negative for anions. By virtue of their

1

charge, ions strongly interact with polar or polarizable molecules and the strength of such an interaction is clearly manifested by the behavior of gaseous ions (see Chapter 2). Even at a very low partial pressure of a solvating agent, and at relatively high temperatures, the solvating molecules become tightly aggregated around a gaseous ion, forming large clusters in less than a microsecond.

In a liquid phase, ions cannot avoid contact with the neighboring molecules that ceaselessly surround them. Thus in any solution ions are constantly in touch with solvent molecules, and in this sense, they are fully solvated. We may prefer, however, to reserve the term "solvation" for those ions that interact strongly and perhaps even specifically with their neighbors and to refer to the weakly interacting ions as "nonsolvated" species. Although this classification is not rigorous, and the borderline is fuzzy and ill defined, it seems desirable to maintain this terminology for didactic reasons.

As pointed out by Born [1], transfer of a charged sphere of radius r from vacuum into a medium of dielectric constant D decreases the free energy of the system by $(e^2/2r)(1 - 1/D)$, where e denotes the charge residing on the sphere. This relation, derived from classic electrostatics, does not invoke any molecular model that needs to be considered in the calculation. Indeed, we are concerned here with a structureless continuous medium endowed with an electric property manifested through its dielectric constant.

The dielectric constant appears in this expression as a purely empirical parameter. Its temperature dependence, if any, is therefore again empirical, but whenever D is temperature dependent the thermodynamic formalism leads to the equation

$$\Delta S = \left\{ \frac{e^2}{2rD} \right\} \frac{\partial \ln D}{\partial T}$$

for the entropy change arising from the transfer process.

Born applied the preceding relation for calculating the free energy of ions' solvation. Representing a univalent ion by an equivalent sphere with radius 2 or 3 Å we find the free energy of solvation in a medium of dielectric constant 10 to be -74 kcal/mole and -50 kcal/mole, respectively. The corresponding values for ΔS are negative if $\partial \ln D/\partial T$ is negative. For example, taking a reasonable, although rather low value of -3×10^{-3} for $\partial \ln D/\partial T$, we find ΔS of solvation to be about -25 e.u. and -16 e.u. when univalent ions of radius 2 and 3 Å, respectively, are immersed in a medium of dielectric constant 10.

Born's approach provides a useful starting point for calculations based on the "sphere in continuum" model. However, several problems call for clarification when the computations are attempted. Thus we should know the radii of ions in solutions, provided they are spherical, or the radii of equivalent spheres, whenever the shape of the ions is less symmetric. The present uncertainties about their size are not negligible. For example, the values recommended by Pauling [2] for the radii of alkali ions have been seriously challenged by Gourary and Adrian [3]. The discrepancy between both sets of data may be appreciated by inspecting Table 1. Furthermore,

Table 1 Radii of Some Simple Ions

	$r(Å)$	
Ion	Pauling	Gourary and Adrian
Li^+	0.60	0.94
Na^+	0.95	1.17
K^+	1.33	1.49
Rb^+	1.48	1.63
Cs^+	1.69	1.86
F^-	1.36	1.16
Cl^-	1.81	1.64
Br^-	1.95	1.80
I^-	2.16	2.05

the "radius" of an ion depends on its environment; for example, it is expected to be larger in vacuum than in a polar solvent [4]. The dielectric constant of the solvent is not the same throughout the medium, but it decreases in the vicinity of an ion due to dielectric saturation. The finite size of solvent molecules and their imperfect packing leads to some voids in the medium, especially in the vicinity of ions, and to account for the increased free volume of the solution the ions were treated as if they were larger than expected [5]. However, this effect is somewhat balanced by the electrostriction, which decreases the total volume of the system.

The most important difficulties of the "sphere in continuum" model arise from the fact that the size of solvent molecules is comparable to, and often even larger than, that of the ion. Furthermore, the interaction between a solvent molecule and an ion is not described correctly by a charge-point dipole force. The dipole of a solvent molecule arises from the presence of some atoms or groups in its framework and consequently its interaction with an *adjacent* ion depends on the position of that ion with respect to the nuclear framework. This is clearly shown by the following, perhaps slightly oversimplified model:

$$\left(+\right) \quad O \overset{CH_2-CH_2}{\underset{CH_2-CH_2}{\longleftarrow +}} \qquad \left(-\right) \overset{CH_2-CH_2}{\underset{CH_2-CH_2}{\longleftrightarrow +}} O$$

A positive ion tends to be located close to the oxygen atom of a tetrahydro-furan molecule, thus taking advantage of the strong attractive interaction with its lone pair of electrons. Such an interaction leads to repulsion of a negative ion and hence the anion is preferentially located on the other end of the ether molecule. Therefore tetrahydrofuran solvates cations strongly but anions weakly.

A more realistic approach to solvation problems calls for molecular models that describe the preferential orientation of solvent molecules adjacent to the ions. This approach was developed first by Bernal and Fowler [6] and improved by Eley and Evans [7]. Subsequently, it was utilized by other investigators with variable degrees of success. In principle it is a fruitful model and eventually it should lead to correct results. However, it suffers from numerous practical difficulties: the polarity and polarizibility of solvating molecules have to be described in great detail, their shapes and the allowed conformations fully outlined, the repulsive forces acting between the molecules packed in the solvation shell, as well as those acting between the ion and the adjacent solvent molecules, should be well understood, etc. Furthermore, in calculation of the free energy of solvation the mutual interaction of solvent molecules in the bulk of the liquid must be accounted for, because the insertion of a solvated ion into a solvent requires formation of a "hole" in the investigated medium, and this may distort the structure, if any, of the unperturbed solvent. This effect is of great importance for understanding the behavior of aqueous solutions.

It may be interesting to compare the problems of solvation of ions in the liquid and gaseous phase. In a liquid phase the abundance of solvent molecules leads to a complete hierarchy of solvation shells. The packing of the first shell, formed by the nearest neighbors to the ion, is affected by the insertion of the next-neighbor solvent molecules in the voids created by the molecules of the first layer. There is no shortage of solvent molecules and therefore there is no need for an undesirable empty space. On the other hand, a given cluster of solvent molecules surrounding a gaseous ion involves a fixed number of solvating species. Hence situations may be encountered where the packing is looser in order to accommodate a greater number of nearest neighbors.

A solvation shell surrounding an ion affects its physical and chemical properties. It increases the size of the ion and therefore decreases its mobility and diffusion coefficient. The mobility of an ion, Λ_0^+ or Λ_0^-, may be given in terms of its Stokes radius r_S; this is the radius of a sphere with identical hydrodynamic behavior as the solvated ion. The precise mathematical relation between the mobility, Λ_0 (+ or −), and r_S is still in dispute, although it is frequently assumed that the Stokes-Einstein equation,

$$r_S = \frac{0.819}{\Lambda_0 \eta}$$

where η denotes the viscosity of the solvent in which the mobility is measured, is valid. The numerical coefficient appearing in the foregoing equation is derived from the hydrodynamic theory of motion of a relatively large sphere in a continuous medium. There is evidence [8] that its value should be increased when the dimension of the sphere approaches that of solvent molecules. For example, the conventional Stokes radius of $N(CH_3)_4^+$ ion is 1.7 times smaller than the radius calculated from atomic models, whereas for $N(C_5H_{11})_4^+$ ions both radii seem to be identical.

The simple "sphere in continuum" model requires some further modification to account for various relaxation phenomena associated with the motion of a charged particle through polar molecules. This problem has been studied extensively, and much theoretical and experimental work is now available in this field. The interested reader may consult numerous texts dealing with this topic [9].

In spite of all the uncertainties and ambiguities, calculations of Stokes radii are useful because they provide a valuable diagnostic method of differentiation between "solvated" and "bare" ions. For example, the conductance studies of alkali tetraphenyl borides in tetrahydrofuran [10] showed that the Stokes radius of Na^+ is about 4.2 Å, substantially larger than that of the "bare" sodium ($\sim$1.2 Å), whereas for the Cs^+ ion the values of both radii are comparable, 2.4 and 1.9 Å, respectively. On the basis of these results it has been concluded that Na^+, but not Cs^+, is solvated by tetrahydrofuran molecules. Similar studies of dimethoxyethane solutions [10b] revealed that both cations are "solvated" by dimethoxyethane, a result which has been satisfactorily rationalized in terms of molecular structure of both solvents.

Solvation should affect other physical properties of ions, for example, the chemical shift of sodium nucleus is affected by solvation of Na^+ cation [11]. However, not much information is available yet on this subject. The strong interaction between some ions (particularly the cations of transition metals) and solvent are described in terms of ligand theory and may be reflected in a substantial change of their electronic spectra. For example, the anhydrous Cu^{2+} ion is colorless, whereas the hydrated ion is blue.

Solvation is expected to reduce the reactivity of an ion because the reagent and solvent molecules have to compete for the vacant site on the ion. For example, Cl^- ion is a much more powerful base in dimethylformamide—a medium which poorly solvates anions—than in water.

The solvating power of various aprotic solvents does not depend on their dielectric constant, contrary to past belief, but on their ability to donate electrons to the cations (cation solvation) or accept electrons from anions (anion solvation). The donor property is measured by solvent's "donicity" [12] and for the sake of illustration the donicity number and the dielectric constant of several aprotic solvents are given in Table 2. In protic solvents,

Table 2   Donicities and Dielectric Constants of Some Aprotic Solvents

Solvent	Donicity Number	D
Sulfurylchloride	0.1	10.0
Thionylchloride	0.4	9.2
Acetylchloride	0.7	15.8
Nitromethane	2.7	35.9
Nitrobenzene	4.4	34.8
Aceticanhydride	10.5	20.7
Benzonitrile	11.9	25.2
Acetonitrile	14.1	38.0
Sulfolane	14.8	42.0
Benzylcyanide	15.1	18.4
Ethylenesulphite	15.3	41.0
Propionitrile	16.1	27.7
Acetone	17.0	20.7
Diethylether	19.2	4.3
Tetrahydrofuran	20.0	7.6
Dimethylacetamide	27.8	38.9
Dimethylsulfoxide	29.8	45.0
Dimethylformamide	30.9	∼35.
Pyridine	33.1	12.3
Hexamethylphosphortriamide	38.8	30.

as is well known, the ability of solvent molecules to form hydrogen bonds with the ion play an important role in solvation of anions.

An interesting situation is encountered in relatively poorly solvating media containing small amounts of powerfully solvating agents. In such systems two or more thermodynamically distinct ions may be observed. For example sodium ions exist in tetrahydropyrane (THP) containing a small amount of tetraglyme (TG) as $Na^+(THP)_n$ and $Na^+(THP)_m(TG)$, and in the presence of triglyme (TrG), species such as $Na^+(THP)_k(TrG)_2$ have been observed [13].

2. THE CONCEPT OF ION PAIRS

The concept of ion pairs was introduced in 1926 by Bjerrum [14]. It was known in those days that ionophores—compounds built up of ions and not neutral molecules—are completely dissociated in aqueous solution, and it was expected that they should behave in the same way in other suitable solvents. It therefore came as a surprise when Krauss [15] reported that sodium chloride, a typical ionophore, behaves like a weak electrolyte, an ionogene, when dissolved in liquid ammonia. The electric conductance of such a solution is given by the law governing the conductance of aqueous solutions of acetic acid indicating that only a small fraction of the dissolved salt is dissociated into free ions. To account for these observations Bjerrum proposed that in liquid ammonia and in other nonaqueous solvents the oppositely charged ions are associated into neutral ion pairs which do not contribute to the electric conductance.

A question arises as to how to differentiate between the free ions and ion pairs. Bjerrum, who was occupied with the phenomenon of conductance and the importance of Coulombic forces, tackled the problem by asking what form of the radial distribution function $F(r)$ gives the concentration of counterions surrounding a reference ion. Taking into account the geometric factor, given by $4\pi r^2\, dr$, and the Bolzmann factor, $\exp(-e^2/rDkT)$, involving the Coulombic interaction only, he found $F(r)$ to have a minimum for $r = r_c = e^2/2DkT$. Subsequently, he assumed that two ions separated by a distance greater than the critical distance r_c are free and contribute to the electric conductance of the solution, whereas the ions separated by a distance smaller than r_c form the nonconducting ion pairs.

The approach of Bjerrum was refined by other investigators and the developed theories were successfully confirmed by numerous experiments. The whole subject has been critically reviewed and discussed by the present writer elsewhere [16] and the interested reader may find there the references to the original papers. However, in spite of its elegancy and originality, Bjerrum's approach does not seem to be satisfactory. There is no a priori reason to associate the minimum in the distribution function with the critical distance r_c differentiating between the free ions and ion pairs. Furthermore, the concept of a critical distance is incompatible with molecular approaches and deprives an ion pair of a molecular status. It is shown in the following chapters that ion pairs are well defined chemical species characterized by their own physical properties. Their own specific chemical behavior, distinguishing them from free ions, will be discussed in Volume II. Hence the equilibrium established in the free-ions–ion pairs system should be treated by the conventional thermodynamic methods and not by techniques based on the distribution functions.

The thermodynamic approach was attempted first by Denison and Ramsey [17], who envisaged only two situations: (1) either the two oppositely charged ions are in a binding distance, or (2) they are infinitely far from each other. Hence the change in the electrostatic free energy of the system resulting from such a dissociation is given by

$$\Delta F'_e = \frac{Ne^2}{(r_1 + r_2)D}$$

where r_1 and r_2 denote the radii of the ions which, for the sake of simplicity, are visualized as hard spheres immersed in a continuous medium having a dielectric constant D. The dissociation may be accomplished by the following sequence of steps:

1. Transfer of an ion pair from its solution into evacuated space.
2. Separation of the ions in vacuum.
3. Immersion of the two isolated ions into solvent, while still keeping them far apart.

In the first step, the ion pair becomes desolvated and the free energy increases by $-\Delta F_{sol.\ ion\ pair}$. The free energy of the system increases again in the second step, $\Delta F_v = Ne^2/(r_1 + r_2)$. In the third step, the gaseous ions become solvated and this decreases the free energy of the system by $\Delta F_{sol.\ free\ ions}$. Hence the difference in the free energy of solvation of a free ion, separated ion, and that of an ion pair is

$$\Delta F_{sol.\ free\ ion} - \Delta F_{sol.\ ion\ pair} = -\left(\frac{Ne^2}{r_1 + r_2}\right)\left(1 - \frac{1}{D}\right)$$

The free ions are more strongly solvated than the ion pairs, and the increase in the degree of solvation provides the driving force for the dissociation process. Therefore the dissociation in well-solvating liquids is more extensive than in the gas phase.

3. THERMODYNAMICS OF DISSOCIATION OF ION PAIRS INTO FREE IONS

The thermodynamic formalism embodied in the equation $\Delta F_e = Ne^2/(r_1 + r_2)D$ permits us to calculate the effect of electrostatic interactions upon the entropy and enthalpy of dissociation. Thus

$$\Delta S'_{diss} = \left[\frac{Ne^2}{(r_1 + r_2)D}\right]\left(\frac{\partial \ln D}{\partial T}\right)_p$$

and

$$\Delta H'_{diss} = \left[\frac{Ne^2}{(r_1 + r_2)D}\right]\left(1 + \frac{\partial \ln D}{\partial \ln T}\right)$$

$\Delta S'_{diss}$ represents only the entropy decrease arising from the change in the degree of physical solvation — that is, the entropy change of the surrounding medium. The fact that one particle (one ion pair) dissociates into two is not accounted for. The coefficient $(\partial \ln D/\partial \ln T)_p$ is always negative. For many liquids its absolute value exceeds unity, as shown in Table 3, and hence $(1 + \partial \ln D/\partial \ln T)_p$ is usually also negative. This simple treatment thus accounts for the exothermicity of the dissociation process observed for most solvent systems.

Table 3 Temperature Dependence of the Dielectric Constants of Common Solvents

Solvent	Temperature Range (°C)	$\partial \ln D/\partial \ln T$	D at 0°C	Reference
Tetrahydrofuran	−70 to +25	−1.16	8.23	a
Tetrahydropyran	−40 to +25	−0.97	6.12	b
Dimethoxyethane	−70 to +25	−1.28	8.00	a
Diethyl ether	−70 to +25	−1.33	4.88	b
2-Methyltetrahydrofuran	−70 to +25	−1.125	6.92	b

[a] Ref. 10b.
[b] D. Nichols, C. A. Sutphen, and M. Szwarc, *J. Phys. Chem.*, 72, 1021 (1968).

The formation of two particles (free ions) from one ion pair increases the entropy of the system by a term ΔS_t, its value being determined mainly by a greater translational freedom of two ions when compared with that of the ion pair. Hence ΔF_{diss} should be given by

$$\Delta F_{\text{diss}} = \frac{Ne^2}{(r_1 + r_2)D} - T \Delta S_t$$

leading to

$$-\ln K_{\text{diss}} = \frac{e^2}{(r_1 + r_2)DkT} - \frac{\Delta S_t}{R}$$

The term $T \Delta S_t$ introduces concentration units; in fact, its omission left the original expression of Ramsey independent of units. This error was corrected in his later paper [17b]; K_{diss} was given by the equation

$$-\ln K_{\text{diss}} = -\ln K_{\text{diss}}^0 + e^2/(r_1 + r_2)DkT$$

where K_{diss}^0 denotes the dissociation constant of an "uncharged" ion pair, an associate of two fictitious species differing from the real ions only by the lack of charge. K_{diss}^0 accounts for the change in translational entropy as well as for other energy and entropy contributions for example, those resulting from the van der Waals interactions between the ions of a pair.

The thermodynamic approach of Ramsey was elaborated further by Gilkerson [18], who pointed out that the simple treatment predicts, contrary to experimental findings, that ΔF_{diss} and K_{diss} should remain identical in a series of solvents having the same dielectric constant. Gilkerson's treatment starts with Kirkwood's zero approximation to the partition function of a particle in solution [19]. Thus

$$f = \left(\frac{2\pi mkT}{h^2}\right)^{3/2} gv_F \, \delta \exp\left(\frac{-E_0}{RT}\right)$$

where v_F is the available free volume per particle, g the partition function accounting for the internal degrees of freedom, and δ a constant slightly larger than unity. The dissociation constant is therefore

$$K_{\text{diss}} = \left(\frac{2\pi\mu kT}{h^2}\right)^{3/2} \overline{gv\,\delta} \exp\left(\frac{E_s}{RT}\right) \exp\left[\frac{-e^2}{(r_1 + r_2)DkT}\right]$$

where $\mu = m_+ m_-/(m_+ + m_-)$ is the reduced mass of the pair

$$\overline{gv\,\delta} = \frac{g_+ g_- v_+ v_- \delta_+ \delta_-}{g_\pm v_\pm \delta_\pm}$$

and E_s denotes the difference of the specific interaction energies of ions and ion pairs with the dipoles of the nearest solvent molecules. Gilkerson explicitly assumed that the energy needed to separate the charges in the solution is given by $e^2/(r_1 + r_2)D$ where D is the *macroscopic* dielectric constant of that solvent. This is a too drastic an approximation for a treatment that attempts to account for the specific properties of solvents.

The influence of the solvent on K_{diss} appears now in two terms: the one concerned with the free volume $\bar{V} = v_+ v_-/v_\pm$, and the other involving the interaction energy E_s. The relative effect of the solvent's dipole moment upon the E_s value was calculated, and the relation obtained was confirmed by the data derived from experiments performed in three solvents having approximately the same bulk dielectric constants but different dipole moments. Although the usefulness of the calculations may be doubted, Gilkerson's approach correctly emphasizes the importance of free volume and of specific solvation in calculating the equilibrium constants of ionic dissociations.

The main binding energy of ion pairs arises from Coulombic interactions* but other forces may still substantially contribute to their stability. For example, nonspherical ions may possess a dipole moment, as well as a charge. Dissociation of ion pairs formed from such species was considered by Accascina, D'Aprano, and Fuoss [20], who included an appropriate term in the Ramsey equation:

$$-\ln K_{\text{diss}} = -\ln K_{\text{diss}}^0 + \frac{e^2}{(r_1 + r_2)DkT} + \frac{\mu e^2}{d^2 DkT}$$

Here K_{diss}^0 denotes again the dissociation constant for the "uncharged" ion pair, μ is the dipole moment of the unsymmetrical ion, and d is the distance between the center of its dipole and the center of the charge of the symmetrical counterion.

---

* In aqueous solution, Coulombic interaction between a positive and a negative ion is not sufficient to secure the stability of the pair. In this medium, the contributions due to other interactions must be of paramount importance.

The contribution of the dispersion forces, which bind the two ions, is included in the term K_{diss}^0. For polarizable, colored ions, these interactions may be sufficiently strong to contribute significantly to the stability of the resulting ion pair. For example, Grunwald [21] recently compared the dissociation of acids forming colored ions, such as picric acid, with those yielding colorless anions. To account for the experimental data, it was necessary to assume that dispersion forces contribute as much as 4 kcal/mole to the energy of hydration of colored anions, and semiquantitative calculations confirmed these results. A similar problem was discussed by Noyes [22], who presented a treatment demanding a substantial contribution of nonelectrostatic forces to the free energy of hydration of inorganic ions. The dispersion forces stabilize also the ion pairs of radical anions derived from aromatic hydrocarbons, the respective binding energy being probably 2–3 kcal/mole.

In some systems the destruction of the solvent structure caused by the presence of free ions contributes to the stabilization of ion pairs. This factor is often important in aqueous solution, because liquid water possesses a well defined and relatively stable structure. A recent example illustrating such an effect, which contributes to the stabilization of ion pairs in methanol, was discussed by Kay et al. [23]. Probably a similar effect is responsible for a larger degree of association of ions in D_2O than in H_2O [24]. Another manifestation of this phenomenon may be found in the action of hydrophobic bonds advocated by Scheraga [25].

4. THE ROLE OF SOLVENT IN THE FORMATION OF ION PAIRS

Let us consider once again the problem of ion-pair dissociation. For a hypothetical *continuous* and *structureless* solvent the dielectric constant D is temperature *independent*.* Ions or ion pairs immersed in such a medium do not induce in it any degree of order and hence the dissociation constant of a pair in this liquid is

$$-RT \ln K_{diss} = \Delta E_{diss} - T \Delta S_t$$

where $\Delta E_{diss} \approx \Delta H_{diss}$ equals the sum of Coulombic energy needed for the separation of ions and the binding energy BD arising from other kinds of interaction. Note that the Coulombic energy is now independent of temperature. The entropy term ΔS_t represents the change of the entropy resulting merely from the formation of *two* free ions from *one* ion pair.

---

* Solvents having static dielectric constant equal to n^2 (n = the refractive index) show a behavior closely similar to that of the hypothetical solvent.

Real solvents have discrete molecular structure and often dipolar properties. Their molecules perform an endless Brownian dance, its vigor increasing with temperature. The Brownian motion destroys, at least partially, their expected orderly arrangements around the ions or ion pairs, as well as an order induced by an external electric field, and this makes the dielectric constant of a real solvent temperature *dependent*.

The dissociation process may be visualized in mechanical terms as a movement of a representative point on a potential energy hypersurface or curve, and in the latter case the distance a separating the ions is the reaction coordinate. In this model the potential energy is given by the sum of Coulombic energy, e^2/aD_{eff}, and the binding energy, $BD(a)$, both D_{eff} and BD being functions of a. On the other hand, the thermodynamically derived $\Delta H \approx \Delta E$ is inappropriate and even misleading for such a model because the free energy stored in the medium $(-T \Delta S_{electrostat})$ is then disregarded. Note however that the above discussed potential energy hypersurface, or curve, is affected by the nature of solvent and depends on temperature, since both factors determine the values of D_{eff} and BD.

Let us restate the last argument. In a solvent the energy of two ions separated by distance a depends on the *average* configuration of the surrounding molecules. The *average* varies with temperature and thus the potential energy curve also becomes a function of temperature. Indeed, a solvent at two different temperatures provides, after all, two different surroundings for ions or ion pairs. We may say with some justification that the contents of a bottle filled with a solvent varies with its temperature and only the label on the container remains the same.

5. ION PAIRS AND COVALENT MOLECULES

It has been shown in the preceding section that ion pairs are thermodynamically distinct entities coexisting in equilibrium with the free ions:

$$\underset{\text{ion pair}}{A^+, B^-} \quad \leftrightarrows \quad \underset{\text{free ions}}{A^+ + B^-} \qquad K_{diss}$$

This species should be distinguished from a covalent molecule AB, provided such a molecule exists. The distinction between an ion pair A^+, B^- and a covalent molecule AB requires some clarification. Chemical bond linking the two fragments A and B may have a polar character and its polarity may vary from 0 (a purely covalent bond) to 1 (a purely electrovalent bond), depending on the nature of the bonded fragments. Could we visualize two distinct molecules, both composed of A linked to B, but one possessing a high covalent character while the other was more polar? In answering this question, it is desirable to differentiate between molecules AB or A^+, B^- in the gas phase and those in solution.

It appears, at first glance, that gaseous molecules exist entirely in one or the other form, and apparently both forms are never simultaneously observed in any system. For example, the gaseous HCl undoubtedly represents a covalently bonded molecule, whereas the gaseous sodium chloride should be classified as an ion pair. However, suitably electronically excited HCl could be treated as an ion pair, and hence it should be possible to have both species, HCl and H^+, Cl^- simultaneously present in a system at equilibrium with each other. The problem is purely practical. The energy gap between the two forms is large and therefore the proportion of H^+, Cl^- would be too low to permit its observation at any reasonable temperature. Consider, however, some charge-transfer complexes. In such a system the excitation energy needed for conversion of a predominantly covalent species into its polar "isomer" may be low, and thermal excitation could then maintain an appreciable concentration of the isomer. Thus both forms would coexist then in equilibrium in the gaseous phase.

In conclusion, the coexistence of AB with A^+, B^- in the gaseous phase is in principle possible, although for most molecules the concentration of one form may be vanishingly small under conventional conditions. The geometry of the two forms would be different; for example, the A-B distance could be smaller in the covalent AB molecule than in the A^+, B^- ion pair, and of course each form would be described by a different electronic ψ-function.

Although for a gaseous system the energy of the polar form A^+, B^- is usually higher than that of the covalent AB, in solution this relation may be reversed because the A^+, B^- pair is stabilized by solvation. Indeed, in many polar solvents the reaction AB $\rightarrow$ A^+, B^- is exothermic and it proceeds with a decrease in the entropy of the system since the solvent molecules become immobilized around the ion pairs. A similar phenomenon is observed on condensation of some vapors. For example, the condensation of the covalently bonded N_2O_5 yields ionic crystals NO_2^+, NO_3^- [26], the ionic form being stabilized in the solid phase by lattice energy.

The donicity of the solvent plays the major role in the solution reaction AB $\rightarrow$ A^+, B^-, whereas its dielectric constant is of lesser importance. The donation of electrons of the donor-solvent to the fragment A facilitates the transfer of the σ A-B electrons to the moiety B and eventually this leads to the heterolytic fission of the A-B bond. On the other hand, a high dielectric constant of the solvent increases the degree of dissociation of the resulting ion pair into free A^+ and B^- ions.

Let us close this section with a discussion of an interesting example of an equilibrium between an ion pair and its nonpolar form. Solvated electrons $(e^-)_S$ may be generated in a solvent containing alkali ions [27, 28] and under appropriate conditions an equilibrium

$$(e^-)_S + Na^+ \rightleftharpoons (e^-)_S, Na^+$$

may be established. In fact, we have succeeded recently in determining that equilibrium constant in tetrahydrofuran [28] and also have shown that the unimolecular reaction forming the neutral sodium atoms,

$$(e^-)_S, Na^+ \rightarrow Na^0$$

may be observed in this system, its rate constant being $\sim 10^4$ sec^{-1}. Hence a solution of alkali atoms should be in equilibrium with their ion pairs and the equilibrium constant of the pair formation is expected to increase with increasing solvating power of the solvent. This seems to be the case, since we have found [29] the collapse of the $(e^-)_S$, Na$^+$ pair to be much faster in tetrahydropyrane (a relatively poor solvent) than in tetrahydrofuran (a relatively good solvent).

6. LIMITATION OF THE CONCEPT OF ION PAIRS

The concept of ion pairs, like many other concepts of chemistry, is justified and profitable within some range of temperatures and concentrations and loses its utility beyond these boundaries. Molecules "exist" at those temperatures at which kT is smaller than their binding energy. At sufficiently high temperatures their lifetime becomes comparable to the time of molecular collisions, and under these conditions the significance of molecules is lost. The largest contribution to the binding energy of an ion pair comes from Coulombic forces, the respective term being $z_1 z_2 e^2 / aD$, where $z_1 e$ and $z_2 e$ are the charges of the bonded ions and a the distance separating them. Hence the concept of ion pairs is valid only at temperatures lower than $z_1 z_2 e^2 / aDk$. Consider, for the sake of illustration, an ion pair composed of bulky ions 10 Å apart immersed in a medium of dielectric constant 80. Such ion pairs lose their stability above 200°K and then it becomes ambiguous and perhaps even meaningless to discuss this system in terms of ion pairs, because such species cannot be treated as thermodynamically distinct from free ions.

Similarly, the concept of ion pairs is useless when the concentration of ions is too high. For example, it would be impossible to differentiate between free ions and ion pairs in a fused sodium chloride. Of course, each sodium ion has some chloride ions as its nearest neighbors, and vice versa, but it is impossible and unprofitable to assign two oppositely charged ions to a lasting pair. Such a system is better described by a suitable distribution function and not by an equilibrium between free ions and ion pairs.

Some solutions acquire new properties when the concentration of ions becomes too high. Let us consider a rather esoteric example, a solution of an alkali metal in a suitable solvent. This is not at all a simple system, but at low concentrations of the metal the electric conductance of the resulting solutions could be accounted for by postulating an equilibrium between solvated electrons and alkali cations. In accordance with this model the specific conductance of such solutions decreases with increasing concentration of the solute. However, as the concentration of the metal increases the specific conductance eventually reaches its minimum and then rapidly increases. At that stage the character of the conductance changes and it seems that the previous ionic conductance acquires the feature of metallic conductance. Hence the concentrated solutions of alkali metals should not be described as the equilibrium between the free ions and ion pairs (or still higher aggregates), but an entirely new model is needed to account for their behavior.

7. DIFFERENT TYPES OF ION PAIR

In 1954 Fuoss [30] and Winstein [31] simultaneously and independently suggested that ion pairs may exist in two distinct forms. We refer to them as the loose and tight ion pairs. The argument used by Fuoss to establish the existence of two types of ion pair is interesting and instructive, and therefore is presented here in a somewhat modified version.

An ion surrounded by a tight solvation shell may approach a counterion without hindrance until its solvation shell contacts the partner. Thereafter, either the associate maintains its structure of a solvent-separated, loose ion pair, or the solvent molecules separating the partners are squeezed out and then a tight-contact ion pair is formed. Such two-step association of ions has been revealed by various relaxation techniques [32].

This model implies that loose ion pairs may exist only in those solvents in which at least one of the ions possesses a tight solvation shell, or, according to our nomenclature, only when it is solvated. Whenever the interaction of both ions with the solvent is weak, that is, both ions are "bare," the association process produces tight ion pairs only; in other words, no loose pairs exist in such a medium. On the other hand, the solution contains exclusively loose ions when their interaction with solvent is very strong, especially if the pair involves a large counterion [33]. To clarify the last point, consider an ion of radius r_1 surrounded by a solvation shell of Δr thickness which combines with a counterion of radius r_2. The collapse of the initially formed loose pair into a tight one releases Coulombic energy approximately given by

$$\left(\frac{e^2}{D}\right)\frac{\Delta r}{(r_1 + r_2 + \Delta r)(r_1 + r_2)}$$

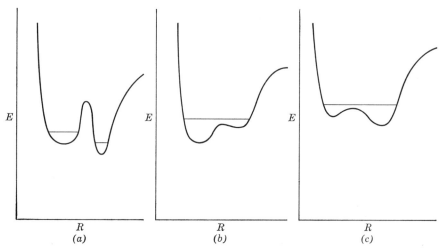

Figure 1. Potential energy E of any ion pair as function of interionic distance R. (a) The concept of two kinds of thermodynamically distinct ion pairs is justified; (b) and (c) The concept of two distinct ion pairs is not justified.

For a constant r_1 and Δr this "driving force" decreases with increasing r_2, and thus for a sufficiently large r_2 it becomes too small to cause a collapse.

The discrete molecular structure of the solvent affects the shape of the pseudo-potential energy curve describing the dissociation process. As the two bare ions are pulled apart, the resulting empty space cannot accommodate even one solvent molecule until the separation becomes sufficiently large, because the solvent molecules have a finite size [34]. The potential energy curve therefore may acquire the shape shown in Fig. 1a, the second minimum appearing at the distance at which a solvent molecule may be squeezed in. The two minima correspond to two distinct species coexisting in equilibrium.

This description, although useful, is oversimplified [35]. As has been pointed out previously, the pseudo-potential curve is temperature dependent; the *average* configuration of the surrounding solvent molecules affects the energy of a pair separated by distance a. The steep maximum shown in Fig. 1 may appear at one temperature but not at another. In brief, this model, originally proposed by Grunwald [34], assumes that the potential energy is uniquely determined by the interionic distance a. This is the usual situation when we deal with isolated gaseous molecules, whereas in solution the ion pairs are imbedded in a fluctuating environment whose properties vary with temperature.

For deep and narrow potential wells, the concept of contact and solvent-separated pairs as two thermodynamically distinct species is still justified within a relatively wide temperature range. However, for wide and shallow wells shown in Figs. 1*b* and 1*c*, the distinction may become meaningless [36]. The well may change its shape; it may appear like 1*b* at low temperature but acquire a shape like 1*c* at higher temperatures. The pairs are then gradually transformed from a contact into a solvent-separated type as the temperature rises, and the concept of two *thermodynamically* distinct species does not apply to these systems. The model of the tight-loose ion pairs illustrated in Figs. 1*b* and 1*c* is referred to as the static model in contrast to the dynamic model invoking a genuine equilibrium between two distinct types of pair. An experimental attempt to distinguish between these two models has been reported recently by de Boer [37] (see also Chapter 8 of this book).

Alternatively, the differentiation between the static and dynamic model of ion pairs can be achieved by utilizing the appropriate distribution functions. This approach has been described by this writer elsewhere [38].

The concept of different types of ion pair can be presented in a more general way. Ions of a pair and the neighboring solvent molecules form a class of variable patterns, the variations being due to Brownian motion. Under some conditions, two or more nonoverlapping groups of patterns may retain their characteristic configuration for a time longer than 10^{-10} sec, the correlation time of Brownian motion. Such groups of patterns may be treated as thermodynamically distinct species whenever the intermediate configurations are improbable [40, 41]. The tight and loose ion pairs fall into this category.

Variation of temperature affects the lifetime of a pattern as well as the range of configurations included in its variations. Thus two patterns that are nonoverlapping at lower temperatures may coalesce at higher temperatures; put another way, although two distinct types of ion pairs are coexisting at lower temperatures, the distinction between them is lost at higher temperatures.

It is hoped that these comments clarify the meaning of different types of ion pair and under what conditions the distinctions are or are not justified.

8. SOLVATION OF ION PAIRS BY EXTERNAL AGENTS

A suitable agent E added to a solution of ion pairs may produce new ionic species and the equilibria such as

$$A^+, B^- + E \rightleftharpoons A^+, B^-, E$$
$$A^+, B^-, E + E \rightleftharpoons A^+, B^-, 2E$$
$$\cdot$$
$$\cdot$$
$$\cdot$$

and

$$A^+ + E \rightleftarrows A^+, E$$

$$A^+, E + E \rightleftarrows A^+, 2E$$

.

.

.

are then established. The agent E may react with a tight ion pair in two ways: forming a tight A^+, B^- pair coordinated on its periphery with E, or forming a loose A^+, E, B^- pair in which E separates the ions. Hence in such a system we may encounter an isomerization equilibrium,

$$A^+, B^-, E \qquad \rightleftarrows \qquad A^+, E, B^-$$

a tight ion pair
externally coordinated
with E

a loose ion pair
separated by E

This phenomenon of isomerization was first reported by Slates and Szwarc [36], who carried out spectrophotometric studies of an equilibrium established between an electron-acceptor (biphenyl) and metallic sodium. Other examples of such an isomerization are discussed in Chapter 3.

A coordinating agent usually replaces a solvent molecule associated with an ion pair. For example, the addition of tetrahydrofuran (THF) to a solution of tight sodium napthhalenide ion pairs ($N^{\cdot -}$, Na^+) in diethylether (DEE) leads to reactions

$$(N^{\cdot -}, Na^+)(DEE)_n + THF \rightleftarrows (N^{\cdot -}, Na^+)(DEE)_{n-1}(THF)$$

and

$$(N^{\cdot -}, Na^+)(DEE)_n + 2THF \rightleftarrows (N^{\cdot -}, Na^+)(DEE)_{n-2}(THF)_2$$

forming tight ion pairs externally coordinated with THF [39]. Such processes result in a negligible change of the entropy of the system because the solvent molecule gains its freedom by enslaving a molecule of the coordinating agent.

In systems where ion pairs are in equilibrium with nonionic molecules, for example,

$$\text{metallic sodium} + \text{biphenyl} \rightleftarrows \text{sodium}^+ \text{ biphenylide}^-$$

the addition of a coordinating agent increases the total concentration of ion pairs. In fact, the change in the total concentration of ion pairs resulting from the addition of a coordinating agent permits us to study the coordination phenomena [36, 39]. Similarly, if an equilibrium is established between tight and loose ion pairs,

$$A^+, B^- \rightleftarrows A^+, S, B^-$$

tight   loose

S denoting a solvent molecule, the addition of a coordinating agent E may lead to an increase in the fraction of loose pairs if the reaction proceeds according to equation

$$\underset{\text{tight}}{A^+, B^-} + E \rightleftarrows \underset{\text{loose}}{A^+, E, B^-}$$

Studies of such systems provide information about the equilibrium constant of exchanges:

$$A^+, S, B^- + E \rightleftarrows A^+, E, B^- + S$$

9. PROPERTIES OF DIFFERENT TYPES OF ION PAIR

Properties of ion pairs depend on their structure and tight and loose ion pairs are often very different in this respect. In fact, it will be shown in the following chapters how the electronic spectra of these pairs change with their structure (Chapter 3), how the structure affects the ESR coupling constant of pairs involving paramagnetic ions (Chapters 5 and 8), etc. The relation between the characteristic time of observation and the lifetime of species has been clarified by quantum-mechanical considerations, and it is well understood now that two species in a system may reveal themselves as distinct in respect to one kind of measurement, although another kind of observation does not distinguish between them and the measured property has an average value only.

The properties of tight and loose ion pairs, although different for each kind, frequently are independent of solvent. For example, the optical spectra of loose pairs are virtually the same in various solvents. An interesting case illustrating this principle has been reported recently [42]. Studies of equilibria,

sodium biphenylide (Na^+, B^-) + naphthalene (N)

$$\rightleftarrows \text{sodium naphthalenide } (Na^+, N^-) + \text{biphenyl (B)}$$

demonstrate its dependence on solvent. However, closer examination of the data showed that the equilibria

$$(Na^+, B^-)_{\text{tight}} + N \rightleftarrows (Na^+, N^-)_{\text{tight}} + B$$

and

$$(Na^+, B^-)_{\text{loose}} + N \rightleftarrows (Na^+, N^-)_{\text{loose}} + N$$

seem to be independent of solvent. The observed dependence of the overall equilibrium constant on the solvent's nature arises from the solvent dependence of the equilibrium of transformation,

$$\text{tight pair} \rightleftarrows \text{loose pair}$$

which in turn affects the proportion of both kinds of pair present in the system.

The preceding rule is not rigorously valid. For example, the reactivity of loose pairs of living polystyrene depends on whether THF molecules or tetraglyme molecules separate the pair [43].

10. CONDUCTANCE OF ION PAIR SOLUTIONS

Conductance studies provided the first evidence for ion-pairing, but only recently have such investigations been extended to cover a wide temperature range, thus permitting a direct determination of heat and entropy of dissociation [10b, 33, 35, 44–47]. For the sake of illustration, the van't Hoff plots giving log K_{diss} versus $1/T$ are shown in Figs. 2 and 3. The pronounced curvature shown by some lines needs stressing.

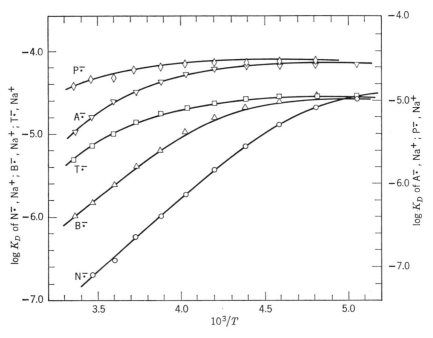

Figure 2. The overall dissociation constant K_D for sodium salt of aromatic radical anions in tetrahydrofuran given by the van't Hoff plot. The curvature results from changes in the structure of ion pairs, which are tight at higher and loose at lower temperatures. N—naphthalene, B—biphenyl, T—triphenylene (left scale); A—anthracene, P—perylene (right scale).

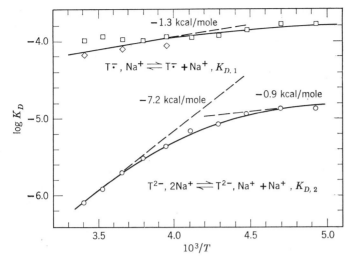

Figure 3. Van't Hoff plot of dissociation processes T^-, $Na^+ \rightleftarrows T^{\bar{\cdot}} + Na^+$ and T^{2-}, $2Na^+ \rightleftarrows T^{2-}$, $Na^+ + Na^+$ in tetrahydrofuran. $T^{\bar{\cdot}}$—radical anion of tetraphenylethylene; T^{2-} its dianion $T^{\bar{\cdot}}$, Na^+ is a loose pair, whereas T^{2-}, $2Na^+$ is a tight pair (at higher temperature).

Whenever two or more kinds of ion pair are present in a solution, their dissociation into free ions is governed by three or more interrelated equilibria. For example, consider a system composed of tight and loose ion pairs. Their dissociation is then determined by

$$(A^+, B^-)_{tight} \rightleftarrows A^+ + B^- \qquad K_{diss,t}$$

$$(A^+, B^-)_{loose} \rightleftarrows A^+ + B^- \qquad K_{diss,l}$$

and

$$(A^+, B^-)_{tight} \rightleftarrows (A^+, B^-)_{loose} \qquad K_{conv}$$

The foregoing equilibrium constants are correlated by the equation

$$\frac{K_{diss,t}}{K_{diss,l}} = K_{conv}$$

and the apparent equilibrium constant of dissociation, defined through

$$K_{ap} = \frac{[A^+][B^-]}{[\text{all ion pairs}]}$$

is given by

$$K_{ap} = \frac{K_{diss,t}}{1 + K_{conv}} = \frac{K_{diss,l}}{1 + K_{conv}^{-1}}$$

Hence $K_{ap} \approx K_{diss,t}$ when $K_{conv} \ll 1$ and it approaches $K_{diss,l}$ when $K_{conv}^{-1} \ll 1$. Since the conversion of tight into loose pairs is exothermic, K_{ap} may be identified with $K_{diss,t}$ at sufficiently high temperatures and with $K_{diss,l}$ at sufficiently low temperatures. The apparent heat of dissociation, determined by the tangent to the experimentally obtained van't Hoff plots, is given by $f\Delta H_{diss,t} + (1 - f)\Delta H_{diss,l}$ where f and $1 - f$ are the mole fractions of the tight and loose pairs, respectively. Of course, $(1 - f)/f = K_{conv}$ and $f\Delta H_{diss,t} + (1 - f)\Delta H_{diss,l} = \Delta H_{conv}$. These relations explain the observed curvature of the van't Hoff plot and show how ΔH_{conv} may be calculated from conductance data. The results of such calculations seem to agree with ΔH_{conv} determined from spectrophotometric studies [42, 47, see also 48].

The situation is slightly more complex when the system contains a coordinating agent E, because then not only is a new kind of ion pair formed but a new type of free ion is formed as well. To illustrate the point, consider a system composed of tight ion pairs A^+, B^- and a solvating agent E that gives rise to A^+, E, B^- pairs and A^+, E ions. The following equilibria are then pertinent:

$$A^+, B^- \rightleftarrows A^+ + B^- \qquad K_{diss,t}$$

$$A^+, E, B^- \rightleftarrows A^+, E + B^- \qquad K_{diss,E}$$

$$A^+, B^- + E \rightleftarrows A^+, E, B^- \qquad K_{E,p}$$

$$A^+ + E \rightleftarrows A^+, E \qquad K_{E,i}$$

and the condition of electroneutrality demands

$$[A^+] + [A^+, E] = [B^-]$$

The apparent equilibrium constant of dissociation, K_{ap}, defined as

$$K_{ap} = \frac{[B^-]^2}{\{[A^+, B^-] + [A^+, E, B^-]\}}$$

is given by

$$K_{ap} = \frac{K_{diss,t}(1 + K_{Ei}[E])}{1 + K_{Ep}[E]}$$

As expected, $K_{ap} = K_{diss,t}$ for $[E] = 0$ and approaches $K_{diss,E}$ as the concentration of E increases. Conductance of such a system has been studied recently by Shinohara [49].

11. AGGREGATION OF ION PAIRS

Thorough studies of conductance of salt solutions in nonaqueous media indicated the formation of triple ions at higher salt concentrations [50]:

$$A^+, B^- + A^+ \rightleftarrows A^+, B^-, A^+ \qquad K_{+-+}$$

and

$$A^+, B^- + B^- \rightleftarrows B^-, A^+, B^- \qquad K_{-+-}$$

The formation of triple ions increases the specific conductance of the solution and thus Λ, the specific conductance, seems to be given by $Ac^{-1/2} + Bc^{1/2}$ where c denotes the concentration of the salt. Such a relation leads to a minimum in the Λ-c relation. The original treatment of triple-ion formation by Fuoss and Kraus [50] assumed $K_{+-+} = K_{-+-}$; the generalization to those cases where this equality does not hold are due to Wooster [51] and Dole [52].

An interesting case of *intramolecular* formation of triple ions was reported by Bhattacharyya, Smid, and Szwarc [53]. Anionic polymerization initiated by electron-transfer yields polymers endowed with two terminal carbanions, each associated with a counterion into an ion pair. Dissociation of one pair leaves the resulting free ion attached through the polymeric chain to an ion pair. Subsequently, these two have a high probability to associate into a triple ion. The equilibrium constant of such an association should be independent of the concentration of the partners, that is, A^+, B^-vvvB^-, but it should decrease with increasing length of the polymeric chain. Both conclusions were confirmed by experiments.

The aggregation processes may proceed further. For example, ion pairs formed in the phenanthroquinone radical-anion system dimerize into pairs of ion pairs, and these associate into even larger aggregates [54, 55]. Although dipole-dipole interaction substantially contributes to the bonding energy, other forces seem to be also of paramount importance in stabilizing the aggregate.

The self-association of $PhC \vdots CPh^{2-}$, $2Na^+$ seems to form large aggregates in tetrahydrofuran solution [56]. The evidence is indirect, but nevertheless convincing. Most complex aggregation processes are observed in benzene solution and in solutions of other hydrocarbons. This is particularly noticeable for carbanions coupled with Li^+ cations, a subject reviewed recently by Bywater [57]. Extensive studies of aggregation of alkoxides have been reported by Steiner [58], and examples of other aggregations are revealed in various papers published recently.

12. EVIDENCE FOR THE FORMATION OF ION PAIRS AND TRIPLE IONS

Although the concepts of ion pairs and triple ions developed from studies of electric conductance, a direct evidence for their formation came from ESR studies of paramagnetic radical anions. Similarly, ESR studies conclusively demonstrated the formation of triple ions. The ESR studies are thoroughly reviewed in Chapter 5 and again in Chapter 8.

Indirect evidence is provided by spectrophotometric investigations, especially when *nonconducting* ion pairs are formed from covalently bonded molecules. Such a reaction was first proposed by Ziegler and Wollschitt [59], who studied the behavior of triphenyl-methylchloride in liquid sulfur dioxide. The recent status of that subject is reviewed by Monk [60] and by Gutmann [12].

The effect of pairing and aggregation on chemical reactivity of ionic species provides further evidence supporting the concept of ion pairs and other aggregates as independent chemical species. It will be shown that the reactivity of free ions, loose ion-pairs, tight ion-pairs, and so on may differ greatly. A change in the aggregation of ionic species often profoundly affects the rate and stereochemistry of reactions as well as the position of the relevant equilibria. These topics are thoroughly reviewed in Volume II of this book where the effect of pairing and aggregation on proton-transfer, electron transfer, ionic polymerization, and so on, will be discussed.

REFERENCES

1. M. Born, *Z. Physik*, **1**, 45 (1920).
2. L. Pauling, *The Nature of the Chemical Bond*, 3rd ed., Cornell University Press, Ithaca, N.Y., 1960.
3. B. S. Gourary and F. J. Adrian, *Solid State Physics*, **1k**, 127 (1960).
4. (a) R. H. Stokes, *J. Am. Chem. Soc.* **86**, 979 (1964); (b) D. V. S. Jain, *Indian J. Chem.*, **4**, 466 (1965). See, however, Y. C. Wu and H. L. Friedman, *J. Phys. Chem.*, **70**, 501 (1966).
5. (a) E. Glueckauf, *Trans. Faraday Soc.*, **60**, 572 (1964); see also, (b) W. M. Latimer, *J. Chem. Phys.*, **23**, 90 (1955); (c) U. Sen and J. W. Cobble, forthcoming.
6. J. D. Bernal and R. H. Fowler, *J. Chem. Phys.*, **1**, 515 (1933).
7. D. D. Eley and M. G. Evans, *Trans. Faraday Soc.*, **34**, 1093 (1939).
8. R. A. Robinson, and R. H. Stokes, *Electrolyte Solutions*, 2nd ed., Academic Press, New York, 1959.
9. See, e.g., R. M. Fuoss and F. Accascina, *Electrolytic Conductance*, Interscience, New York, 1959.
10. (a) D. N. Bhattacharyya, C. L. Lee, J. Smid, and M. Szwarc, *J. Phys. Chem.*, **69**, 608 (1965); (b) C. Carvajal, K. J. Tölle, J. Smid, and M. Szwarc, *J. Am. Chem. Soc.*, **87**, 5548 (1965).
11. (a) E. G. Bloor and R. G. Kidd, *Can. J. Chem.*, **46**, 3425 (1968); (b) J. L. Wuepper and A. I. Popov; *J. Am. Chem. Soc.*, **91**, 4352 (1969); (c) R. H. Erlich, E. Roach, and A. I. Popov, *J. Am. Chem. Soc.*, **92**, 4989 (1970); (d) R. H. Erlich and A. I. Popov, *J. Am. Chem. Soc.*, **93**, 5620 (1971); (e) M. K. Wong, W. J. McKinney, and A. I. Popov; *J. Phys. Chem.*, **75**, 56 (1971).
12. V. Gutmann, *Coordination Chemistry in non-Aqueous Solutions*, Springer-Verlag, Berlin, 1968.

13. M. Shinohara, J. Smid, and M. Szwarc, *J. Am. Chem. Soc.*, **90**, 2175 (1968).

14. N. Bjerrum, *Kgt. dauske Vidensk. Selsk.*, **7**, 9 (1926).

15. R. M. Fuoss and C. A. Kraus, *J. Am. Chem. Soc.*, **55**, 476 (1933).

16. M. Szwarc, *Carbanions, Living Polymers and Electron Transfer Processes*, Interscience, New York, 1968, Chapter V, pp. 218–225.

17. (a) J. T. Denison and J. B. Ramsey, *J. Am. Chem. Soc.*, **77**, 2615 (1955); (b) Y. H. Inami, H. K. Bodeusch, and J. B. Ramsey, *J. Am. Chem. Soc.*, **83**, 4745 (1961).

18. W. R. Gilkerson, *J. Chem. Phys.* **25**, 1199 (1956).

19. J. G. Kirkwood, *J. Chem. Phys.*, **18**, 380 (1950).

20. F. Accascina, A. D'Aprano, and R. M. Fuoss, *J. Am. Chem. Soc.*, **81**, 1058 (1959).

21. E. Grunwald and E. Price, *J. Am. Chem. Soc.*, **86**, 4517 (1964).

22. R. M. Noyes, *J. Am. Chem. Soc.*, **84**, 513 (1962).

23. R. L. Kay, C. Zawoyski, and D. Fennell-Evans, *J. Phys. Chem.*, **69**, 4208 (1965).

24. R. L. Kay and D. Fennell-Evans, *J. Phys. Chem.*, **69**, 4216 (1965).

25. H. A. Scheraga, S. J. Leach, and R. A. Scott, *Disc. Faraday Soc.*, **40**, 268 (1965).

26. E. Grison, K. Eriks, and J. L. de Vries, *Act. Cryst.* **3**, 290 (1950).

27. (a) J. G. Kloosterboer, L. J. Giling, R. P. H. Rettschnick, and J. D. W. van Voorst, *Chem. Phys. Letters*, **8**, 457 (1971); (b) L. J. Giling, J. G. Kloosterboer, R. P. H. Rettschnick, and J. D. W. van Voorst, *Chem. Phys. Letters*, **8**, 462 (1971).

28. M. Fisher, G. Rämme, S. Claesson, and M. Szwarc, *Chem. Phys. Letters*, **9**, 309 (1971).

29. M. Fisher, G. Rämme, S. Claesson, and M. Szwarc, unpublished results.

30. H. Sadek and R. M. Fuoss, *J. Am. Chem. Soc.*, **76**, 5897, 5905 (1954).

31. S. Winstein, E. Clippinger, A. H. Fainberg, and G. C. Robinson, *J. Am. Chem. Soc.*, **76**, 2597 (1954).

32. (a) H. Diebler and M. Eigen, *Z. phys. Chem. (Frankfurt)*, **20**, 299 (1959); (b) M. Eigen and K. Tamm, *Z. Elektrochem.*, **66**, 107 (1962).

33. R. C. Roberts and M. Szwarc, *J. Am. Chem. Soc.*, **87**, 5542 (1962).

34. E. Grunwald, *Anal. Chem.*, **26**, 1696 (1954).

35. P. Chang, R. V. Slates, and M. Szwarc, *J. Phys. Chem.*, **70**, 3180 (1966).

36. R. V. Slates and M. Szwarc, *J. Am. Chem. Soc.*, **89**, 6043 (1967).

37. G. W. Canters, E. de Boer, B. M. P. Hendriks, and A. H. K. Klaassen, *Proc. Coll. Ampere XV*, North-Holland, Amsterdam, 1969, p. 242.

38. M. Szwarc, ref. 16, pp. 263–264.

39. L. Lee, R. Adams, J. Jagur-Grodzinski, and M. Szwarc, *J. Am. Chem. Soc.*, **93**, 4149 (1971).

40. K. Höfelmann, J. Jagur-Grodzinski, and M. Szwarc, *J. Am. Chem. Soc.*, **91**, 4645 (1969).

41. M. Szwarc, *Science*, **170**, 23 (1970).

42. Y. Karasawa, G. Levin, and M. Szwarc, *J. Am. Chem. Soc.*, **93**, 4614 (1971); *Proc. Roy. Soc.*, in press.

43. M. Shinohara, J. Smid, and M. Szwarc, *Chem. Communications, London*, **1969**, 1232.

44. D. Nicholls, C. Sutphen, and M. Szwarc, *J. Phys. Chem.*, **72**, 1021 (1968).

45. T. Shimomura, K. J. Tölle, J. Smid, and M. Szwarc, *J. Am. Chem. Soc.*, **89**, 796 (1967).

46. T. Shimomura, J. Smid, and M. Szwarc, *J. Am. Chem. Soc.*, **89**, 5743 (1967).

47. T. E. Hogen-Esch and J. Smid, *J. Am. Chem. Soc.*, **88**, 307 (1966).
48. N. Hirota, *J. Am. Chem. Soc.*, **90**, 3603 (1968).
49. M. Shinohara, unpublished results.
50. R. M. Fuoss and C. A. Kraus, *J. Am. Chem. Soc.*, **55**, 2387 (1933).
51. C. B. Wooster, *J. Am. Chem. Soc.*, **60**, 1609 (1938).
52. M. Dole, *Trans. Electrochem. Soc.*, **77**, 385 (1940).
53. D. N. Bhattacharyya, J. Smid, and M. Szwarc, *J. Am. Chem. Soc.*, **86**, 5024 (1964).
54. K. Maruyama, *Bull. Chem. Soc. Japan*, **37**, 553 (1964).
55. T. L. Staples and M. Szwarc, *J. Am. Chem. Soc.*, **92**, 5022 (1970).
56. G. Levin, J. Jagur-Grodzinski, and M. Szwarc, *J. Am. Chem. Soc.*, **92**, 2268 (1970).
57. S. Bywater, *Fortschr. Hochpolimer Forsch.*, **4**, 66 (1965).
58. E. Steiner, *Ind. Eng. Chem.* (*Fundamentals*), **9**, 334 (1970).
59. K. Ziegler and M. Wollschitt, *Ann.*, **90**, 479 (1930).
60. C. B. Monk, *Electrolytic Dissociation*, Academic Press, New York, 1961.

2

Ions and Ion-Solvent Molecule Interactions in the Gas Phase

PAUL KEBARLE

Chemistry Department, University of Alberta, Edmonton, Canada

1. ENERGETICS OF HETEROLYTIC ORGANIC REACTIONS FROM PHYSICAL MEASUREMENTS IN THE GAS PHASE

1.1. Introduction

The system of structure-reactivity correlations which represents the body of physical organic chemistry is based primarily on studies of organic reactions in solution. Many of the rules and correlations depend on arguments about the energetics of the reactants, electronic stabilization or destabilization of the activated complex, reaction intermediates, and reaction products. Since the solvent plays an important part in heterolytic reactions by stabilizing the ions, ion pairs, or incipient ions by solvation, it is at times difficult to distinguish whether an increase of reactivity by a given substituent is due to an inherent electronic stabilization of the activated complex or to a favorable change of the solvation energy. In such cases theoretical or semi-empirical arguments, which are not always simple or reliable, must be used. It would be clearly a great advantage if the energies of organic ions, involved as reaction intermediates, were known from gas phase experiments. It would also be an advantage if some of the reactions occurring in solution could be executed in the gas phase as well. Finally, the most attractive possibility would be the study of these reactions in a "semidispersed phase," that is, under conditions where only one or a small and controlled number of solvent molecules surround the reactants.

Substantial advances have been reported since 1960 in the studies of the energetics of gaseous ions and their reactions and partial solvation in the gas phase. In this chapter we review first the energetics—the thermochemistry—of gaseous ions, and then the reported studies of ionic solvation in the gas phase. However, before turning to that task, let us consider one example of how the available gas phase data may be utilized.

The energy change in the gas phase dissociation reaction (1) may be considered as a measure of gas phase acidity:

$$HA = H^+ + A^- \tag{1}$$

This reaction is without counterpart in solution since in a solvent the proton is always accepted by a base. Therefore, the dissociation of an acid in solution should be compared with the gas phase acid base reaction, 2a or 2b:

$$HA + HA = H_2A^+ + A^- \tag{2a}$$

$$HA + B = HB^+ + A^- \tag{2b}$$

The energy changes associated with the gas phase reactions (1) and (2) are obtained from the heats of formation of the reactants. To evaluate the heats of formation of the ionic reactants we need to know bond dissociation energies, ionization potentials, electron affinities, and proton affinities—quantities which are discussed in Section 1.

For example, to calculate the enthalpy changes for the gaseous dissociation of water (3) and methanol (4),

$$2H_2O \rightarrow H_3O^+ + OH^- \tag{3}$$

$$2CH_3OH \rightarrow CH_3OH_2^+ + CH_3O^- \tag{4}$$

we need to know the proton affinities of water and methanol, the electron affinities of OH and CH_3O radicals, and the bond dissociation energies $D(HO\!-\!H)$ and $D(CH_3O\!-\!H)$. The results are instructive. While both reactions are endothermic, it is found that (4) is about 20 kcal/mole less endothermic than (3). This shows that in the gas phase the self-dissociation of methanol is more favorable than that of water. This result could have been expected as a consequence of the stabilizing effect of the methyl group on the $CH_3OH_2^+$ ion. From the standpoint of solution chemistry these energetics appear strange since, as is well known, the self-dissociation of liquid water is more extensive than that of methanol. The results of studies of gas phase ion solvation, described in Section 2, explain this strong solvent effect. They show the gradual change in the energetics of the "solvation" Reactions 5 and 6:

$$H^+ + nH_2O = H^+(H_2O)_n \tag{5}$$

$$H^+ + nCH_3OH = H^+(CH_3OH)_n \tag{6}$$

when more than one molecule of "solvent" is involved. It is found that for small n Reaction 6 is indeed more favorable than Reaction 5. However, the heat of proton's solvation by consecutive "solvent" molecules falls off more rapidly for methanol than for water. At $n \approx 9$ water and methanol solvate equally well and for $n > 9$ water becomes the better solvent. The changes of energetics in the stepwise solvation of the corresponding negative ions,

$$OH^- + nH_2O = OH^-(H_2O)_n \tag{7}$$

$$CH_3O^- + nH_2O = CH_3O^-(CH_3OH)_n \tag{8}$$

have not been studied yet, although such studies are now possible. Thus it is within the power of present technique to determine not only the energetics of the reactions involving the bare ions but also to provide thermal data for the reactions in which a variable number of solvent molecules become attached to the ion. This forms a bridge linking the behavior of gaseous ions with their behavior in solutions.

1.2. Heats of Formation of Positive Molecular Ions. Ionization Potentials

In principle, all the important thermochemical information on reaction of ions in the gas phase is given by the heats of their formation. When these are known the energy change in a given ionic reaction can be evaluated by subtracting the enthalpies of formation of the products from the enthalpies of formation of the reactants. Fortunately, a considerable volume of data on heats of formation of gaseous ions has become available in the past decades, and the quality of these data has been steadily improving. Furthermore, due to the application of advanced methods the amount of accurate data is expected to increase rapidly in the near future.

The most recent and complete compilation of heats of formation of positive ions is available in the monograph *Ionization Potentials, Appearance Potentials and Heats of Formation of Gaseous Positive Ions* prepared by Franklin et al. [1] Another very useful monograph which provides data on bond energies, ionization potentials, and electron and proton affinities was published by Vedeneyev [2]. Finally, a compilation by Blaunstein and Christophorou [92] was published in 1971.

The heats of formation of positive molecular ions in their ground state are computed by adding the heat of formation of the pertinent molecule to its lowest ionization potential:

$$\Delta H_f(M^+) = \Delta H_f(M) + I_p(M)$$

The heats of formation of many neutral molecules are available in standard compilations [3]. The ionization and appearance potentials of some thousand species have been determined, hence reliable heats of formation of several hundred organic molecular ions are now available [1]. The ionization potentials and heats of formation of some representative molecules and corresponding molecular ions are given in Table 1. The methods by which the ionization potentials were determined are described in Section 1.3.

The ionization potential represents the energy of a given orbital. The lowest ionization potential gives the energy of the most weakly bonded outer electron. Inner ionization potentials give the energies of inner orbitals, or orbitals of lower energy. The values given in Table 1 correspond

Table 1 Ionization Potentials[a] of Some Representative Molecules M and Heats of Formation[b] of the Ions M^+ [c]

n-Paraffins	I_p	ΔH_f	i-Paraffins	I_p	ΔH_f	cyclo-Paraffins	I_p	ΔH_f	Olefins	I_p	ΔH_f	Diolefins	I_p	ΔH_f	Acetylenes	I_p	ΔH_f
CH_4	12.7[a]	275[b]	iC_4H_{10}	10.57	212	cycloprop.	10.1	245	C_2H_4	10.4	253	C_3H_4	10.16	280	C_2H_2	11.4	317
C_2H_6	11.5	245	neo Pent.	10.35	199	cyclobut.	10.05	250	C_3H_6	9.74	229	$1,3\text{-}C_4H_6$	9.07	236	C_3H_4	10.36	283
C_3H_8	11.14	231				cyclopent.	10.53	224	$1\text{-}C_4H_8$	9.58	221	$1,3\text{-}C_5H_8$	8.68	219	$1\text{-}C_4H_6$	10.18	274
C_4H_{10}	10.63	215							$1\text{-}C_5H_{10}$	9.5	214						
C_5H_{12}	10.35	204															

Alkyl Chlorides	I_p	ΔH_f	Alkyl Bromides	I_p	ΔH_f	Alkyl Alcohols	I_p	ΔH_f	Ethers	I_p	ΔH_f	Alkyl Amines	I_p	ΔH_f	Thio Alkyls	I_p	ΔH_f
HCl	12.7	272	HBr	11.62	255	HOH	12.6	233	HOH	12.6	233	NH_3	10.15	223	H_2S	10.4	235
CH_3Cl	11.22	239	CH_3Br	10.52	234	CH_3OH	10.85	202	$(CH_3)_2O$	10.0	186	CH_3NH_2	8.97	201	CH_3SH	9.4	212
C_2H_5Cl	10.97	226	C_2H_5Br	10.29	222	C_2H_5OH	10.48	185	$(C_2H_5)_2O$	9.53	161	$C_2H_5NH_2$	8.86	193	C_2H_5SH	9.28	202
$n\text{-}C_3H_7Cl$	10.82	218	$n\text{-}C_3H_7Br$	10.18	216	C_3H_7OH	10.2	172				$(CH_3)_2NH$	8.24	186	$(CH_3)_2S$	8.7	191
												$(CH_3)_3N$	8.12	175			

Aldehydes	I_p	ΔH_f	Ketones	I_p	ΔH_f	Acids-Esters	I_p	ΔH_f	Aromatics	I_p	ΔH_f
CH_2O	10.9	223	$(CH_3)_2CO$	10.0	186	CH_3COOH	10.35	135	Benzene	9.24	233
CH_3CHO	10.2	196	$(C_2H_5)_2CO$	9.32	153	C_2H_5COOH	10.24	127	Toluene	8.8	215
C_2H_5CHO	9.98	181				CH_3COOCH_3	10.2	138	Naphthalene	8.12	220
									Biphenyl	8.27	230
									Anthracene	7.55	228

[a] Ionization potentials in electron volts.
[b] Heats of formation in kcal/mole.
[c] All values are taken from reference 1.

to the lowest ionization potentials leading to the ground state ions with little or no excess vibrational energy. Some simple rules can be established by a quick examination of the table. The ionization potential of paraffins decreases with increasing number of carbon atoms. The ionization potentials reach a low limit of about 10.3 eV, which virtually does not change with further increase of the carbon skeleton. The ionization potentials of olefins are lower than those of paraffins, reflecting the higher energy of the π-orbitals as compared with σ-orbitals. The ionization potentials of conjugated dienes are still lower. The ionization potentials of acetylenes and allenes are higher than those of the olefins. The ionization potentials of the sub-stituted alkyls increase in the following order: $RNH_2 < RSH < ROCH_3 < RBr < ROH < RCl < RF$. Since in these compounds a nonbonding p electron is removed, this order is representative of the energy of the p-orbitals in the corresponding molecules. The presence of alkyl groups leads to a stabilization of the positive ion and thus to a lowering of the ionization potential. This effect is easily noticed on inspection of Table 1. The ionization potentials of aromatic compounds are low and continue to decrease with increase of the aromatic skeleton.

1.3. Experimental Methods for the Determination of Ionization Potentials

The following experimental methods have been used for the determination of ionization potentials of atoms and molecules: optical spectroscopy, electron impact, photoionization, and photoelectron spectroscopy. These methods are described briefly here; more detailed reviews are to be found in references 4 to 6. An excellent review of the recent work utilizing electron impact and photoelectron spectroscopy has been prepared by Berry [7]. Modern studies of interactions of electrons and photons with particles has been described in Volume 7 of *Methods of Experimental Physics* [8].

1.3.1. *Optical Spectroscopy*

Spectroscopy provides the basic method for determining the ionization potential of atoms [9]. In it the ionization potential is deduced from the limit of Rydberg type series of the optical spectrum of the relevant neutral atom. In the simple cases, like those of the alkali atoms which have only one electron in the outer shell, the spectrum contains a well-developed Rydberg series. The series clearly converges to the ionization limit and allows an easy and accurate determination of the ionization potential. This is done by fitting the observed frequencies into Eq. 9 where a and b are constants characteristic of the atoms and n takes on integral values representing different Rydberg lines.

$$v = v_\infty - \frac{a}{(n + b)^2} \tag{9}$$

Thus hv_∞ is equal to the ionization energy. For atoms with a large number of electrons in the outer shell such simple series are not available. Although the interpretation of their spectra is more involved, the difficulties have been resolved so that accurate spectroscopic ionization potentials are available now for all atoms.

In spectroscopic determinations of the ionization potentials of molecules [5] we again rely on the detection of Rydberg type series. The electric field in molecules is generally far from spherically symmetric, and therefore the occurrence of Rydberg series might appear surprising. However, the electron in the highly excited electronic levels is at a considerable distance from the molecule and hence under the influence of an approximately spherically symmetrical, hydrogenlike electric field. Thus the energies of the levels near the ionization limit fit a Rydberg type series.

Transfer of energy to molecules may lead not only to ionization but also to vibrational excitation. Therefore it must be established by analysis of the band systems that the levels selected to fit the series formula correspond to the vibrational ground state of the Rydberg states. Only then does the series limit lead to the adiabatic ionization potential corresponding to the transition from the zero vibrational state of the molecule to the zero vibrational state of the ion. Difficulties in unravelling the often very complicated electronic band spectra have restricted the application of the spectroscopic method to relatively simple molecules in which the electron generally comes from a π- or a nonbonding p-orbital.

1.3.2. *Electron-Impact Technique*

The electron-impact method was the first to be applied for the determination of the ionization potential of virtually any molecule which forms a stable ion, that is, any molecule that does not fall apart after ionization. The principle of this method is simple. The gaseous molecules of the compound under investigation are bombarded in an ionization chamber with electrons accelerated by a known potential. The electron energy is gradually increased. At a certain point (the threshold) the positive molecular ions appear as the result of the electron-impact process:

$$e + M = M^+ + 2e \tag{10}$$

The positive ion current is detected with a mass spectrometer. Further increase of the electron energy usually leads to a rapid increase of the current. The plot of the ion current versus electron energy is called the ionization efficiency curve and examination of its shape near the threshold allows the determination of the ionization potential.

The ionization threshold corresponds to the adiabatic ionization potential only if the spatial configuration of the ion in its vibrational ground state is similar to that of the molecule. This is the case when the removed electron comes from a nonbonding or weakly bonding orbital. Since the ionization process, which involves electronic motion, is of much shorter duration ($\sim 10^{-16}$ sec) than the period of nuclear vibrational motion of the nuclei ($\sim 10^{-13}$ sec), the initial configuration of the produced ion is the same as that of the ionized molecule (Frank-Condon principle). If the ejected electron was a bonding one, then the initial configuration of the resulting ions differs from that of its vibrational ground state because one or more interatomic distances are shorter than their equilibrium values. Such a state corresponds of course to vibrational excitation of the ion. Vibrational excitation is also induced if an antibonding electron is removed since then some atoms are farther apart than they should be. The examination of the ionization efficiency curve near the threshold permits us to decide whether a bonding, antibonding, or nonbonding electron was removed. Only in the last case does the ionization onset (threshold) correspond to the adiabatic ionization potential; otherwise the adiabatic Ip is lower than the onset of ionization. Transitions to higher vibrational levels of the ion increase the slope of the ionization efficiency curve (for a schematic representation see Fig. 1). When a bonding or anti-bonding electron is removed, the ionization efficiency curve shows a very gradual increase and in favorable cases, it is possible to estimate from the changes of its slope the vibrational spacings. The position of the adiabatic ionization potential may be then evaluated on the basis of the Frank-Condon principle.

Most of the early determinations of molecular ionization potentials were done under conditions where the probable error was considerably larger than the reproducibility of the measurement, which was about 0.1 eV (~ 2 kcal/mole). The electron beams used in the early studies were not monoenergetic but had a considerable thermal spread (~ 10 kcal) due to the high temperature (3000°K) of the filament from which they were ejected. The voltages between the filament and the ionization chamber did not correspond exactly to the potential difference by which the impacting electrons were accelerated; this failure arose from the potential drops on the not perfectly conducting surfaces of the electrodes (contact potentials) and potential drops caused by the presence of electric charge between the filament and the molecules to be ionized (space charge). To eliminate these errors the ionization efficiency curves of the unknown molecules and of a gas of known ionization potential (Ar, Kr, or Xe) were determined in an experiment in which both species were present simultaneously in the ion source. By a

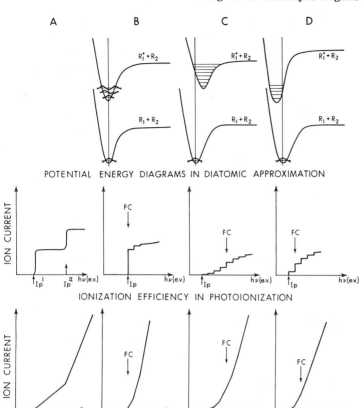

Figure 1. Schematic representation of ionization efficiency curves in photoionization and electron impact. The ion current is proportional to the ionization efficiency. (A) Ionization of an atom leading to the ground state and first electronic excited state of the atomic ion. (B) Ionization of molecule R_1R_2. Geometry of ion $R_1R_2^+$ nearly identical to geometry of the parent molecule. Frank-Condon region of highest transition probability, FC, coincides with adiabatic ionization potential I_p. (C) Ionization involves a bonding electron such that bond distance in the ion is larger than in the present molecule. I_p lower than FC. (D) Ionization involves an antibonding electron such that bond distance in the ion is shorter than in the molecule. I_p lower than FC.

comparison of the two curves, the shifts due to the electron energy spread, contact potential, and space charge could be eliminated. However, this procedure is accurate only if the ionization efficiency curves in the threshold region of the unknown compound and the calibrating gas correspond to transitions to a single isolated energy level. This, however, is practically never the case. Thus, if the ionization of a molecule is observed, ionization in the threshold region leads to various vibrational levels of the molecular ion. Closely lying electronic excited states of the ion of the calibrating gas or of the investigated species introduce further structure in each ion efficiency curve. Since this structure is smeared out by the electron energy spread, a meaningful "comparison" of the ionization efficiency curves of the calibrating and investigated gas cannot be made. Hence the errors depend on the particular pair of investigated substances and on the empirical or semi-empirical procedure by which the ionization efficiency curves were compared [7, 10]. In general, the ionization potentials determined in this manner are higher than the pertinent adiabatic ionization potentials by about 0.1 to 0.3 eV.

The difficulties just described led several research groups to attempt to reduce the electron energy spread. Two principles have been used. The first method, known as the Retarding Potential Difference Method or RPD [11], utilizes a set of electric grids. One of the grids is at a negative (retarding) voltage, V_r, with respect to the electron-producing filament and lets through only the electrons having initial thermal kinetic energies in excess of the retarding voltage. These electrons are then accelerated to the desired ionizing voltage V and conducted into the ion source. The change of ion current arising from a small decrease ΔV_r of the retarding voltage (at constant V) permits us to determine the ionization due to electrons with energy V and a narrow thermal energy spread ΔV_r.

A proper electron energy selector is more advantageous. For example, energy selection can be achieved by means of the electric field produced by the sector of a cylindrical capacitor. The electrons entering the radial field separate according to their velocities; the faster the electrons, the bigger the radii of their trajectories. An electron beam with a narrow spread of energies is then obtained by means of a slit placed at a suitable position on the output end of the selector. The first successful studies utilizing an electron monochromator of this type were reported by Clarke [12] and by Kerwin and Marmet [13]. Considerable improvements in the technique were achieved later so that it is possible now to use monochromatic electrons with an energy spread as narrow as 0.01 eV [6]. The ionization efficiency curves obtained with such electron beams often show clearly the structure due to transitions to vibrationally excited levels of the electronic ground state of the ion.

1.3.3. *Photoionization*

The photoionization method is capable of giving highly accurate ionization potentials. Photons of a given energy are obtained from a suitable light source by means of a monochromator. Molecules of the compound to be investigated are then irradiated in the ion source of a mass spectrometer. When the energy of the photons becomes sufficiently great the molecules are ionized and the appearance of positive ions M^+, produced by the photo-ionization process (11), is observed

$$h\nu + M = M^+ + e \qquad (11)$$

with the mass spectrometer. On further increase of the photon energy the current due to M^+ ions increases too, and thus a photoionization efficiency curve is recorded. Its analysis at and above the threshold can lead to the determination of the ionization potential. A comparison of the ionization efficiency curves obtained by electron and photon impact techniques is given in Fig. 1.

The mass spectrometric photoionization technique was first developed by Hürzeler, Inghram, and Morrison [14]. Since then a large number of photoionization studies have been carried out by several groups [15, 16] and consequently the number of ionization potentials determined by photoionization amounts presently to 10% of those determined by electron impact. The relative importance of the photoionization technique is expected to increase in the coming years, because it is possible to produce photon beams of extremely narrow energy spread and therefore a much better energy resolution can be obtained by photoionization than by the electron-impact technique. The photoionization efficiency curves are steeper than those produced by electron impact (see Fig. 1), and therefore the threshold potential is determined more accurately by the former than by the latter.

In spite of the high resolution achieved with the photon-impact method and the adequate resolution obtained with electron impact, the adiabatic ionization potentials of many molecules cannot be determined satisfactorily by either method. This is the case when the configuration of the ion differs considerably from that of the molecule. The observed photoionization efficiency curves increase then very gradually, the threshold is very poorly defined, and the vibrational structure is smeared out. The threshold gives then the upper limit of the adiabatic Ip, its actual value probably being by some tenths of electron volts lower than the extrapolated threshold.

1.4. Inner Ionization Potentials, Heats of Formation of Electronically Excited Ions

Study of the ionization efficiency curves above the ionization threshold may lead to the determination of the energies of ions in electronically excited states. The energies of these states, relative to the ground state of the molecule, are called inner ionization potentials. The onset of transitions to such states appears as an increase in the slope of the electron-impact ionization efficiency curve and as a higher step in the photoionization curve (see Fig. 1). Thus inner ionization potentials of many important molecules have been determined by examination of breaks in the electron and photon impact curves

However, the occurrence of autoionization processes:

$$hv + M \rightarrow M^* \xrightarrow{M} M^+ + M + e$$

$$e + M \rightarrow M^* + e \xrightarrow{M} M^+ + M + 2e \qquad (12)$$

leads to the appearance of spikes and bumps on the ionization efficiency curves. These are sometimes so numerous that they obscure the underlying structure. An autoionization may be mistaken for an upward break or step due to a transition to a new electronic level and thus lead to an incorrect assignment of nonexisting electronic states.

Best suited for the determination of inner ionization potentials is the photoelectron spectroscopy method introduced by Turner [17], who described his results and their significance in two review articles [18, 19] addressed to nonspecialists; recent work is discussed by Berry [7].

In the photoelectron method the molecules are ionized by a collimated beam of photons of known energy that is higher than the inner ionization potential which is to be determined. The energy of the electrons emitted by the photoionization process is determined by means of suitable retarding grids or by magnetic deflection analysis. Thus a plot referred to as the photoelectron spectrum is obtained by relating the abundance of photoelectrons (electron current) versus their energy, e_k. The transition to higher energy levels of the ion gives rise to steps or peaks in the photoelectron spectrum. The energy of the transition, I (the relevant ionization potential), is related to the kinetic energy, e_k, of the electron at the onset of such steps by equation (13).

$$I = hv - e_k \qquad (13)$$

The occurrence of autoionization does not affect the photoelectron spectrum since the photoelectrons have energies defined by the state of the photoionized ion M^+ which are independent of the path by which M^+ was produced, either direct photoionization or through an autoionization. The photoelectron spectroscopy is thus ideally suited for the determination of inner ionization potentials and is being rapidly applied to a number of organic molecules of current interest [6].

1.5. Appearance Potentials and Heats of Formation of Ionized Radicals

The appearance of fragment ions R_1^+ following the ionization of the molecule R_1R_2 by electrons or photons (Eq. 14) can be observed with a mass spectrometer. The appearance potential abbreviated as $Ap(R_1^+, R_1 \cdot R_2)$ is defined as the lowest electron energy at which the fragment R_1^+ is observed when the molecule $R_1 \cdot R_2$ is bombarded with ionizing radiation of increasing energy. If both fragments R_1^+ and R_2 contain no excess energy (internal or kinetic), then the appearance potential is equal to the energy required to produce R_1^+ and R_2 from the molecule R_1R_2 and then the following simple relationships are valid:

$$Ap(R_1^+, R_1 \cdot R_2) = D(R_1 - R_2) + Ip(R_1 \cdot) \tag{15}$$

$$Ap(R_1^+, R_1 \cdot R_2) = \Delta H_f(R_1^+) + \Delta H_f(R_2 \cdot) - \Delta H_f(R_1 \cdot R_2) \tag{16}$$

$D(R_1 - R_2)$ is the bond dissociation energy. In such a case the determination of the appearance potential gives the ionization potential of the radical R_1 or the bond dissociation energy $D(R_1 - R_2)$, provided one of these quantities is already known. If both are unknown, then it is possible, by combining two carefully chosen appearance potentials, to eliminate one of the unknowns and obtain the thermochemical quantity of interest. This is illustrated in Eqs. 17 to 19. Consider the processes 17 and 18 and the Ap equations associated with them. Subtracting Eq. 17 from 18,

$$e + RH = R^+ + H + 2e; \quad Ap(R^+, RH) = \Delta H_f(R^+) + \Delta H_f(H)$$
$$- \Delta H_f(RH) \tag{17}$$

$$e + R \cdot R = R^+ + R + 2e; \quad Ap(R^+, RR) = \Delta H_f(R^+) + \Delta H_f(R)$$
$$- \Delta H_f(RR) \tag{18}$$

we obtain Relation 19.

$$Ap(R^+, RR) - Ap(R^+, RH) = \Delta H_f(R) - \Delta H_f(RR) - \Delta H_f(H)$$
$$+ \Delta H_f(RH) \tag{19}$$

Since $\Delta H_f(H)$ is known and the heats of formation of the neutral molecules usually are also known, we can calculate $\Delta H_f(R)$ from the two appearance potentials. After finding $\Delta H_f(R)$ it is easy to evaluate $D(R - R)$, $D(R - H)$, $\Delta H_f(R^+)$, and $Ip(R)$ using Equations 20 to 24. The determination of an appearance potential is not time consuming. If the compound and an instrument are available such a task can be completed in twenty minutes or at most in one afternoon. Thus a great wealth of thermochemical data can be accumulated in a relatively short time, provided the products do not contain excess energy.

$$D(R - R) = 2\Delta H_f(R) - \Delta H_f(RR) \tag{20}$$

$$D(R - H) = \Delta H_f(R) + \Delta H_f(H) - \Delta H_f(RH) \tag{21}$$

$$\Delta H_f(R^+) = Ap(R^+, RH) + \Delta H_f(RH) - \Delta H_f(H) \tag{22}$$

$$\Delta H_f(R^+) = Ap(R^+, RR) + \Delta H_f(RR) - \Delta H_f(R) \tag{23}$$

$$Ip(R) = \Delta H_f(R^+) - \Delta H_f(R) \tag{24}$$

There are some theoretical reasons to believe that the appearance potential of ionic fragments resulting from the dissociation of fairly large polyatomic ions (more than 6 atoms) involve only little excess energy. Therefore such data are thermochemically useful [5b]. Indeed let us assume that the molecular ion $R_1R_2^+$, created by electron or photon impact, contains an internal energy E induced by the ionization process. The dissociation into $R_1^+ + R_2$ may occur whenever this energy is larger than the minimum energy $E_0 = D(R_1^+ - R_2)$ required to dissociate the ion. The energy E imparted during the ionization process may be in the form of electronic as well as vibrational excitation. However, if the $R_1R_2^+$ ion is sufficiently complex, it has many electronic states and subsequently many intercrossings of potential energy. Therefore the electronic excitation is expected to be rapidly converted into inner energy (vibration and internal rotation) of the electronic ground state of the ion. According to the quasi-equilibrium theory of mass spectra [5b] the internal energy E distributes itself with a certain statistical probability into the various vibrations and inner rotations of the electronic ground state. Fluctuation of the energy distribution eventually leads to cumulation of sufficient energy into that particular mode, which causes the dissociation into $R_1^+ + R_2$. According to the theory the time required for these processes leading to dissociation decreases rapidly with increasing excess of internal energy $E - E_0$. It is found that "average" lifetime of an ion is substantially shorter than 1 μsec when the difference $E - E_0$ amounts to a few kcal/mole only. The residence time of ions in the ion source is about 1 μsec. Thus appearance potentials of fragments should yield heats of formation that are higher than the true values by only a few kcal/mole. Unfortunately this is not always the case. The conversion of electronic energy may be slow in some cases when compared with a direct dissociation of the electronically and vibrationally excited state of $R_1R_2^+$ into $R_1^+ + R_2$. This would mean that considerable excitation may be present either in R_1^+ or in R_2. The thermal fragments R_1^+ and R_2, if at all produced at lower ionization energy, may be of so much lower relative abundance that they are not detected. Therefore it is always desirable to check whether or not the appearance potential involves appreciable excess energy.

In a most common check the heat of formation of a given radical ion R_1^+ is calculated from as many appearance potentials as possible, each obtained by mass spectrographic study of a different compound, for example, R_1H, R_1R_2, R_1X. Thus, in the example used here, Eqs. 22 and 23 should lead to the same $\Delta H_f(R_1^+)$ if no excess energy is involved in both cases. By collecting such data it is frequently possible to find the same (and lowest) $\Delta H_f(R_1^+)$ from several appearance potentials. It is safe to assume that such lowest $\Delta H_f(R_1^+)$ represents its true value and the other higher "$\Delta H_f(R_1^+)$" were derived from processes involving excess energy. This thermochemical consistency method was systematically applied by Field and Franklin [4].

A technique utilizing photoelectron spectroscopy (absence of interference from autoionization) in which the appearance potentials of pertinent ionic fragments is determined, has been developed by Brehm and Puttkamer [19]. It involves mass spectrometric analysis of the ionic fragments and simultaneous analysis of the energy of the photo-emitted electrons. The simultaneous studies of ions and electrons created in the same event is achieved by introducing a coincidence circuit into the apparatus and by limiting the studies to low ionization rates. Further application of this technique promises to provide highly reliable values of appearance potentials and inner ionization potentials.

The absence of inner excitation energy in the neutral fragment R_2 resulting from the dissociation of $R_1R_2^+$ can be established by a method developed by Beck, Osberghaus, and Niehaus [22]. The neutral fragments R_2 produced by electron-impact ionization near the appearance potential of R_1^+ are ionized by a second electron beam. Variation of the electron energy of the second beam allows the determination of the appearance potential of the fragment R_2. If its value is identical with the adiabatic ionization potential of R_2 (which has to be known), then the fragment R_2 does not possess any appreciable internal energy. The method has been applied to few systems. The technique is difficult since the abundance of neutral R_2 is very low, particularly when working near the appearance potential.

The kinetic energies of the fragments R_1^+ and R_2 resulting from the dissociation of $R_1R_2^+$ may be determined by applying to R_1^+ retarding fields after mass separation [20] or deflection fields before mass separation [21]. The kinetic energy of the neutral fragment and thus the total kinetic energy involved in the process can then be calculated from the law of conservation of momentum. Unfortunately, such determinations are not frequent. The observations reported so far can be summarized as follows: for fairly large molecules R_1R_2 the abundant fragments formed in a simple dissociation process $R_1R_2^+ \rightarrow R_1^+ + R_2$ occurring near the appearance potential of R_1^+ are seldom produced with significant kinetic energy. Fragments of high kinetic energy are often observed in the dissociation of diatomic, triatomic, and other very simple molecules. The fragments resulting from a succession of dissociation processes like the ion C^+ from C_2H_6 generally have a considerable kinetic energy.

In conclusion, the appearance potentials of fragments produced from fairly complex molecules in a single dissociation process usually lead to heats of formation of radical ions only slightly higher than their true values. Photoionization technique leads to the heats of formation higher by 2–3 kcal than the true ones and the electron-impact data yield values which exceed the proper ones by 6–10 kcal.

Heats of formation of a radical ion can be also determined by the direct measurement of the ionization potential of the relevant free radical usually

Table 2 Ionization Potentials of Some Representative Free Radicals R and Heats of Formation of the Corresponding Ions R^+

Radical R	Ionization Potential (eV)	Heat of Formation (kcal/mole)	Ref.
CH_3	9.85	260	a
CH_3CH_2	8.4	219	b
$CNCH_2$	10.81		c
$C_2H_5CH_2$	8.1	209	b
$CNCH_2CH_2$	9.85		c
CH_3CHCH_3	7.5	190	f
CH_3CHCN	9.76		c
Cyclopropyl	8.05	239	d
Cyclobutyl	7.88	213	d
Cyclopentyl	7.79	194	d
Cyclopentadienyl	8.69		e
Cycloheptatrienyl	6.6	217	e
benzyl	7.76	216	f
m-xylyl	7.65		g
o-xylyl	7.61		g
p-xylyl	7.46		g
m-CN benzyl	8.58		h
m-NO_2 benzyl	8.56		h
p-CN benzyl	8.36		h
m-F benzyl	8.18		h
p-Cl benzyl	7.95		h
p-F benzyl	7.78		h
β-Naphthyl methyl	7.56		h
α-Naphthyl methyl	7.35		i
Diphenyl methyl	7.32		i
CH_3	9.85	260	a
CH_2Cl	9.32		j
CH_2Br	9.3	261	j
CH_2F	9.37	209	j
$CHCl_2$	9.3	245	j
CHF_2	9.3		j
CCl_3	8.78	214	k
CN	14.5	430	l
HS	10.5	276	m
CH_3S	8.06	218	n
Phenyl S	8.63	250	o

produced by pyrolysis or photolysis of a suitable compound. This technique, which requires the knowledge of the heat of formation of the radical, has been extensively used by Lossing (see his review article in reference 6). A recent review of this subject by Harrison is published in the monograph edited by McLafferty [5a].

Although it is not feasible to list here all the available thermochemical data on organic ions, it may be desirable to quote some to illustrate the magnitudes involved and their variation with structure. Tables 1 and 2, which give ionization potentials of chosen molecules and radicals in order of decreasing ionization potential, are therefore included in this section.

1.6. Heats of Formation of Protonated Compounds. Proton Affinities

The proton affinity of compound M, called PA(M), is defined as the exothermicity of the gaseous reaction:

$$H^+ + M = MH^+ \tag{25}$$

where $-\Delta H_{25} = PA(M)$. The heat of formation of a protonated species MH^+ is given then by Eq. 26:

$$\Delta H_f(MH^+) = \Delta H_f(M) + \Delta H_f(H^+) - PA(MH^+) \tag{26}$$

References for Table 2.

[a] G. Herzberg and J. Shoosmith, *Can. J. Phys.*, 34, 523 (1956).

[b] F. A. Elder, C. Giese, B. Steiner, and M. Inghram, *J. Chem. Phys.*, 36, 3292 (1962).

[c] R. F. Pottie and F. P. Lossing, *J. Am. Chem. Soc.*, 83, 4737 (1961).

[d] R. F. Pottie, A. G. Harrison, and F. P. Lossing, *J. Am. Chem. Soc.*, 83, 3204 (1961).

[e] A. G. Harrison, A. G. Honnen, H. J. Dauben, and F. P. Lossing, *J. Am. Chem. Soc.*, 82, 5593 (1960).

[f] J. B. Farmer, I. H. S. Henderson, C. A. McDowell, and F. P. Lossing, *J. Chem. Phys.*, 22, 1948 (1954).

[g] J. B. Farmer, F. P. Lossing, D. G. H. Marsden, and C. A. McDowell, *J. Chem. Phys.*, 24, 52 (1956).

[h] A. G. Harrison, P. Kebarle, and F. P. Lossing, *J. Am. Chem. Soc.*, 83, 777 (1960).

[i] A. G. Harrison and F. P. Lossing, *J. Am. Chem. Soc.*, 82, 1052 (1960).

[j] F. P. Lossing, P. Kebarle, and J. B. de Sousa, *Adv. Mass Spectrom.*, 1, 431 (1959).

[k] J. B. Farmer, I. H. S. Henderson, F. P. Lossing, and D. H. Marsden, *J. Chem. Phys.*, 24, 348 (1956).

[l] V. H. Dibeler, R. M. Reese, J. L. Franklin, *J. Am. Chem. Soc.*, 83, 1813 (1961).

[m] T. F. Palmer and F. P. Lossing, *J. Am. Chem. Soc.*, 84, 4661 (1962).

Table 3 Proton Affinities of Selected Compounds

Compound	PA (kcal/mole)	Ref.
He	41	a
H_2	70	b
Kr	92	c
CH_4	126	d
$PA(C_3H_8) > PA(C_2H_6) >$	126	
HBr	140	d
HCl	141	d
HI	145	d
CH_3Cl	164	d
H_2O	166	e, f, g
HCHO	168	h
HCN	170	d
H_2S	174	d, f
HCOOH	179	d
$PA(CH_3COOH) > PA(HCOOH)$		i
CH_3OH	182	d, j, k
$PA(C_2H_5OH) > PA(CH_3OH)$		
CH_3CHO	185	h
CH_3SH	185	d
CH_3OCH_3	187	d
CH_3COCH_3	188	d
PH_3	193	l
NH_3	207	e
CH_3NH_2	211	d
$PA[(CH_3)_2NH]$	>211	
LiOH	241	m
NaOH	248	m
KOH	263	m
CsOH	269	m

a A. A. Evett, *J. Chem. Phys.*, 24, 50 (1956).
b J. H. Simons, C. M. Fontana, E. E. Muschlitz, and S. R. Jackson, Jr., *J. Chem. Phys.*, 11, 312 (1943).
c O. P. Stevenson and D. O. Schissler, *J. Chem. Phys.*, 23, 1353 (1955).
d M. A. Haney and J. L. Franklin, *J. Chem. Phys.*, forthcoming.
e M. A. Haney and J. L. Franklin, *J. Chem. Phys.*, 50, 2028 (1969).
f J. L. Beauchamp and S. E. Buttrill, Jr., *J. Chem. Phys.*, 48, 1783 (1968).
g S. T. Vetchinkin, I. I. Pshenichnov, and N. B. Sokolov, *J. Phys. Chem. Moscow*, 33, 1269 (1959).

A considerable amount of energy is required to form a proton in the gas phase, $[\Delta H_f(H^+) = 366 \text{ kcal/mole}]$, because the ionization potential of the H atom is so high $[Ip(H) = 313.6 \text{ kcal/mole}]$. The hydrogen ion is thus an intensely electrophylic reagent and attaches itself strongly to any neutral or negative species, that is, the proton affinities of virtually all neutral or negative species are positive and large. The greater the ability of a compound to release an electron pair, the larger its proton affinity. Taking only neutral species under consideration, we would expect that amines, ethers, alcohols, and compounds containing carbonyl and carboxyl groups will have high proton affinities since they contain nonbonding p electrons. On similar grounds we would expect that the presence of carbon-carbon double or triple bonds should lead to high proton affinities. The proton affinity of a compound containing electron pairs which are easily donated should be further increased by the presence of electron-donating substituents. Thus we would expect that $PA(HOH) < PA(CH_3OH) < PA(CH_3OCH_3)$ because of the electron-releasing ability of the alkyl groups [24]. The proton affinities listed in Table 3 confirm the agreement with these general expectations.

1.7. Determination of Proton Affinities

The proton affinities of simple molecules like H_2 and CH_4 have been obtained by theoretical calculations. Thus Hirschfelder's [25] calculations lead to $PA(H_2) = 81$ kcal/mole, whereas a somewhat more elaborate approach by Eyring and Barker [26] gave $PA(H_2) = 77$ kcal/mole. The calculations for H_2 and other simple systems such as He are probably reliable to about 10 kcal/mole. Calculations for more complicated systems probably would be unreliable and often unfeasible. Therefore we must rely on experimental proton affinity determinations for all but the simplest molecules.

Two principle methods have been used in studies of proton affinities in the gas phase, both relying on mass spectrometric detection of the ions. The first method utilizes the appearance potentials produced by a rearrangement

References for Table 3.   Continued

[h] K. M. A. Rafacy and W. A. Chupka, *J. Chem. Phys.*, 48, 5205 (1967).
[i] E. W. Godbole and P. Kebarle, *Trans. Farad. Soc.*, 58, 1897 (1962).
[j] M. S. B. Munson, *J. Am. Chem. Soc.*, 87, 2332 (1965).
[k] P. Kebarle, R. N. Haynes, and J. G. Collins, *J. Am. Chem. Soc.*, 89, 5753 (1967).
[l] T. C. Waddington, *Trans. Farad. Soc.*, 61, 2652 (1965).
[m] S. K. Searles, I. Džidić, and P. Kebarle, *J. Am. Chem. Soc.*, 91 2810 (1969).

process of the protonated species. For example, Van Raalte and Harrison [27] have studied the appearance potential of the H_3O^+ ion observed in the mass spectra of compounds like ethanol and propyl alcohol. The H_3O^+ ion observed in the ethanol mass spectrum probably originates from the following process:

$$e + CH_3CH_2OH \rightarrow H_3O^+ + C_2H_3 \qquad (27)$$

The appearance potential of H_3O^+ is related to the minimum energy required for Reaction 27. The energy of this

$$Ap(H_3O^+) \geq D(CH_3CH_2 - OH) + D(CH_2CH_2 - H) + D(C_2H_3 - H)$$
$$+ Ip(H) - D(H - OH) - PA(H_2O) \quad (28)$$

process* is thus expressed by the following steps: dissociation of the C—OH bond in ethanol, creation of two H atoms by dissociation of two C—H bonds of the ethyl radical such that a vinyl radical and two H atoms are produced; addition of one H atom to OH, ionization of the other, and addition of the resulting proton to H_2O; values for all thermochemical quantities except the proton affinity are available. Therefore by measuring the appearance potential and assuming the equality sign (i.e., no excess energy present in the products), we can calculate a proton affinity for water. Since the process leading to the observed H_3O^+ may not be the one assumed but might involve further fragmentation of the C_2H_3 radical or considerable excess energy in the products, it is advisable to determine the appearance potential of H_3O^+ from a number of compounds and then select only the proton affinity values that are in agreement with each other. In practice the number of compounds that can be used is limited by the lack of required thermochemical data.

Alternatively, studies of ion-molecule reactions involving proton transfer,

$$M_xH^+ + M_i \rightarrow M_x + M_iH^+ \qquad (29)$$

may lead to evaluation of the required proton affinities.

The reaction conditions in conventional ion-molecule experiments are such that only fast reactions can be observed. It follows that all the observed reactions must be exothermic, which for process 29 means: $PA(M_i) > PA(M_x)$. Studies of the exchange reaction 29 involving various compounds M_x and M_i lead to a series of consecutive inequalities $PA(M_x) < PA(M_x') < \cdots < PA(M_i)$. If some of these proton affinities are known, the values of others may be estimated.

The above method was first introduced by Talroze [28] and then used

---

* Equation 28 implies the structure of C_2H_3=CH_2:CH. This need not be the case. For example, C_2H_3=CH_3—C.

extensively by Beauchamp [29], Franklin [30], and others for the determination of proton affinities of several important compounds such as H_2O, H_2S, and CH_2O. A detailed discussion of this subject is given in a recently published article by Haney and Franklin [30].

1.8. Heats of Formation of Negative Ions. Electron Affinities

The electron affinity is defined as the exothermicity of process 30.

$$e + M = M^- \tag{30}$$

The heat of formation of a negative ion can be obtained from the heat of formation of the neutral species and its electron affinity by Eq. 31.

$$\Delta H_f(M^-) = \Delta H_f(M) - EA(M) \tag{31}$$

The electron affinity of neutral species arises from the imperfect screening of the nuclear charge by the electrons present in the compound. For example, in free radicals, and in some atoms, a low-lying half-occupied orbital is available. These species therefore have higher electron affinities than neutral molecules, and the respective negative ions, such as Cl^-, NO_3^-, ClO_4^-, RCO_2^-, are stable in solution.

Singlet molecules with extensive delocalized π electron systems can have vacant orbitals which are sufficiently low to lead to a positive electron affinity. This is particularly so if carbonyl, nitro, or cyano groups are present because these stabilize the corresponding negative ion. Quinones, cyano, and nitro aromatics are examples. The resulting radical-ions have been extensively studied in solution by a variety of techniques (see, e.g., reference 33).

SF_6, BF_3, etc., form an interesting group of molecules having positive electron affinity. The negative charge of the resulting ions is probably spread out over all the halide atoms. Part of the interaction seems to be due to electrostatic attraction by the polarization of the outer atoms induced by the electron [31].

Some electron affinities are listed in Table 4, the respective species being arranged in order of decreasing electron affinity.

1.9. Determination of Electron Affinities

Early studies of electron affinities were reviewed by Pritchard [32], who also described the pertinent methods. A recent review of this field and a compilation of available electron affinities is given by Vedeneyev [2] and by Szwarc [33]. A brief description of the newer methods was reported also by Page [31].

Table 4   Electron Affinities of Selected Compounds

Radical Compound	EA (kcal/mole)	Ref.
ClO_4	133	a
ClO_3	91	a
ClO_2	79	a
ClO	64	a
Cl	83	b
F	79.5	b
Br	77.5	b
I	70.5	b
CN	88.1	b
SH	53	c
OH	42	d
CCl_3	48	e
Allyl	48	a
Benzyl	41	u
Methyl	24	f
H	17	a

Molecular Compound	EA (kcal/mole)	Ref.
Chloranil, bromanil	∼60	g
Benzoquinone	∼32.46	g
Anthracene	27	g
Benzaldehyde	10	g
Naphthalene	∼10	g

[a] N. S. Buchelnikova, *Usp. Fiz. Nauk*, **65**, 351 (1958).
[b] R. S. Berry, *Chem. Rev.*, **69**, 533 (1969).
[c] B. Steiner, *J. Chem. Phys.*, **49**, 5097 (1968).
[d] L. M. Branscomb, *Phys. Rev.*, **148**, 11 (1966).
[e] R. K. Curran, *J. Chem. Phys.*, **34**, 2007 (1961).
[f] H. O. Pritchard, *Chem. Rev.*, **52**, 529 (1953).
[g] G. R. Freeman, *Radiation Res. Rev.*, **1**, 1 (1968).
Note: References a, b, f, and g are review articles.

The most accurate and promising method for the determination of electron affinities is the electron detachment method introduced by Branscomb [34]. Negative ions created by some suitable means, for instance, by an electric discharge, are extracted from the ionization region and mass analyzed. The ion beam of the desired mass is illuminated by monochromatic light of known and gradually increasing frequency. At the photodetachment threshold production of electrons begins:

$$M^- + h\nu = M + e \qquad (32)$$

The photoelectrons are collected by suitable electrodes and registered as electron current. Analysis of the dependence of the current on the light frequency at and above the photodetachment threshold leads to the relevant electron affinity. Thus if the geometry of M^- is similar to that of the molecule M, the threshold gives the electron affinity. However, if the geometries are different, vibrational structure may be visible on the photodetachment curve. The analysis of this structure is in principle similar to that performed on the ionization efficiency curves for positive ion production (see the previous sections). The photodetachment method has been applied so far only to relatively simple species, such as O^-, OH^-, I^-, S^- [34] and SH^-, $OH^- \cdot H_2O$ [35]. Steiner intends to apply this very promising technique to a wide range of molecular negative ions (reference 8, page 195).

Another spectroscopic method for the determination of electron affinities was reported by Berry [36]. Negative ions are produced by thermal energy created in a shock tube. The photodetachment continuum is then observed spectroscopically. The experiments described dealt with the halide ions. Obviously only thermally stable species can be studied by this method.

Electron affinities can be determined by mass spectrometric measurement of the appearance potentials of negative fragments produced by electron or photon impact. The pertinent method is described by McDowell [6]. Under the influence of electron impact either dissociative electron capture (Eq. 33) or formation of a pair of ions (Eq. 34) may take place:

$$R_1R_2 + e \rightarrow R_1^- + R_2 \tag{33}$$

$$R_1R_2 + e \rightarrow (R_1R_2)^* + e \rightarrow R_1^+ + R_2^- + e \tag{34}$$

The appearance potentials are related to the electron affinities by Eqs. 33a and 34a. KE and IE denote kinetic and internal energy of the fragments.

$$Ap(R_1^-, R_1R_2) = D(R_1R_2) - EA(R_1) + KE + IE \tag{33a}$$

$$Ap(R_2^-, R_1R_2) = D(R_1R_2) + Ip(R_1) - EA(R_2) + KE + IE \tag{34a}$$

The amount of kinetic energy may be determined by momentum analysis of the ions. The presence of internal energy can be also inferred by comparison of the results with available thermochemical data. Examples of determinations of electron affinities by the electron-impact technique in which the kinetic energy analysis of the fragment ion was performed are given in reference 37. Other electron-impact determinations of electron affinities are described in references 38 and 39. On the whole, the mass spectrometric appearance potential method has been less successful in determinations of electron affinities than ionization potentials.

Two more gas phase methods should be mentioned briefly. One, known as the magnetron method, has been extensively used by Page [31]. Unfortunately this technique appears to be unreliable, judging by the cases where the electron affinities were determined also by some other methods. For example, EA(OH) = 65 kcal/mole according to Page [40], whereas EA(OH) = 42 kcal/mole according to Branscomb [34]. The major weakness of the magnetron method lies in the fact that the identity of the negative ions, whose electron affinity is measured, is simply *assumed* on the basis of the reactants used. However, this procedure is not reliable, as has been shown by Herron [41], who detected with a mass spectrometer ions different from those Page assumed to be present.

Becker and Wentworth [42a] have developed a method for determining electron affinities of stable molecules which involves studies of the thermal equilibrium:

$$c \mid M \rightleftharpoons M^-$$ (35)

The compound M is carried by an inert carrier gas (Ar). Electrons are produced by the radioactive induced ionization of argon (a tritiated foil being used as the ionizing source). The position of the equilibrium (Eq. 35) is determined by collecting the remaining free electrons by means of pulsed electric fields. The method can be used for thermally stable compounds which are electron acceptors. The larger aromatic compounds fulfill the preceding requirements and were among the first studied [42]. The identity of the negative ions is again assumed on the basis of the reactants used. However, because the system studied is simpler than that established in the magnetron, this assumption is fairly justified. Nevertheless, minute impurities, or the products of thermal reactions, may remove electrons by forming negative ions and thus interfere with the observation of the assumed equilibrium. Other shortcomings of this method are discussed in reference 33.

2. ION-SOLVENT MOLECULE INTERACTIONS IN THE GAS PHASE

2.1. Introduction

The nature of the interactions of the ion with the solvent is central to the study of ionic solutions. In assessing the effect of the solvent it is natural to consider first the strong interactions of the ion with the nearest solvent molecules and then examine the structure modifications and energy effects arising at longer distances. It is therefore desirable to have experimental methods which allow us to study the interaction of the ion with a varying number of "inner sphere" molecules in the absence of bulk of the solvent. Such experiments have the additional advantages of eliminating the ions of opposite charge whose presence generally complicates the interpretation of the data obtained from studies of solutions.

To a worker in the field of ionic solutions, studies of isolated ion-solvent molecule complexes may seem impossible at first glance. However, we need reflect only a little to realize that ion-solvent molecule "clusters" held by ion-dipole forces should exist in the gas phase and therefore they can be "prepared" and studied.

Before considering how ion-molecule clusters can be studied in the gas phase it will be worthwhile to recall some of the early work on gaseous ions. Investigations of electrical discharges in rarefied gases started about 1750 and led, through the work of Hittdorf, G. Goldstein, W. Wien, and most notably J. J. Thompson to recognition of the existence of electrons and the presence of positive and negative ions in the gas phase. Thompson, through his studies of gas discharges and the effect of electric and magnetic fields on the trajectories of the ions, was led to construct the first mass spectrograph. For various technical reasons the pressure in the ion source of Thompson's instrument was higher than that used in modern analytical instruments. At higher pressure, the primary ions created by the ionizing medium (ionization is most often obtained by electron impact) may collide with neutral molecules and react with them. A possible reaction sequence following the primary ionization of a water molecule by a fast electron is shown in reaction (a) to $(n-1, n)$.

$e + H_2O \rightarrow H_2O^+ + 2e$	primary ionization	(a)
$e + H_2O \rightarrow OH^+ + H + 2e$	primary ionization	(b)
$H_2O^+ + H_2O \rightarrow H_3O^+ + OH$	ion-molecule reaction	(c)
$OH^+ + H_2O \rightarrow H_3O^+ + O$	ion-molecule reaction	(d)
$H_3O^+ + H_2O \rightarrow H^+(H_2O)_2$	forward clustering reaction	(1, 2)
$H^+(H_2O)_{n-1} + H_2O \rightleftarrows H^+(H_2O)_n$	clustering equilibrium	$(n-1, n)$

Reactions of the type shown above did occur in Thompson's instrument (at least up to the step c) [43] and consequently the product ions of such reactions were observed by him.

Ion-molecule reactions obviously interfere with the analytical uses of the mass spectrometer. Thus, for example, the determination of isotope ratios would be greatly complicated if the ions were to participate in a variety of ion-molecule reactions. For this reason Aston, Dempster, Nier, and other pioneers of the mass spectrometric technique made all efforts to reduce the pressure in the ion source to eliminate these undesirable ion-molecule reactions. The technical problems were soon solved and eventually instruments were developed which operate at pressures lower than 10^{-5} torr, that is, under conditions when most ions can leave the ion source without colliding and reacting with other molecules.

For many years interest in mass spectrometry centered on the analytical applications and the importance of investigating ion-molecule reactions was not appreciated. Therefore, in spite of the early discovery of such reactions, their systematic study was initiated only around 1952 by the work of Talroze [44] rapidly followed by Stevenson [45], Hamill [46], and Field, Franklin, and Lampe [47]. Since then most valuable information on the reactivity of gaseous ions has been provided by many laboratories. Reactions of gaseous ions are significant for various fields of research, for example, in elucidating radiation chemistry, for understanding gaseous electronics, ionic processes in the upper atmosphere, and ionic reactions in solution.

For the purposes of the present discussion, ion-molecule reactions can be divided into two types: proper chemical reactions in which chemical bonds are formed and broken and clustering reactions where association occurs mainly due to the operation of "physical" forces, that is, ion-dipole and ion-induced dipole attractions. A chemical process is exemplified by the addition of ethylene ion to ethylene molecule yielding a butene ion (reaction 36), whereas reactions 37 and 38 represent typical clustering reactions.

$$C_2H_4^+ + C_2H_4 = C_4H_8^+ \tag{36}$$

$$Cs^+ + nAr = Cs^+(Ar)_n \tag{37}$$

$$Cs^+ + nH_2O = Cs^+(H_2O)_n \tag{38}$$

Most ion-molecule investigations have dealt with reactions of the chemical type; clustered ions were observed rather incidentally and not as the results of deliberate studies. For example, the clustered ion $K^+(H_2O)$ was detected by Chupka [48] in a mass spectrometric study of ions emerging from a heated Knudsen cell containing potassium chloride. The ions $H^+(H_2O)_n$ were observed by Knewstubb and Sugden [49] in flames, by Knewstubb and Tickner [50] in gas discharges and by Beckey [51] in field emission, but these observations were not pursued further. Systematic investigations of clustering reactions were started in 1964 [53] after an incidental observation [52] of $H^+(H_2O)_n$ as an impurity ion in gases irradiated at high pressure by ionizing radiation. These investigations were initiated with the belief that unique information can be obtained from gas phase ion-solvation studies. However, a number of instrumental obstacles had to be overcome first. The instrumental requirements are therefore discussed in the following sections together with the principles underlying this study.

2.2. Principles of the Mass Spectrometric Method of Gas Phase Solvation Studies

The mass spectrometric method permits us to determine the concentrations of clustered ions formed in reactions represented by equations $(n - 1, n)$

$$A^+S_{(n-1)} + S \rightleftarrows A^+S_n$$

or $\qquad\qquad\qquad\qquad\qquad (n = 1, 2 \cdots) \qquad\qquad (n - 1, n)$

$$B^-S_{(n-1)} + S \rightleftarrows B^-S_n$$

where S is a molecule of the "solvent" and A^+ and B^- are some positive or negative ions. The system is assumed to be in equilibrium and the knowledge of concentrations of the various ionic species at fixed temperature and pressure of the "solvent" vapor allows us to calculate the equilibrium constants of formation of the clusters A^+S_n. Naturally, the relative concentration of a cluster is a measure of its stability, that is, whenever a particular structure with a given content of n solvent molecules is exceptionally stable, its equilibrium concentration is relatively high. The equilibrium constant $K_{n-1,n}$ for the reaction $(n - 1, n)$ is given by Eq. 39:

$$K_{n-1,n} = \frac{P_n}{P_{(n-1)}P_s} \tag{39}$$

P_n and P_{n-1} are the partial pressures of the clusters A^+S_n and A^+S_{n-1}, respectively, and P_s is the partial pressure of the "solvent." The partial pressures P_n and P_{n-1} of the ions are measured by continually bleeding the gas out of the reaction chamber (ion source) into a vacuum system. There the ions are captured by appropriate electric fields, focused, and subjected to mass analysis and electric detection. Thus

$$K_{n-1,n} = \frac{I_n}{I_{(n-1)}P_s} \tag{40}$$

where I_n/I_{n-1} is the intensity ratio, measured directly in the experiment and assumed equal to P_n/P_{n-1}. The pressure of the "solvent" in the reaction chamber is also measured directly by suitable manometers.

The equilibrium constants $K_{n-1,n}$ determined at different reaction temperatures T allow the evaluation of $\Delta H^0_{n-1,n}$ from van't Hoff plots. Hence the free energy changes $\Delta G^0_{n-1,n}$ and the entropy changes for the individual steps may be calculated:

$$-RT \ln K_{n-1,n} = \Delta G^0_{n-1,n} \tag{41}$$

$$\Delta G^0_{n-1,n} = \Delta H^0_{n-1,n} - T \Delta S^0_{n-1,n} \tag{42}$$

Determination of the relative equilibrium concentrations of the clusters leads therefore to detailed knowledge of the thermochemistry of ion-solvent molecule interactions. Such studies may be eventually extended to mixtures of "solvent" vapors. Thus the competitive solvation and formation of mixed solvent aggregates also may be investigated.

2.3. Experimental Requirements and Apparatus

In order to observe the equilibrium concentrations of the ion clusters we must create reaction conditions at which the clustering and dissociation of the ions A^+S_n are faster than any other reaction in which they are involved. For example, ions in the gas phase gradually disappear either by diffusion and discharge on the wall of the apparatus or by recombination with ions of opposite charge. Thus the reactions leading to the clustering equilibrium should be faster than the charge neutralization. The clustering is speeded up by high "solvent" pressures and by the addition, if necessary, of an inert gas.

Experimental studies [54] have shown that the half-life of a typical clustering reaction is less than 10 μsec at solvent pressures over 0.1 torr and total pressures over 1 torr. It is therefore necessary that the time required for the charge neutralization reactions be considerably longer than 10 μsec. The time of the positive-negative ion recombination can be made long by keeping the concentration of ions sufficiently low. The time for diffusion to the wall may be prolonged by the relatively high pressure established in the reaction chamber. Thus the relatively high pressure offers a twofold advantage—it speeds up the clustering and slows down the diffusion to the walls. Finally, the ions have to be formed relatively far from the sampling leak to assure their presence in the reaction chamber for a time longer than 10 μsec before they escape through the leak into the analyzing device. This is achieved by using an ionizing beam which passes some 4 mm above the sampling leak.

All these requirements are met in the apparatus shown in Fig. 2. It consists of three major parts: (1) a reaction chamber (ion source) where the ions are created (by electron impact) and where they react with the solvent vapor; (2) an electron gun producing a focused electron beam which enters through an electron entrance slit into the ion source; (3) a mass analysis section in which the ions escaping through the ion exit slit of the source are mass analyzed and detected. The three major components are connected through three ports with a large pumping chamber linked to a high-capacity pumping system. The high-capacity pumping system maintains a pressure differential of about 10^5 between the ion source and the vacuum chamber even though gas is continually escaping from the electron entrance slit and the ion exit slit of the ion source. Typical operating conditions are 5 torr pressure in the ion source and about 10^{-4} torr pressure in the vacuum chamber. The mass analysis section and the electron gun operate fairly well at and below 10^{-4} torr but begin to fail at higher pressures.

Pulsing of the electron beam allows us to observe the history of the ions during the time between the pulses. The pulsing method, described in greater detail elsewhere [55], allows us to study the kinetics of approach to the equilibrium as well as to measure the equilibrium concentrations. Figure 3

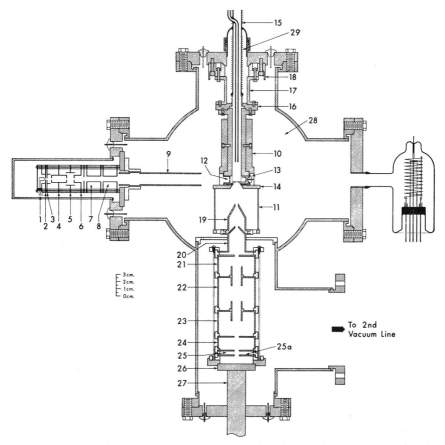

Figure 2. Electron-beam high-pressure ion source. (1) Electron filament. (2–6) Electrodes for electrostatic focusing of electron beam. (7, 8) Deflection electrodes. (9) Magnetic and electrostatic shielding of electron beam. (10) Ion source with copper heating mantle. (11) Electrostatic shielding of ions. (12) Electron entrance slit. (13) Electron trap. (14) Copper lid holding ion exit slit flange. (15) Gas in and outlet. (16–18) Ion source supports and insulation. (19–26) Ion beam acceleration and focusing. (27) Mass spectrometer tube to 90° magnetic sector.

illustrates the concentration-time dependence observed in the hydration reactions occurring in nitrogen gas which contains small amounts of water vapor. The N_2^+ ions are the primary products of the electron impact since nitrogen predominates in the investigated mixture. This ion reacts with more N_2 to give N_4^+. The N_4^+ ion would have been the final product in the absence of water vapor. However, in the presence of water vapor it may pick up an electron from a water molecule (charge transfer) and then the following

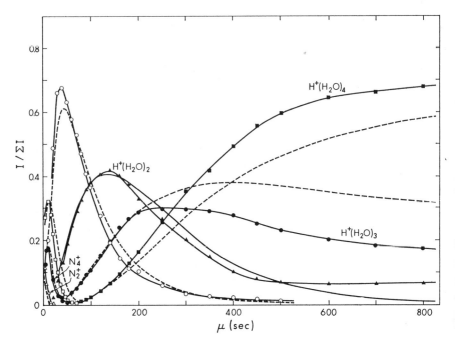

Figure 3. Normalized ion intensity curves for ions in moist nitrogen. $P_{N_2} = 2$ torr, $P_{H_2O} = 1.6 \times 10^{-3}$ torr, $300°K$. Successive intensity maxima indicate sequence $N_2^+ \rightarrow N_4^+ \rightarrow H_2O^+ \rightarrow H_3O^+ \rightarrow H^+(H_2O)_2 \rightarrow H^+(H_2O)_3 \rightleftarrows H^+(H_2O)_4$. Dashed lines represent theoretical curves calculated from integrated rate equations for the consecutive reversible reactions using the average rate constants determined in reference 53. In experiments where only the ultimate equilibrium was studied the concentration of water was higher; hence the equilibrium was established in less than ~ 20 μsec.

reaction sequence develops:

$$N_2^+ + 2N_2 \rightarrow N_4^+ + N_2$$

$$N_4^+ + H_2O \rightarrow 2N_2 + H_2O^+$$

$$H_2O^+ + H_2O \rightarrow H_3O^+ + OH$$

$$H_3O^+ + H_2O + N_2 \rightarrow H^+(H_2O)_2 + N_2$$

$$H^+(H_2O)_{n-1} + H_2O + N_2 \rightleftarrows H^+(H_2O)_n + N_2$$

The decay of N_4^+, the buildup and decay of the intermediate ions, and the establishment of stationary (equilibrium) concentrations of the final ions $H^+(H_2O)_3$ and $H^+(H_2O)_4$ are clearly seen in Fig. 3. Detailed analysis of such data allows the determination of the rate constants of all the relevant reactions, the prediction of the time required for attaining the equilibrium at various conditions, and the determination of the pertinent equilibrium constants.

In studies of the equilibrium it is convenient to suppress the ion detection for the period needed to establish the equilibrium conditions and to detect the ions thereafter. With proper choice of the reaction conditions, the delay period may be reduced to 10–50 μsec.

The pulsed electron beam apparatus described here was developed only recently. Most of the experiments described in the subsequent sections were carried out with a less sophisticated apparatus under conditions of steady primary ionization. In those experiments the bulk of primary ions was produced at a certain minimum distance from the ion exit slit. Thus the ions that diffused to the exit slit had sufficient time to attain the clustering equilibrium before being detected.

A critical point in measurement of the clustering equilibrium is the assumption that the ratios of the ion intensities, as determined by the mass spectrometer, are equal to the concentration ratios of the ions at equilibrium. Various factors may render this assumption invalid; some of them have been discussed in previous publications [56, 57]. Although detailed discussion of this subject is not warranted at this point, it may be stressed that reliable and meaningful results can be obtained if sufficient care is taken to avoid the more serious causes of ion beam adulteration.

2.4. Hydration of Spherically Symmetrical Ions. The Positive Alkali and Negative Halide Ions

The alkali and halide ions have played an important role in the studies of ionic hydration and solvation. Alkali halides form simple salts, soluble in water, their ions have spherical symmetry and sufficiently vary in size to permit studies of the dependence of solvation parameters on the ionic diameter. However, the formation of alkali ions in water vapor at pressures of a few torr presents some experimental difficulties, which were overcome only recently with the aid of a thermionic ion source [58, 59]. The relative ion intensities of the sodium ion hydrates at various partial pressure of water at 300°K are shown in Fig. 4. It is assumed that equilibrium is established in the investigated systems and that the relative intensities are proportional to the relative stabilities of the clusters. The results indicate that three or even four different types of clusters may coexist at comparable concentrations. Furthermore, the number of ligands is not restricted to any fixed "coordination" number like 4 or 6 or 8; the higher the partial pressure of water the larger the number of ligands and no cluster appears to be much more stable than any other.

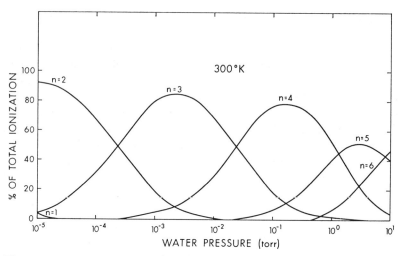

Figure 4. Relative intensities of sodium hydrates $Na^+(H_2O)_n$ at $300^\circ K$ and variable water pressure.

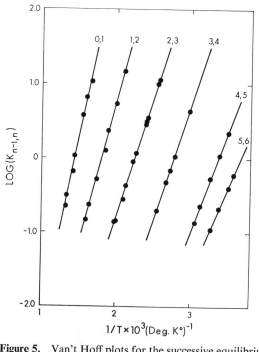

Figure 5. Van't Hoff plots for the successive equilibrium constants $K_{n-1,n}$ of the reactions $Na^+(H_2O)_{n-1} + H_2O = Na^+(H_2O)_n$.

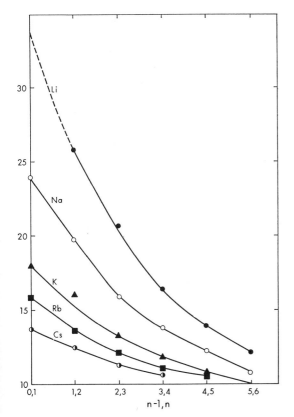

Figure 6. $-\Delta H_{n-1,n}$ for the successive hydrations of alkali ions: $M^+(H_2O)_{n-1} + H_2O = M^+(H_2O)_n$.

These findings could have been predicted on the basis of simple electro-static and statistical thermodynamics calculations; nevertheless, they appear somewhat surprising since it is generally believed that certain structures with a discrete number of ligand molecules are the most stable.* For example, it is commonly assumed that 4, 6, or 8 (generally even numbers) of ligands form stable complexes but, 3, 5, or 7 never do. This attitude apparently arose from the experience gained in the studies of crystal structure. Their three-dimen-sional extention imposes symmetry requirements which makes only certain structures allowed. Such restrictions do not apply to an isolated $M^+(H_2O)_n$ ion formed in the gas phase.

---

* The existence of an unoccupied site is plausible for a gaseous, partially "solvated" ion but highly improbable in solution.

The symmetry requirements of the solid state are greatly relaxed in the liquid phase. Hence we might expect that the inner solvent sphere of an ion in a liquid could easily change the number of ligands by one or more units, similarly to the gaseous complex.

The equilibrium constants $K_{n-1,n}$ determined at different temperatures yield quantitative thermochemical data for the clustering reactions. Figure 5 shows typical Van't Hoff plots obtained for the various equilibrium constants of the $Na^+(H_2O)_n$ formation. The slopes and intercepts of these plots lead to the relevant $\Delta H^0_{n-1,n}$ and $\Delta S^0_{n-1,n}$ values. The thermodynamic data obtained in this manner for other alkali hydrates and the halide hydrates are summarized in Tables 5 and 6.

In Figs. 6 and 7 the variation of $\Delta H_{n-1,n}$ with n is shown for the alkali and halide ion hydration reactions. For each alkali ion $-\Delta H_{n-1,n}$ decreases with increasing n, whereas for a constant n the enthalpy change is the largest for Li^+

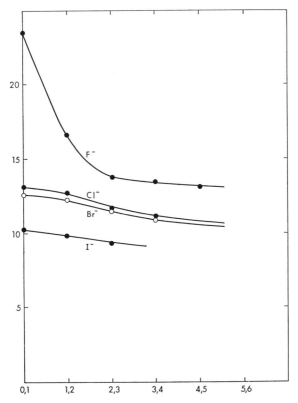

Figure 7. $-\Delta H_{n-1,n}$ for the successive hydrations of halide ions: $X^-(H_2O)_{n-1} + H_2O = X^-(H_2O)_n$.

and the smallest for Cs^+, intermediate values being found for other alkali ions. These results could be expected. Similar trends are found for the halide ions.

Experimental findings were compared with theoretical calculations of the potential energies of the clusters $M^+(H_2O)_n$ and $X^-(H_2O)_n$ [59–63]. In these calculations the potential energy is expressed as a sum of terms resulting from the ion-dipole, ion-induced dipole and Van der Waals' attractive interactions and terms arising from the ion-water electron cloud repulsions and dipole-dipole repulsions. The absolute values of the calculated energies are greatly affected by the assumed value of the constant A appearing in the ion-water electron cloud repulsion term: A/R^{12}. The constant A could not be calculated from the first principles; therefore its value was so adjusted as to give the best agreement with the experimental $\Delta H_{0,1}$. Such an A was then used for the calculation of the energies of the higher clusters of the same ion. The degree of agreements between the calculated and observed values is shown in Fig. 8. For the large Cs^+ ion the agreement is good. However, the calculated and experimental curves show distinctly different shapes for the two small ions Na^+ and Li^+. The initial rapid fall-off of the experimental energies in Li^+

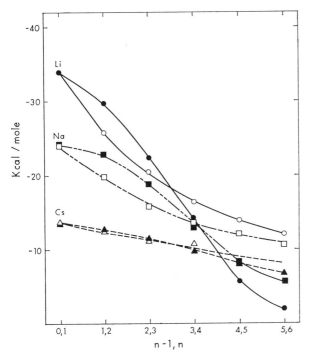

Figure 8.   Comparison of the calculated $\Delta E_{n-1,n}$ (open circles squares and triangles) and the experimental $\Delta H_{n-1,n}$ (shaded points).

(and to a lesser extent Na^+) should be attributed to specific chemical bonding interactions in complexes of small n which become saturated. Probably a dative bond between the lone pair of oxygen and the lowest empty orbital of the alkali ion accounts for the high values of $\Delta H_{0,1}$. The effect of the dative bonding should be the largest with Li^+ and the weakest with Cs^+, in agreement with the observation.

Similar comparisons between calculated energies and measured enthalpy values can be also made for the halide hydrates [62, 63]. However, the problem is more complex since different orientation of the water molecules toward the ion must be also considered. Two basic orientations seem to be important: (1) the oxygen atom and ion lie on the bisector of the HOH angle: and (2) the ion, one hydrogen atom, and the oxygen atom lie on one line. The electrostatic calculations indicate that the first orientation gives lower energy and is preferred for small n. The alternative orientation allows a larger number of water molecules to be placed around the ion. Growth of

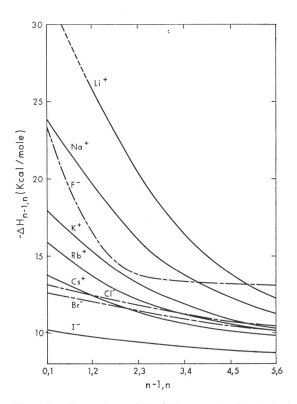

Figure 9.   Comparison of the $\Delta H_{n-1,n}$ values for the hydration of alkali and halide ions.

the cluster should therefore lead to a probable change from the former symmetric to the latter nonsymmetric orientation of water molecules [62, 63].

The very high $-\Delta H_{0,1}$ in $F^-(H_2O)$ (see Fig. 7) probably is due again to chemical bonding, viz.

$$\begin{array}{ccc} \text{H} & \text{H} & \text{H} \\ {}^-\text{F} \diagup\diagdown \text{O} \leftrightarrow \text{F} \diagup\diagdown \text{O}^- \leftrightarrow \text{F} \diagup\diagdown \text{O}_- \\ \text{H} & \text{H} & \text{H} \end{array}$$

The tendency to form a covalent bond falls off in the halides with increase of atomic number. This is reflected in the results presented in Fig. 9.

2.5. Comparison of the Hydration Energies of the Alkali and Halide Ions

The correlation of the present results with the total single-ion hydration enthalpies is interesting. The single-ion hydration enthalpies ΔH_h correspond to the enthalpy changes for processes in which one mole of ions, say M^+, is transferred from the gas phase into aqueous solution. The calculation of single-ion hydration enthalpies is accomplished through Born cycles. Some of the steps of the Born cycle involve enthalpy changes for processes concerned with the salt and not with the individual cation or anion. The determination of the hydration energy of a single ion depends therefore on assumptions by which the total contribution is apportioned between the cation and the anion. Thus, for example, Latimer, Pitzer, and Slanski [64] and Verway [65] derived hydration energies of single ions by utilizing the Born equation. Their values were accepted until a few years ago when a measurement of EMF of a galvanic halfcell coupled to a mercury metal electrode through the gas phase allowed Randles to construct Born cycle for a single ion [66]. The single-ion hydration energies thus deduced differ from those advocated by Latimer et al. It will be shown that the present experimental results support Randles data.

According to Latimer the $-\Delta H_h$ for the negative ion is substantially larger than that for the positive isoelectronic ion. Thus $-\Delta H_h(\text{Br}^-) = 81.4$ kcal/mole, whereas $-\Delta H_h(\text{Rb}^+) = 69.2$ kcal/mole, contradicting the results quoted in Table 5. Two basically different explanations of Latimer's data were proposed. One school [67] assumes that the arrangement of water molecules around the negative ion and (or) in the transition from the hydrated ion to the bulk of the liquid is more favorable than that around a positive ion. Alternatively, it has been suggested that water molecules possess an electrical quadrupole moment which is of such a sign that it leads to an attraction with negative and a repulsion with positive ions [68]. The present

Table 5 Enthalpy Changes for Gas Phase Reactions $M^+H_2(O)_{n-1} + H_2O = M^+(H_2O)_n$ and $X^-(H_2O)_{n-1} + H_2O = X^-(H_2O)_n$

$$-\Delta H_{n-1,n}$$
(kcal/mole)

$n-1, n$	H⁺	Li⁺	Na⁺	K⁺	Rb⁺	Cs⁺	OH⁻	F⁻	Cl⁻	Br⁻	I⁻
0, 1	165	34	24	17.9	15.9	13.7	22.5	23.3	13.1	12.6	10.2
1, 2	36	25.8	19.8	16.1	13.6	12.5	16.4	16.6	12.7	12.3	9.8
2, 3	22.3	20.7	15.8	13.2	12.2	11.2	15.1	13.7	11.7	11.5	9.4
3, 4	17.	16.4	13.8	11.8	11.2	10.6	14.2	13.5	11.1	10.9	
4, 5	15.3	13.9	12.3	10.7	10.5		14.1	13.2			
5, 6	13	12.1	10.7	10.0							
6, 7	11.7										
7, 8	10.3										

References: H⁺, [57], alkali ions, [59], OH⁻, [76], halide ions, [62].

64

results, which lead to higher energies of interactions for the positive ions, seem to discredit the quadrupole theory since the effect of the quadrupole should be particularly large at close range, that is, for gaseous clusters. It should be stressed that water molecules may possess a quadrupole moment but its magnitude cannot be as large as required to explain the Latimer data.

Randles data lead to a slightly higher $-\Delta H_h$ for F^- than for Na^+ but for the larger ions the $-\Delta H_h$ of the positive isoelectronic ions are somewhat larger than those of the negative ions—a reverse relation to that deduced from Latimer's data. A correlation of the gaseous hydration data with single-ion hydration energies of Latimer and of Randles is given in Fig. 10, which shows plots of $[-\Delta H_{0,n}(M^+)] - [-\Delta H_{0,n}(X^-)]$ for isoelectronic pairs. The plots are based on the experimental $\Delta H_{n-1,n}$ whenever these were available (low n). For higher n the values extrapolated from Fig. 9 were used.

We can summarize the observations on positive and negative isoelectronic pairs as follows. For isoelectronic ions the gaseous anion is larger than the corresponding cation due to its smaller nuclear charge. In the initial hydration interactions (n small) the positive ion gives higher energies of interaction, as would be expected from its smaller size. As the cluster grows interactions in the negative ion become gradually more favorable (Figs. 9 and 10). This is

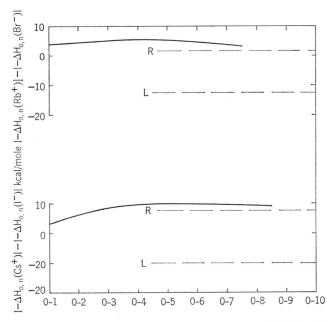

Figure 10. Comparison of $\Delta H_{0,n}$ of the hydration of alkali and halide ions with the total enthalpies of hydration of Latimer, Pitzer, and Slanski (L) and of Randles (R).

probably due to the ability of the negative ion to pack the water molecules without too large water-water repulsions. Extrapolation of the data to moderately high n seems to lead to an asymptotic approach to the Randles data. The decisive interactions determining the total heats of hydration occur during the attachment of the first 8–12 molecules. The further path toward formation of a liquid solution may be considered as a single step in which the large clusters are fitted into the bulk of the liquid. The present results indicate that the hydration energy for this final "step" is similar for a positive or negative cluster containing the corresponding isoelectronic ion.

2.6. The Hydrated Proton in the Gas Phase

The state of the proton in liquid water is of great interest. A widely accepted model [69] assumes that the hydrated proton forms the symmetrical structure $H_3O^+(H_2O)_3$ shown in Fig. 11. The question arises whether the cluster containing one proton and four molecules of water has an exceptional stability. The behavior of the proton in the gas phase [57] may be significant in answering this problem. The pertinent results are shown in Fig. 12, which gives the equilibrium concentrations of the clusters $(H_{2n+1}O_n)^+$, determined mass spectrometrically at different pressures of water at 300 and 400°K. They clearly demonstrate that the cluster $H_9O_4^+$ is not exceptionally stable, since its equilibrium concentrations are not prominent but fit the general concentration distribution of the clusters. The enthalpy changes for the reactions $(n - 1, n)$, shown in Fig. 13, support this conclusion. The figure shows that the exothermicity of the clustering reactions decreases in a regular manner, with no evident discontinuities. Similar gradations are observed for the entropy changes $\Delta S^0_{n-1,n}$, as summarized in Table 6.

Thus the gas phase data do not support the proposed exceptional stability

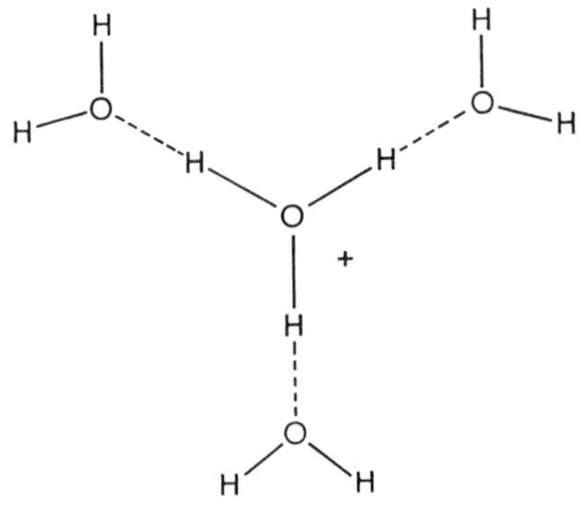

Figure 11. The symmetric structure of $H_3O^+(H_2O)_3$ ion.

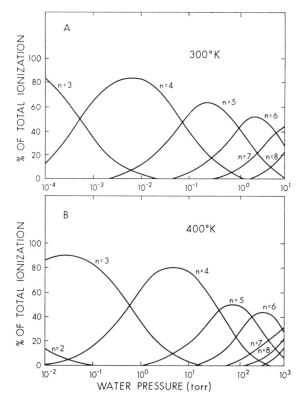

Figure 12. Equilibrium distributions of clusters $H^+(H_2O)_n$ derived from the experimental data. (A) 300°K. (B) 400°K.

of the structure shown in Fig. 11. On the other hand, a preference for the cluster $NH_4^+(NH_3)_4$ is indicated by the results shown in Fig. 14.

2.7. Competitive Solvation of H^+ by Water and Methanol and by Water and Ammonia Molecules

In liquid-solvent mixtures the immediate neighborhood of a given ion may have composition different from that of the bulk of the liquid; that is, the immediate neighbors of the ion seem to be selected by some specific inter-actions between the ion and the solvent molecules. Predictions based on dipole moment, polarizability, dielectric constant, acidity, basicity, etc., of the solvent molecules are not always reliable.

The mass spectrometric cluster method can give some interesting insights into the situation arising in a mixed solvent. Two examples will be considered.

Table 6 Entropy Changes for Gas Phase Reactions $M^+(H_2O)_{n-1} + H_2O = M^+(H_2O)_n$ **and** $X^-(H_2O)_{n-1} + H_2O = X^-(H_2O)_n$

$n-1, n$	H^+	Li^+	Na^+	K^+	Rb^+	Cs^+	OH^-	F^-	Cl^-	Br^-	I^-
						$-\Delta S_{n-1,n}$ (e.u.)					
0, 1		23	21.5	21.6	21.2	19.4	19.1	17.4	16.5	18.4	16.3
0, 1 (calculated)		23	22.7	21.6	22	21.2		22.5	20.9	20.0	17.7
1, 2	33.3	21.1	22.2	24.2	22.2	22.2	19.3	18.7	20.8	22.9	19.0
2, 3	29	24.9	21.9	23.0	24.0	23.7	24.8	20.4	23.2	24.8	21.3
3, 4	28.3	29.9	25.0	24.7	24.8	25.4	29.5	36.9	25.8	26.8	
4, 5	32.6	31.4	28.1	25.2	25.7		33.2	30.7			
5, 6	30.3	32	26.0	25.7							
6, 7	29.6										
7, 8	27										

References: H^+, [57], alkali ions, [59], OH^-, [76], halide ions, [62].

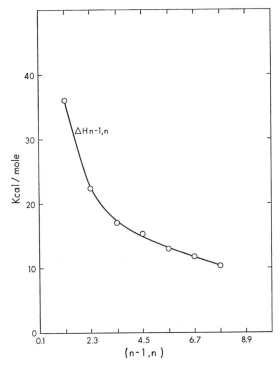

Figure 13.   Plot of the $\Delta H_{n-1,n}$ versus n for the reaction $H^+(H_2O)_{n-1} + H_2O = H^+(H_2O)_n$.

The water-methanol system is a classical example of a mixed solvent and has received considerable attention in studies of liquid solutions. Since the proton affinity of methanol is by 10–20 kcal higher than that of water, whereas the dielectric constant of water is much higher than that of methanol, the following questions may be interesting. Is water or methanol preferentially taken up in small clusters? What are the changes in the cluster composition as it grows? How does the situation evolve when a "macrocluster" is transferred into a liquid solution?

The mass spectrometric studies [70] which provide some information on these questions were performed with mixtures of water and methanol vapor. The observed distribution of the clusters containing varying amounts of methanol molecules is shown in Fig. 15. The average mole % of methanol in a given cluster $H^+(L)_n$ (L is a ligand) is shown in Fig. 16. It is evident from Figs. 15 and 16 that methanol is preferentially taken up by the clusters. For example, for a vapor containing only 5 mole % methanol, the cluster groups L_4H^+, L_5H^+, and L_6H^+ contain an average of 80, 65, and 55 % methanol molecules. Let us define a preference factor γ_n as the ratio of

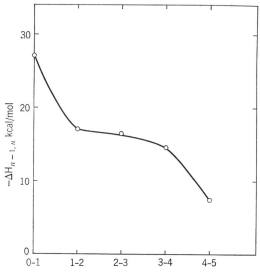

Figure 14. Plot of the $\Delta H_{n-1,n}$ for the reaction $NH_4^+(NH_3)_{n-1} + NH_3 = NH_4^+(NH_3)_n$.

methanol to water molecules in the clusters L_nH^+ over the ratio of methanol to water molecules in the gas phase. Denoting by μ_n the mole fraction of methanol in cluster n, we can express γ_n by Equation 43

$$\gamma_n = \frac{\mu_n P_W}{(1 - \mu_n) P_M} \tag{43}$$

where P_W and P_M denote the partial pressures of water and methanol vapor. Figure 17 shows γ_n for various n as a function of mole % of methanol in the vapor. One finds that for a constant n, γ_n is essentially constant. It is also observed that γ decreases as n increases, for example, while $\gamma_2 > 1000$, $\gamma_9 \approx 1$ (see Fig. 18). We would therefore expect the preference factor for methanol to be smaller than unity for clusters greater than 9, and thereafter water will be taken up preferentially.

The preferential uptake of methanol in clusters of small size is not surprising. It has been established (see Table 3) by mass spectrometric studies that the reaction

$$H_3O^+ + CH_3OH \rightarrow CH_3OH_2^+ + H_2O$$

is exothermic by some 1–20 kcal. The exothermicity of this reaction, which represents the difference between the proton affinities of methanol and water, must be due to the positive inductive effect of the CH_3 group which leads to a

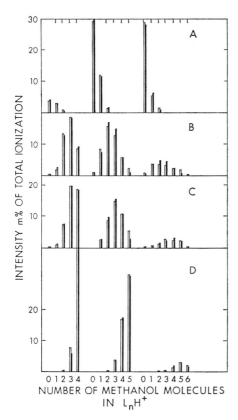

Figure 15. Ion intensities of clusters observed in water-methanol vapor mixtures at 5 torr total pressure and 50°C. L denotes the ligand, i.e. either a water or methanol molecule. (A) Traces of methanol. (B) 2.3 mole % of methanol. (C) 5 mole % of methanol. (D) 20 mole % methanol. □ observed values; ■ calculated values. The calculated values were obtained by fitting the probability distributions to the experimental data (see text). The data show that methanol is preferentially taken up by the cluster, but the preferential uptake of methanol decreases with the size of the cluster. Uptake of methanol shifts the intensity distributions to lower ligand number n.

stabilization of the $CH_3OH_2^+$ ion. The attachment of a second molecule of methanol to $CH_3OH_2^+$ should again be favored over water since the electron-releasing property of the methyl group in the second methanol molecule stabilizes any reasonable structure of the M_2H^+ ion. The gradual decrease of the preference for methanol probably is due to two factors: (1) the inductive effect of the methyl group rapidly falls off as the distance between the proton and the solvating molecule increases; and (2) the bulkier methanol molecules reduce the number of nearest neighbors around the ion. This "volume" effect is clearly seen in Fig. 19, which shows the relative intensities of clusters L_nH^+ at constant total pressure and increasing relative concentration of methanol. It can be seen that the preferential uptake of methanol by clusters tends to reduce their size. Whereas the L_7H^+ clusters form 10% of the clusters in pure water vapor, their abundance becomes negligible after addition of 5% of methanol to the vapor. Under the same conditions the abundance of L_6H^+ decreases from 39 to 10% while the smaller clusters

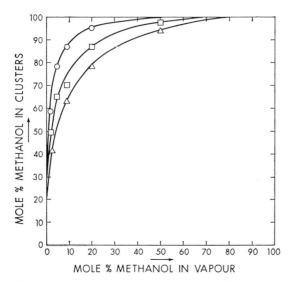

Figure 16. Methanol content of clusters L_nH^+, $\bigcirc = L_4H^+$, $\square = L_5H^+$, $\triangle = L_6H^+$.

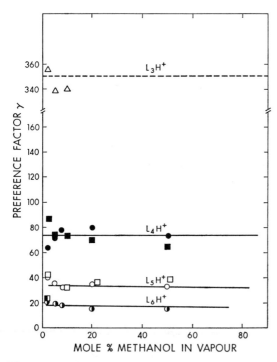

Figure 17. Plot of preference factor γ for methanol versus mole % methanol in vapor; total pressure $\square$, 2.5 torr; $\bigcirc$, 5 torr.

72

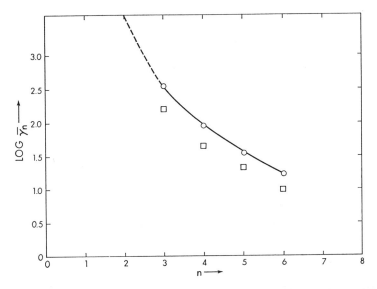

Figure 18.   Plot of the factor $\bar{\gamma}_n$ (which expresses the preference for methanol in a cluster group L_nH^+) versus n. $\bar{\gamma}_n$ is the average value of γ_n given in Fig. 17. ○, the results obtained on an α particle mass spectrometer for a total pressure of 5 or 2.5 torr; □, data from proton beam mass spectrometer obtained at 0.23 torr total pressure. Extrapolation of these data leads to preference for water above $n = 9$.

L_4H^+ and L_3H^+ become more abundant. Apparently three or four molecules of water are expelled from the cluster when two or three molecules of methanol are taken up.

The intensities of different clusters $M_mW_wH^+$ where M and W denote methanol and water molecules respectively and $m + w = n = $ constant can be calculated from the previously determined statistical distributions. Thus the relative intensities of $W_3H^+ : W_2MH^+ : WM_2H^+ : M_3H^+$ are found close to the values of the binomial expansion terms of $(\omega + \mu)^3 = \omega^3 + 3\omega^2\mu + 3\omega\mu^2 + \mu^3$ where $\omega = 1 - \mu$. The statistically calculated distributions are shown in Fig. 15.

A radically different situation is observed for the protonated mixed water-ammonia [56, 71] clusters. The distributions can be fitted only if it is assumed that three groups of clusters should be distinguished: a one-ligand molecule "group" in which there is an extremely large preference for ammonia, a four-ligand molecule group in which there is still a preference for ammonia but only by a factor of about 20, and finally a group containing all the larger clusters in which there is a preference for water by a factor of 20–30. These findings are interpreted in the following way. The "group" of

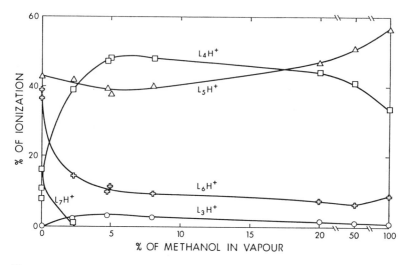

Figure 19. Plot of relative abundance of clusters L_nH^+ versus methanol vapor concentration. Increase of methanol content in clusters leads to decrease of the number of ligands L (i.e., to a decrease of n).

one ligand molecule is the ammonium ion NH_4^+; the group of four ligand molecules represents an inner shell where there is still preference for ammonia; and the last group represents molecules located outside the inner shell.

The results can be understood if we recall that ammonia has a much greater proton affinity than water [PA(H_2O) = 166, PA(NH_3) = 207 kcal/mole]. One ammonia molecule therefore holds the proton as an ammonium ion and the other molecules share partially the charge through the four hydrogen atoms of the ammonium ion. In the outer shell where the molecules are at a considerable distance from the charge, water is taken up preferentially, because the higher permanent dipole of water leads to much stronger interaction.

The feasibility of fitting the methanol-water clusters by a single distribution—without having to divide them into groups—indicates that the proton is not closely associated with one given molecule and that no clear-cut inner and outer shell structure exists in these clusters.

2.8. Solvation of Negative Ions by Various Solvent Molecules. Correlation with Acidity

In the previous sections we discussed systems in which water, ammonia, and methanol molecules were the solvating species of positive ions. The solvation of the negative Cl⁻ and OH⁻ ions by several different molecules was also examined [72].

The Cl⁻ was selected for comparative solvation studies because it is a simple, spherical, easily produced ion. Water, methanol, t-butanol, aniline, phenol, and benzene were used as the solvating agents (S1). The equilibria of the reactions involving the attachment of one or two ligands were studied

$$Cl^- + S1 = Cl^-(S1)$$

$$Cl^-(S1) + S1 = Cl^-(S1)_2$$

at different temperatures. The free energy, enthalpy, and entropy changes determined in this manner are summarized in Table 7.

Table 7 Enthalpy, Entropy and Free Energy Changes for the Solvation of Cl⁻ by Various Molecules [72]

Solvent Molecule	$-\Delta H_{0,1}$ (kcal/mole)	$-\Delta H_{1,2}$	$-\Delta S_{0,1}$ (e.u.)	$-\Delta S_{1,2}$	$-\Delta G_{0,1}$ (kcal/mole)	$-\Delta G_{1,2}$
HOH	12.8	12.0	15.7	18.9	8.1	6.4
CH₃OH	14.2	13.0	14.8	19.5	9.8	7.2
t-C₄H₉OH	14.2	13.4	10.3	19.2	11.1	7.7
Aniline	17.3	15.0	18.4	23.3	11.0	8.1
Phenol	19.4	18.5	15.5	24.5	14.8	11.3
Benzene	~5.6		~7.3		~3.4	

When molecules such as HOH or CH₃OH, or generally RH endowed with a dipole moment imparting a partial positive charge on the hydrogen atom, become attached to the Cl⁻ ion, the hydrogen is directed toward the negative ion. The presence of the negative ion will induce a further shift of electrons away from the H atom. Therefore we may consider that in the complex R⁻H⁺ ⋯ Cl⁻ a partial proton donation to the negative ion has occurred. The process resembles the neutralization of the Bronsted base Cl⁻ by the acid HR. The stronger the proton-donating ability of HR, the stronger its "gas phase acidity" and the stronger the interaction with Cl⁻. Therefore the $-\Delta H_{0,1}$ and $-\Delta G_{0,1}$ should increase with the acidity of HR. Accepting this premise, we can construct the following gas phase acidities series using the data listed in Table 7: benzene < H₂O < methanol < t-butanol < aniline < phenol. The same order of "gas phase acidities" was found by Tiernan [73] and by Brauman [74]. These authors studied the gas phase ion molecule reactions:

$$R'O^- + R''OH = R'OH + R''O^-$$

and established that the ease with which ROH loses a proton increases in the following order: $R = H < CH_3 < C_2H_5 < iC_3H_7 < t$-butyl.

An even clearer correlation between the magnitude of the enthalpy change $\Delta H_{0,1}$ for the addition of the first solvent molecule to the negative ion and the acidity of the negative ion is observed [75] for the reaction series

$$OH^- + HR = (HOHR)^- \qquad (\Delta H_{0,1})$$

where $R = OH, F, Cl, Br, I$. In Fig. 20 the relevant $\Delta H_{0,1}$ values [76] are plotted against the values of $D(R^- - H^+) = D(H - R) - EA(X) + I(H)$. This expression corresponds to the heterolytic dissociation energy of HR, that is, to the energy required for the process

$$HR = H^+ + R^-$$

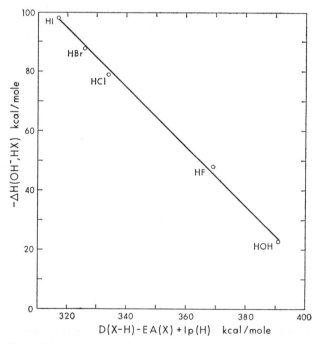

Figure 20. Linear relationship between energy for reaction $(0, 1)$: $OH^- + HX = (OH_2X)^-$ and the heterolytic dissociation energy of HX.

This energy can be considered as a measure of the acidity of HR in the gas phase, the acidity being greater the smaller the heterolytic dissociation energy. As can be seen from the figure, an almost linear correlation is obtained between the $-\Delta H_{0,1}$ and this energy. We observe a gradual increase of $-\Delta H_{0,1}$ and increase of the gas phase acidity in the order HOH < HF < HCl < HBr < HI. For the halogens this order is determined not by the electron affinities, which do not change much, but by the bond dissociation energies $D(H - X)$, which decrease from HF to HI [76]. The very low acidity of water is due to both a relatively low electron affinity of OH and a very high bond dissociation energy $D(H - OH)$.

2.9. Ion Pairs and Higher Aggregates in the Gas Phase

Salts which form ionic lattices in the solid state exist in the vapor mostly as molecules or higher aggregates composed of ions. The investigation of the alkali halide molecules in the vapor is particularly facile because the vapor pressure of these salts is relatively high. Studies of the electronic [77, 78], vibrational [79], and rotational spectra [4] of MX in the gas phase led to determination of the internuclear distances [80], potential energy diagrams [79], and bond dissociation energies [2, 78] of these molecules. Theoretical calculations of Rittner [81] have shown that these properties are well accounted for by a purely ionic model similar to that proposed by Pauling [82].

In addition to MX, higher aggregates like dimers $(MX)_2$ and trimers $(MX)_3$ can exist in the gas phase. The presence of higher aggregates was first demonstrated by Friedman [83], who determined with the aid of a mass spectrometer the composition of a molecular beam effusing from a heated oven containing solid lithium iodide. Temperature dependence of the partial pressures of the various aggregates permitted calculation of the heat of sublimation, dimerization, and trimerization of lithium iodide.

Similar results were reported by Miller and Kusch [84], who analyzed the velocity distribution of the species effusing from an oven containing solid alkali halides. The approximate composition of the vapor formed above heated alkali halides was determined by Berkowitz and Chupka [85, 86], who demonstrated again the existence of $(MX)_2$ dimers and of still higher aggregates.

Datz, Smith, and Taylor [87] were able to study the association equilibria in alkali halide vapors by determining the temperature dependence of the apparent molecular weights of the gaseous NaCl, NaBr, NaI, KCl, KI, RbCl, CsCl. The apparent molecular weight was calculated from the equation

$$PV = \frac{M}{W} RT$$

where M is the apparent molecular weight and W is the weight of the evaporated salt. At the selected temperatures the concentration of the trimer was negligible. Under these conditions, the decrease of the apparent molecular weight with increasing temperature is due to the dissociation of the dimer $(MX)_2$. From the results the dissociation enthalpy and entropy were calculated and these are listed in Table 8, which also includes later data reported by Hagemark, Blander, and Luchsinger [88].

The dissociation energies of the dimers are quite large; they decrease systematically as the radius of the cation or the anion increases. For example, the dissociation energy of $(NaCl)_2$ is 48 kcal/mole, whereas that of $(CsCl)_2$ is only 34.7 kcal/mole.

The bond energies of the dimers and some trimers were calculated by Milne and Cubicciotti [89], who modified the method of Pauling [82]. Their values are included in the table and show good agreement with the experimental data. A dimensional analysis developed by Blander [90] allows us to estimate unknown equilibrium constants and dissociation energies for a given member of a series, such as the alkali halides, if the values for some other member are known.

The vapor phase, in equilibrium with the heated salt MX, may contain, in addition to the monomer MX and the dimer $(MX)_2$, charged species like $(M_2X)^+$, $(MX_2)^-$, and $(M_3X_2)^+$. Chupka [48, 85] detected the presence of such species in mass spectrometric studies of ions escaping from a heated

Table 8 Dissociation Energies of Gas Phase Alkali Halide Species at $1300°K$

Reaction	ΔE_{exp}	ΔE_{cal}
$(NaCl)_2 = 2NaCl$	48	47.9
$(NaBr)_2 = 2NaBr$	43	46.3
$(NaI)_2 = 2NaI$	40	43.1
$(KCl)_2 = 2KCl$	41	42.2
$(KBr)_2 = 2KBr$	38	
$(KI)_2 = 2KI$	35	38.3
$(RbCl)_2 = 2RbCl$	39	40.1
$(CsCl)_2 = 2CsCl$	35	36.9
$Na_2Cl^+ = Na^+ + NaCl$	39.8	
$K_2Cl^+ = K^+ + KCl$	38.6	
$K_2Br^+ = K^+ + KBr$	38.2	
$Rb_2Cl^+ = Rb^+ + RbCl$	25.5	

References: neutral experimental values, [87, 88] calculated values, [89]; ion values, [48].

oven containing the salts MX. He studied the equilibria

$$(M_2X)^+ = MX + M^+$$

at different temperatures and calculated the enthalpies of the dissociation for the systems involving NaCl, KCl, and KBr. These data are included in Table 8, and they show that the dissociation of a dimer requires more energy than the dissociation of the corresponding M_2X^+. Calculations in which the ions are treated as rigid spheres and only the coulombic forces due to the permanent charges are considered lead to the same conclusion. The ratio of the dissociation energies given by the calculation is 1.17. This is quite close to the experimental ratios which can be obtained from the data given in Table 8.

2.10. Future of Studies of Ion-Ligand Interactions in the Gas Phase

The experimental results on ion-solvent molecule clusters $M^+(S)_n$, $X^-(S)_n$, ion pair aggregates $(MX)_n$, and charged ion pair aggregates $M_{n+1}X_n^+$ and $M_nX_{n+1}^-$ discussed in Section 2 of this chapter were all based on measurements of equilbria and their temperature dependence. The extension of the equilibrium measurements to a wider variety of solvent molecules and organic ions is an obvious task for future workers in this field.

The thermodynamic data obtained from the equilibrium studies are not sufficient for the full understanding of the problems discussed in this chapter. Thus structural information can be obtained only indirectly by comparison of the enthalpy and entropy changes, or by comparison of the experimental data with values obtained by calculation of the energies and entropies of clusters having some assumed structures. Obviously it would be desirable to apply spectroscopic methods in studies of charged and neutral aggregates in the gas phase. A simple calculation shows that such measurements would be very difficult. The concentration of the clusters $M^+(S)_n$ in the reaction chamber of a mass spectrometer is of the order of 10^6 ions/cm³. This corresponds to a 10^{-15} molar concentration or an ionic partial pressure of 4×10^{-11} torr. Concentrations many orders of magnitude higher would be required to obtain optical spectra and there is no conceivable way to produce such a high concentration of ion clusters. Somewhat limited and indirect spectroscopic information can be obtained for negative ion clusters. The photodetachment method for determination of electron affinities was mentioned in this chapter. Steiner [35] used this method to study the hydroxyl hydrate $H_3O_2^-$. The photodetachment threshold observed for this species led to an estimate of the energy for the reaction $H_3O_2^- \rightarrow H_2O + OH^-$. More detailed studies of the photodetachment efficiency curve of this species might also give information on vibrational spacings in the $H_3O_2^-$ ion. Similar studies of other negative ion clusters could be possible. A variant of the photodetachment method, laser-beam produced photodetachment [91] with energy analysis of the photoelectrons, might provide an even more promising avenue for studies of negative ion clusters.

Neutral clusters of the type $(MX)_2$ and $(MX)_3$ can be produced in concentrations much higher than those of the ion clusters. Therefore it is likely that optical spectra of such species could be obtained with special methods (laser) of high sensitivity. The optical spectroscopic analysis of such ion pair aggregates and aggregates of solvent molecules like $(H_2O)_2$ and $(CH_3OH)_2$ might be fruitful fields of the future.

Further studies and characterization of ionic and neutral aggregates in the gas phase is a challenge which, if accepted, is likely to provide knowledge of great importance to liquid phase phenomena.

REFERENCES

1. J. L. Franklin, J. G. Dillard, H. M. Rosenstock, J. T. Herron, K. Draxl, and F. H. Field, *Ionization Potentials, Appearance Potentials and Heats of Formation of Gaseous Positive Ions*, National Standards Reference Data System, U.S. Government Printing Office, Washington, D.C., 1969.

2. V. I. Vedeneyev, L. V. Gurvich, Y. N. Kondratiev, V. A. Medvedev, and Y. L. Frankevich, *Bond Energies, Ionization Potentials and Electron Affinities*, Arnold, London, 1966.

3. National Bureau of Standards Circular 500, U.S. Government Printing Office Washington, D.C., 1952; *JANAF Thermochemical Tables*, The Dow Chemical Company, Midland, Mich.; *Selected Values of Physical and Thermodynamic Properties of Hydrocarbons and Related Compounds*, Carnegie Press, Pittsburgh, Pa., 1953.

4. F. H. Field and J. L. Franklin, *Electron Impact Phenomena and Properties of Gaseous Ions*, Academic Press, New York, 1957.

5. (a) M. Krauss and V. H. Dibeler, "Appearance Potential Data of Organic Molecules," in *Mass Spectrometry of Organic Ions*, F. W. McLafferty, (Ed.), Academic Press, New York, 1963; (b) A. L. Wahrhaftig, "The Theory of Mass Spectra and the Interpretation of Ionization Efficiency Curves," in *Application of Mass Spectrometry to Organic Chemistry*, R. I. Reed, (Ed.), Academic Press, New York, 1966.

6. "The Ionization and Dissociation of Molecules," in *Mass Spectrometry*, C. A. McDowell (Ed.), McGraw-Hill, New York, 1963.

7. R. S. Berry, *Ann. Rev. Phys. Chem.*, **20**, 357 (1969).

8. *Atomic and Electron Physics in Methods of Experimental Physics*, Vol. 7, Part A, B. Benderson and W. L. Fite (Eds.), Academic Press, New York, 1968.

9. G. Herzberg, *Molecular Spectra and Molecular Structure*, Vol. III, Van Nostrand, Princeton, N.J., 1966.

10. F. H. Dorman, J. D. Morrison, and A. J. C. Nicholson, *J. Chem. Phys.*, **31**, 1335 (1959).

11. R. E. Fox, W. M. Hickam, D. J. Grove, and T. Kjeldaas, *Rev. Sci. Instr.*, **26**, 1101 (1955).

12. E. M. Clarke, *Can. J. Phys.*, **32**, 764 (1954).

13. P. Marmet and L. Kerwin, *Can. J. Phys.*, **38**, 787 (1960).

14. H. Hürzeler, M. G. Inghram, and J. D. Morrison, *J. Chem. Phys.*, **28**, 76 (1958).

15. D. C. Frost, D. Mak, and C. A. McDowell, *Can. J. Chem.*, **40**, 1064 (1962).

16. V. H. Dibeler and S. K. Liston, *J. Chem. Phys.*, **49**, 482 (1968).

17. M. I. Al-Joboury and D. W. Turner, *J. Chem. Soc.*, 4434 (1964).

18. D. W. Turner, in *Physical Methods in Advanced Inorganic Chemistry*, M. A. Hill and P. Day (Eds.), Interscience, London, 1968, Chapter 3.

19. D. W. Turner, *Chem. Brit.*, **4**, 435 (1968).

20. H. D. Hagstrum, *Rev. Mod. Phys.*, **23**, 185 (1951).

21. H. Ehrhardt and A. Kresling, *Z. Naturforsch.*, **22a**, 2036 (1967).

22. A. Niehaus, *Z. Naturforsch.*, **22**, A690 (1967), D. Beck and O. Osberghaus, *Z. Phys.*, **160**, 406 (1960).

23. E. W. Godbole and P. Kebarle, *Trans. Farad. Soc.*, **58**, 1897 (1962).

24. M. S. B. Munson, *J. Am. Chem. Soc.*, **87**, 2332 (1965).

25. J. O. Hirschfelder, H. Diamond, and H. Eyring, *J. Chem. Phys.*, **5**, 659 (1937); J. O. Hirschfelder, **6**, 795 (1938).

26. R. S. Barker and H. Eyring, *J. Chem. Phys.*, **22**, 2072 (1954).

27. D. Van Raalte and A. G. Harrison, *Can. J. Chem.*, **41**, 3118 (1963).

28. V. L. Talroze and E. L. Frankevich, *Zh. fiz. khim.*, **33**, 955 (1959).

29. J. L. Beauchamp and S. E. Buttrill, Jr., *J. Chem. Phys.*, **48**, 1783 (1968).

30. M. A. Haney and J. L. Franklin, *J. Phys. Chem.*, **73**, 4328 (1969).

31. F. M. Page and G. C. Goode, *Negative Ions and the Magnetron*, Wiley-Interscience, London, 1968, p. 79.

32. H. O. Pritchard, *Chem. Rev.*, **52**, 529 (1953).

33. M. Szwarc, *Carbanions, Living Polymers and Electron Transfer Processes*, Interscience, New York, 1968, pp. 300–344.

34. L. M. Branscomb, D. S. Burch, S. J. Smith, and S. Geltman, *Phys. Rev.*, **111**, 504 (1958); L. M. Branscomb, *Phys. Rev.*, **148**, 11 (1966).

35. B. Steiner, *J Chem. Phys.*, **49**, 5097 (1968), S. Golub and B. Steiner, *J. Chem. Phys.*, **49**, 5191 (1968).

36. R. S. Berry, C. W. Reimann, and G. N. Spokes, *J. Chem. Phys.*, **37**, 2278 (1962); **35**, 2237 (1961).

37. H. D. Hagstrum, *Rev. Mod. Phys.*, **23**, 185 (1951).

38. W. M. Hickam and R. E. Fox, *J. Chem. Phys.*, **25**, 642 (1956).

39. R. K. Curran, *Phys. Rev.*, **125**, 910 (1962).

40. F. M. Page, *Disc. Faraday Soc.*, **19**, 87 (1955).

41. J. T. Herron, H. M. Rosenstock, and W. R. Shields, *Nature*, **206**, 611 (1965).

42. (a) R. S. Becker and W. E. Wentworth, *J. Am. Chem. Soc.*, **85**, 2210 (1963); (b) W. E. Wentworth, E. Chen, and J. Lovelock, *J. Phys. Chem.*, **70**, 445 (1966); (c) R. S. Becker and E. Chen, *J. Chem. Phys.*, **45**, 2403 (1966).

43. J. J. Thompson, *Rays of Positive Electricity*, Longmans, Green, London; M. M. Mann, A. Hustrulid, and J. T. Tate, *Phys. Rev.*, **58**, 340 (1940).

44. V. L. Tallroze and A. K. Lubimova, *Dokl. Akad. Nauk. SSSR*, **86**, 909 (1952).

45. D. P. Stevenson and D. O. Schissler, *J. Chem. Phys.*, **23**, 1353 (1955).

46. G. G. Meisels, W. H. Hamill, and R. R. Williams, *J. Chem. Phys.*, **25**, 790 (1956).

47. F. H. Field, J. L. Franklin, and F. W. Lampe, *J. Am. Chem. Soc.*, **79**, 2419 (1957).

48. W. A. Chupka, *J. Chem. Phys.*, **30**, 458 (1959).

49. P. F. Knewstubb and T. M. Sugden, *Proc. Roy. Soc. (London)*, **A255**, 520 (1960).

50. P. F. Knewstubb and A. W. Tickner, *J. Chem. Phys.*, **36**, 674 (1962).

51. H. D. Beckey, *Z. Naturforsch.*, **15a**, 822 (1960).

52. P. Kebarle and E. W. Godbole, *J. Chem. Phys.*, **39**, 1131 (1963).

53. P. Kebarle and A. M. Hogg, *J. Chem. Phys.*, **42**, 798 (1965).

54. A. Good, D. A. Durden, and P. Kebarle, *J. Chem. Phys.*, **52**, 212 (1970).

55. D. A. Durden, P. Kebarle, and A. Good, *J. Chem. Phys.*, **50**, 805 (1969).

56. A. M. Hogg, R. M. Haynes, and P. Kebarle, *J. Am. Chem. Soc.*, **88**, 28 (1966).

57. S. K. Searles and P. Kebarle, *J. Phys. Chem.*, **72**, 742 (1968).

58. P. Kebarle, S. K. Searles, A. Zolla, J. Scarborough, and M. Arshadi, *J. Am. Chem. Soc.*, **89**, 6393 (1967).

59. S. K. Searles and P. Kebarle, *Can. J. Chem.*, **47**, 2619 (1969).

60. I. Džidić and P. Kebarle, *J. Phys. Chem.*, **74**, 1466 (1970).

61. I. Džidić, Ph.D. Thesis, Chemistry Department, University of Alberta (1970).

62. M. Arshadi, R. Yamdagni, and P. Kebarle, *J. Phys. Chem.*, **74**, 1475 (1970).

63. M. Arshadi, Ph.D. Thesis, Chemistry Department, University of Alberta (1969).

64. W. M. Latimer, K. S. Pitzer, and C. M. Slanski, *J. Chem. Phys.*, **7**, 108 (1939).

65. E. T. Verwey, *Chem. Weekblad.*, **37**, 530 (1940).

66. D. A. Desnoyers, "Hydration Effects and Thermodynamic Properties of Ions," in *Modern Aspects of Electrochemistry*, Vol. 5, J. O. M. Bockris (Ed.), Plenum Press, New York, 1969.

67. E. T. Verwey, *Rec. Trav. Chim.*, **61**, 127 (1942); D. R. Rosseinsky, *Chem. Rev.*, **65**, 467 (1965).

68. A. D. Buckingham, *Disc. Faraday Soc.*, **24**, 151 (1957).

69. F. Vaslow, *J. Phys. Chem.*, **67**, 2773 (1963).

70. P. Kebarle, R. N. Haynes, and J. G. Collins, *J. Am. Chem. Soc.*, **89**, 5753 (1967).

71. A. M. Hogg and P. Kebarle, *J. Chem. Phys.*, **43**, 449 (1965).

72. R. Yamdagni and P. Kebarle, *J. Am. Chem. Soc.*, forthcoming.

73. T. O. Tiernan and B. Mason Hughes, *J. Am. Chem. Soc.*, forthcoming.

74. J. I. Braumann and L. K. Blair, *J. Am. Chem. Soc.*, **90**, 5636, 6561 (1968); **92**, 5986 (1970).

75. M. Arshadi and P. Kebarle, *J. Phys. Chem.*, **74**, 1483 (1970).

76. L. V. Gurvich, G. A. Khachkurusov, V. A. Medvedev, I. V. Veyts, G. A. Bergman, V. S. Yungman, N. P. Rtishcheva, and L. F. Kuratova, *Thermodynamic Properties of Individual Substances*, Izd. AN, USSR, 1962. USSR, 1962.

77. R. F. Barrow and A. D. Caunt, *Proc. Roy. Soc. (London)*, **A219**, 120 (1953).

78. G. Herzberg, *The Spectra of Diatomic Molecules*, 2nd ed., Van Nostrand, Princeton, N.J., 1950.

79. W. Klemperer and S. A. Rice, *J. Chem. Phys.*, **26**, 618 (1957); S. A. Rice and W. Klemperer, *J. Chem. Phys.*, **27**, 573 (1957).

80. A. Honig, M. Mandel, M. L. Stitch, and C. H. Townes, *Phys. Rev.*, **96**, 629 (1954).

81. E. S. Rittner, *J. Chem. Phys.*, **19**, 1030 (1951).

82. L. Pauling, *Proc. Natl. Acad. Sci. India*, **25**, Sec. A, Pt. I, 1 (1956).

83. L. Friedman, *J. Chem. Phys.*, **23**, 477 (1955).

84. R. C. Miller and P. Kusch, *J. Chem. Phys.*, **25**, 860 (1956).

85. W. A. Chupka, Dissertation, Department of Chemistry, University of Chicago (1951).

86. J. Berkowtiz and W. A. Chupka, *J. Chem. Phys.*, **29**, 653 (1958).

87. S. Datz, W. T. Smith, Jr., and E. H. Taylor, *J. Chem. Phys.*, **34**, 558 (1961).

88. K. Hagemark, M. Blander, and E. B. Luchsinger, *J. Phys. Chem.*, **70**, 276 (1966).

89. T. A. Milne and D. Cubicciotti, *J. Chem. Phys.*, **29**, 846 (1958). T. A. Milne and D. Cubicciotti, *J. Chem. Phys.*, **30**, 1625 (1959).

90. M. Blander, *J. Chem. Phys.*, **41**, 170 (1964).

91. B. Brehm, M. A. Gusinow, and J. L. Hall, *Phys. Rev. Letters*, **19**, 737 (1967).

92. R. P. Blaunstein and L. G. Christophorou, *Radiation Research Rev.*, **3**, 69 (1971).

3

Spectrophotometric Studies of Ion-Pair Equilibria

JOHANNES SMID

Department of Chemistry, State University College of Forestry at Syracuse University, Syracuse, New York

1. INTRODUCTION

The behavior and structure of ion pairs and ion-pair complexes in solution have been extensively investigated since 1960 by a variety of techniques such as electron and nuclear magnetic resonance, electrolytic conductance, ultraviolet, visible, infrared, and Raman spectroscopy. Electron spin resonance studies probably have yielded the most detailed description of ion pair structures. For example, the position and motion of one ion with respect to its paired counterion could be investigated as well as the dynamic phenomena involving association of ions or solvation of ion pairs. On the other hand, spectrophotometric methods, particularly the ultraviolet and visible spectroscopy, although less informative, are attractive due to the simplicity of execution and of analysis of the data. Moreover, these techniques are not limited to paramagnetic molecules, like the ESR methods, and do not require high concentrations of reagents, which are often imperative in the NMR studies.

Warhurst et al. [1] were perhaps the first to observe shifts in optical spectra arising from ion pairing. They reported that the absorption peaks of alkali and alkaline earth salts of ketyls of benzophenone, fluorenone, and other ketones shift toward longer wavelength when the radius of the cation increases. Similar effects were later observed by Hogen Esch and Smid [2] for salts of fluorenyl and other carbanions and by Zaugg and Schaefer [3] for alkali salts of phenols and enols.

Under the prevailing experimental conditions, the preceding salts form tight ion pairs and the observed bathochromic shifts most likely result from the greater destabilization of the ground state, as compared to that of the excited state, when the cation is enlarged. In the ground state of an ion pair the cation is located close to the atom possessing the highest electron density. A redistribution of charge occurs on excitation but, as expected from the Frank-Condon principle, the cation has no time to readjust its position. Consequently, the excited state is less stabilized by the counterion than the ground state, and the smaller the cation the larger becomes the increase in the transition frequency. As pointed out by Warhurst et al. [1], the perturbation of the molecular orbital levels of an organic anion due to the field of the cation associated with it depends on the electrostatic interaction energy. Therefore its magnitude is approximately proportional to the reciprocal of the interionic distance given by the sum of the cation radius r and a constant representing the corresponding contribution of the anion. For ketyls, as shown in Fig. 1 [1], the frequency ν_m of the absorption maximum was found to be linearly correlated with the inverse of $r + 2$ (in Å). For enols [3] and fluorenyl salts [2] a straight-line relationship was observed by plotting ν_m against the reciprocal of the cation radius r.

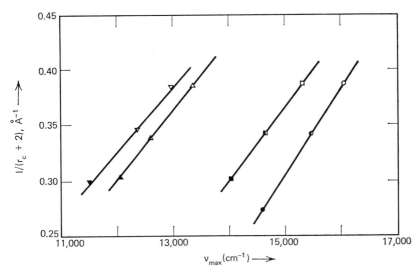

Figure 1. Dependence of the wave number of the principal absorption maximum of ketyls on cation radius. ○, benzophenone in dioxane; □, benzophenone in THF; △, di-biphenylketone in dioxane; ▽, di-biphenylketone in THF; open, half-closed, and closed symbols refer to the lithium, sodium, and cesium salts, respectively.

The interaction between anion and cation is decreased not only by increasing the size of the cation but also by its increased degree of solvation. Therefore increasing the polarity of the solvent or lowering the temperature of the solution often causes bathochromic shifts. This was observed not only for ketyls and carbanions but also for salts of various aromatic hydrocarbon radical anions studied by Hoijtink et al. [4]. Solvation (or complexation with an additive) can lead either to dispersion of cationic charge in the contact ion pair or to the formation of a new species, a solvent-separated or loose ion pair. In the latter case the cation is virtually surrounded by solvent molecules and appreciably separated from the anion. Consequently, the spectrum of this anion resembles that of the free anion.

There are, of course, other ways by which a solvent can affect the optical spectrum of a neutral species or of an ion pair. The various causes of blue or red shifts arising from a change in solvent polarity have been fully discussed by Bayliss and McRae [5]. For example, a blue shift is often observed when the transition of a solute, dissolved in a polar solvent, decreases its dipole or changes its direction. The oriented solvent dipoles destabilize then the Frank-Condon excited state. By the same argument, an increase in the dipole of the solute causes a red shift of its spectrum. Here the polarizability of the solvent, rather than its polarity, is the important factor responsible for the bathochromic shift.

Large dipole changes result from transitions involving charge transfer. For example, Kosower [6] investigated spectra of pyridinium iodides, which are extremely sensitive to the variation in solvent polarities. The relevant optical transition is associated with the transfer of negative charge from the iodide anion to the pyridinium cation and indeed the dissociation of these ion pairs eliminates the charge-transfer band. The interaction of an ion pair with the oriented molecules of solvent, which stabilizes the ground state, is lost therefore on excitation.

Similar effects were noted by Symons [7], who investigated the spectra of other iodides. In this case, the charge-transfer transition is attributed to electron transfer from the iodide anion to solvent molecules oriented around the ion. Ion-pairing affects the solvation shell surrounding the I^- and therefore modifies the position of the respective band. Both systems will be discussed in more detail later.

An interesting solvent effect has recently been observed in polarizable solvents [8]. The optical spectrum of retinylpyrrolidinium perchlorate appreciably shifts to the red in polarizable solvents such as pyridine and dichlorobenzene, but its shape is not affected by a change in the solvent polarity in nonpolarizable media (e.g., the replacement of dioxane by ethanol produces no effect). The excited state of this conjugated chromophore is greatly delocalized and hence, although the slow solvent-solute reorganization is not allowed in the transition, stabilization of the excited state (and therefore a bathochromic shift) may occur when the neighboring solvent molecules are polarizable.

Another interesting effect of ion pairing on optical spectra was observed by Feichtmayr and Schlag [9], who studied some triphenylmethyl dyes in polar and nonpolar media. The spectrum of 4,4′,4″-ethylamino-trinaphthylcarbonium chloride (victoria blue) in water varies with the concentration of the dye. Two absorption maxima at 615 and 555 nm were observed, the first ascribed to monomeric and the second to dimeric species. The dimerization was attributed to strong dispersion forces acting between the *free* carbonium ions. However, concentration-dependent peaks (at 550 and 630 nm) were also observed in nonpolar media, with the peak ratio drastically affected by excess chloride ions. These spectral changes were attributed to ion-pair formation, and the shift itself was explained in terms of an asymmetrical charge distribution in the contact ion pair. Although the charge is equally distributed over the whole molecule in the free carbonium ion, enough coulombic energy is apparently gained by localizing the positive charge close to one of the auxochromic groups when a contact ion pair is formed. The presence of an excess of chlorine ions restores then the symmetry of charge distribution.

In this chapter attention will be focused on the spectra of carbanions and radical anions which, when present as free ions are little, if at all, affected by solvent polarity. Any spectral change observed in the ion pairs can therefore be attributed to perturbations caused by a variation in the type of cation or by its solvation. Such systems are advantageous in studies of ion-pair solvation, the structures of ion-pair complexes, and generally in investigations of the interaction of cations with solvents or complexing agents. The negative ion functions primarily as a probe, revealing the degree of such interactions, although frequently the nature of the anion affects the solvation of the ion pair. The salts of fluorenyl carbanion are perhaps the best examples of such systems. They are distinguished by sharp optical absorption bands, the spectrum of the free anion is essentially solvent independent, and the association with cation, for example, with lithium, to contact ion pair causes a large hyposochromic shift (6.2 kcal/mole for the lithium pair).

In discussing ion-pair behavior we have used the terms contact and solvent separated ion pairs or tight and loose ion pairs. The latter terminology may indicate that more than two kinds of ion pairs can be visualized. A contact ion pair may have a different average interionic distance, depending on the nature of its environment and the temperature. Similarly, solvent-separated ion pairs of a particular salt may possess different structures in different media, depending on the size and geometry of the solvating or complexing molecules. Moreover, for a system the presence of two different species may be revealed by one experimental technique but not by another. For example, electron spin resonance measurements on the triphenylenesodium radical anion show that in 2-methyltetrahydrofuran the sodium coupling disappears below $-50°C$, implying a change to a loose ion-pair structure.* On the other hand, optical measurements in the same region indicate that the ion pairs retain their tight structure even at $-80°C$, a change to a loose structure being observed below this temperature [10]. This example stresses the necessity of applying a variety of techniques in studies of ion-pair behavior and ion-pair structures.

2. SPECTROSCOPIC STUDIES OF SOLVATION AND ION PAIRING OF IODIDES

Some of the earliest spectrophotometric studies of ion pairs and ion-pair–solvent interactions were reported by Symons and his co-workers [7, 11].

---

* Alternatively, this may indicate that the positive and negative contributions to the cation's hyperfine coupling constants cancel each other, see Chapter 8.

For a variety of iodides they observed marked shifts in the position of the first electronic absorption band arising from changes of solvent polarity, temperature, nature of the counterion, or from the addition of electrolytes. The spectral band under consideration has been attributed to a transition involving charge transfer to solvent, the energy of the excited state being affected by the orientation of the adjacent solvent molecules. In fact, the excited electron has been assumed to move in a centrosymmetric orbital encompassing the solvent molecules which are oriented by the anion. The process of excitation is depicted, therefore, as follows:

Electron and iodine atoms in the same cavity. The dotted line symbolizes the orbit of the delocalized electron.

This model led to an equation relating the energy maximum E_m of the first absorption band to the radius r of the solvent cavity:

$$E_m = I_p + \frac{h^2}{8mr^2}$$

where I_p denotes the ionization potential of the gaseous iodide ion [12, 13]. In spite of the simplifying assumptions introduced in the calculations, their results appear to be consistent with most of the experimental findings. For example, the radius of the solvent cavity is expected to increase with temperature for a constant composition of the solvation shell. This accounts for the observed negative value of dE_m/dT, which, although fixed for each solvent, increases in its absolute value with decreasing E_m at a given temperature [12].

In terms of the foregoing model, an increase of E_m implies a decrease in the diameter of the solvent cavity available to the excited electron. Since the solvent cavity can be diminished by applying hydrostatic pressure to the solution, a hypsochromic shift of the spectrum is then expected. This conclusion was confirmed by experiments [14] which showed that E_m varies linearly with pressure, dE_m/dp being characteristic for each solvent.

Numerous iodides were studied in order to determine the effect of ion pairing on E_m for various solvents and cations [15]. The results, shown in Table 1,

Table 1   Effect of Cation on the Energy Maximum E_m (cm^{-1}) of the Charge Transfer Band of Iodides in Solvents of Low Polarity [15]

Cation	CCl$_4$	CH$_2$Cl$_2$	1.4 Dioxane	THF
R$_4$N$^+$	34,500	40,900	42,500	40,600
Me$_4$N$^+$	insoluble	41,700		
Et$_4$N$^+$	34,600	41,000		
n-Pr$_4$P$^+$	34,500	41,200		
n-Bu$_4$P$^+$	35,000	41,350		40,800
Ph$_4$As$^+$	34,600	40,600		
PhMe$_3$N$^+$	35,300	41,900		
PhMe$_3$P$^+$	35,600	42,800	43,400	
BzMe$_3$N$^+$	35,200	41,800		
PhEtMe$_2$N$^+$	34,900	41,400		
(Cyclo-Hex)$_2$H$_2$N$^+$		45,000		43,300
n-Cetyl, Me$_3$N$^+$		41,550	43,050	
Me$_3$S$^+$		42,100		
Na$^+$				42,700
Rb$^+$				42,800
Cs$^+$				42,600

lead to the following ramifications. For tetraalkylammonium ions E_m is independent of the size of the alkyl group as long as none of them is smaller than n-propyl. Substitution of an alkyl group by CH$_3$ increases E_m due to the decreased interionic distance in the pair. No change is observed when ammonium cation is replaced by phosphonium, provided the alkyl groups are large. Neither does substitution of a large alkyl group by phenyl affect E_m. Marked changes are noted, however, when alkali cations replace the ammonium cations, although the resulting shifts are almost identical whether Na$^+$, Rb$^+$, or Cs$^+$ salts are examined.

These observations suggested to the authors that the orbitals of the cation do not contribute to the orbital of the excited electron because the transition energy is independent of the electronic structure of the cation. However, the size of the cation affects the energy of the ground state; its stabilization increases with decreasing radius of the cation.

The shifts of the iodide charge transfer band, although solvent dependent, do not correlate well with Kosower's solvent polarity parameter Z [16]. These shifts are deduced from studies of charge transfer to cation transitions which cause a large change in the dipole of the ion pair. In solvated iodide no change in dipole occurs during the transition and therefore no correlation with Z values is expected.

The low-energy band of iodides observed in carbontetrachloride is attributed to a contact ion pair. Its maximum is practically independent of temperature and pressure, and the authors ascribe it to a transition involving charge transfer to cation [17], in contrast to the higher energy transition attributed to transfer of charge to solvent. Symons et al. [15] conclude that the iodides exist in most solvents of low dielectric constants, for example, dioxane, methylene chloride, dioxolane, as solvent shared ion pairs—pairs of ions linked by one solvent molecule (the term solvent-separated ion pair was restricted by Griffith and Symons [7] to pairs of ions separated by more than one solvent molecule). In this writer's view, the experimental evidence does not appear to be sufficient to support this conclusion. Salts such as sodium or cesium iodide would be expected to dissociate considerably in solvents like THF if the ion pairs were separated by a solvent molecule. However, conductance studies demonstrated that the dissociation constants of most inorganic salts in solvents having dielectric constants comparable to that of THF are very low [18]. Moreover, the optical spectra of fluorenyl sodium or cesium show that when salts of this, charge delocalized, carbanion are dissolved in dioxane, dioxolane, or THF the tight ion pairs are the dominant species [2]. Conductance studies of tetraalkylammonium tetraphenylboron also show that even these salts mainly exist as contact ion pairs in THF [19]. It would be desirable to investigate the possibility of aggregation of the various iodide ion pairs in media such as CCl_4, dioxane, or THF. Aggregation could be quite extensive, especially for the alkali iodides, and the charge transfer band might be appreciably affected by profound changes in the composition of the solvation shells surrounding the iodide ions.

The spectra of some iodides are concentration dependent. For example, the E_m for n-hexylammonium iodide in dichloromethane decreases at lower concentrations [15]. This dependence was attributed to the equilibrium

$$(n - \text{hexyl})_4 N^+, \text{I}^- \rightleftharpoons (n - \text{hexyl})_4 N^+ + \text{I}^-$$

The dissociation constant K_d was calculated from the concentration dependent change in optical density at a fixed frequency, 41,000 cm^{-1}, using the relationship

$$\mathscr{E}_f = \mathscr{E}_p + \frac{2(OD_i - \mathscr{E}_p C_i)}{-K_d + (K_d^2 + 4K_d C_i)^{1/2}}$$

Here $\mathscr{E}_f$ and $\mathscr{E}_p$ denote the molar extinction coefficients of the free ion and ion pair, respectively, at 41,000 cm^{-1}, while OD_i denotes the observed optical density at concentration C_i. Thus the dissociation constant K_d of this salt in CH_2Cl_2 ($D = 9.08$) was calculated to be $10^{-3} M$—an improbably high value, especially when compared with the dissociation constant of Bu_4N^+, ClO_4^- in the same solvent, which was found conductometrically [20] to be $2.8 \times 10^{-5} M$ at 20°C, or with a K_d of $10^{-5} M$ found for Bu_4N^+, I^- in $C_2H_4Cl_2$ ($D = 10.36$). Closer analysis of the spectral data for the tetra-n-hexyl-ammonium iodide reveals that the molar extinction coefficients of the free ion and ion pair at $41,000 \ cm^{-1}$ differ by 12% only. The change in the optical density/concentration ratio, 1.55, 1.52, and 1.47, is therefore exceedingly small for the reported concentrations, which varied by a factor of 100. Small spectral variations due to impurities (e.g., triiodides) or to formation of triple ions could appreciably affect the calculations of K_d.

Addition of dicyclohexylammonium perchlorate to a CH_2Cl_2 solution of tetra-n-hexylammonium iodide increases E_m from 41,000 to $45,000 \ cm^{-1}$. Apparently $(cyclohexyl)_2H_2N^+$ ion has a greater affinity for I^- than the $(hexyl)_4N^+$ ion [15], probably because the interionic distance is shorter in the first pair. Calculation of the equilibrium constant of the reaction

$$(cyclohexyl)_2H_2N^+ + (n - hexyl)_4N^+, I^- \rightleftarrows (cyclohexyl)_2H_2N^+, I^- + (n - hexyl)_4N^+$$

led to a value of 10. However, in this calculation it was assumed that the perchlorates are completely dissociated. This is unlikely in view of the low K_d value found for Bu_4N^+, ClO_4^- in CH_2Cl_2.

Symons and his associates investigated the spectral changes of iodide solutions in mixed solvents. Large variations were observed [17, 21, 22] for mixtures of CCl_4 with CH_3OH, CH_2Cl_2, or CH_3CN, and these were interpreted in terms of a variable structure of the ion pairs. In the case of methanol, infrared studies showed that only one CH_3OH molecule coordinates with the iodide salts even when the CH_3OH/iodide ratio is about 100. The complex is described as an externally solvated contact ion pair, but in this writer's view the separation of ion pairs should not be excluded even in CCl_4. The separation process is not much affected by the dielectric constant of the medium; for example, fluorenyl salts form separated ion pairs even in toluene or dioxane when small quantities of cation coordinating reagents such as dimethylsulfoxide or hexamethylphosphoramide are added to their solution [2].

An interesting and most valuable spectrophotometric method of determining solvent-ion pair interactions was developed by Kosower [6, 16, 23], who discovered that the pyridinium iodides possess electronic transitions which are extremely sensitive to changes in solvent polarity. A convenient substance for such studies is 1-ethyl-4-carbomethoxy pyridinium iodide which is soluble in a wide variety of solvents. Basically, the electronic transition involves the transfer of an electron from the iodide ion to the pyridinium ring. The extent of such a transfer was demonstrated by flash photolysis of pyridinium iodide, which generates pyridinyl radicals [16]. Other evidence for the charge transfer is based on the observation that electron-withdrawing substituents in the pyridine ring decrease the transition energy of the charge transfer band, whereas the electron donating groups increase it. For example, replacing the $COOCH_3$ substituent in the 4 position by a CN group changes the absorption maximum from 448.9 to 491.2 nm, whereas a CH_3 group in the 4 position decreases the λ_{max} to 359 nm.

Kosower explained the solvent sensitivity of the charge transfer band by postulating a "dipole flip" accompanying the transition. In the ground state, the I^- ion is believed to be located above the pyridine ring in the neighborhood of the N^+, and consequently the resulting dipole is perpendicular to the plane. The strong interaction of this dipole with the surrounding solvent molecules leads to their orientation (Kosower refers to it as a "cybotactic region") and hence to stabilization of the ground state. On excitation the negative charge moves to the pyridine ring, causing the dipole to "flip" into the plane of the ring. Moreover, the dipole moment substantially decreases, although not necessarily to zero. According to the Frank-Condon principle, the excitation time is too short to allow for solvent rearrangement, and the electrostatic solvent-solute interaction does not stabilize the excited state. In fact, the excited state is even less stabilized in a polar than in a nonpolar solvent, since in a polar solvent the cybotactic region surrounding the excited state is unstable in respect to a nonorganized solvent. Both factors—stabilization of the ground state and destabilization of the excited state—cause a large increase in the transition energies when the polarity of the solvent is increased.

This striking property of pyridinium iodides was used by Kosower to establish a scale of solvent polarities. A solvent polarity parameter Z was introduced and defined as the transition energy E_T in kcal/mole of the longest wavelength absorption band of 1-ethyl-4-carbomethoxy pyridinium iodide dissolved in that solvent. Some Z-values are given in Table 2.

The charge transfer band in water and some other highly polar solvents is obscured by the strong π-π* transition of the pyridinium ion. This difficulty can be overcome by using mixtures of water with, for example, acetone or alcohol, and the resulting Z-values are then compared with the well known Y-values of Winstein and Grunwald referring to the same mixture. The Z-value of water is then obtained by extrapolation. Ion-pair clustering may be encountered in solvents of very low polarity in addition to difficulties caused by solubility problems (which can be circumvented by using the more soluble salt 1-ethyl-4-carbobutoxy pyridinium iodide). This leads to higher Z-values because the associated ion pairs stabilize each other. It is doubtful whether extrapolation to zero concentration is sufficiently reliable to give the correct Z-value for such systems because the aggregation in solvents like hexane and benzene may persist even at extremely low concentrations.

Table 2 Z Values[a]

Solvent	Z-Value (kcal/mole)
Water	94.6
Methanol	83.6
Ethanol	79.6
t-Butylalcohol	71.3
Methylene chloride	64.2
Acetonitrile	71.3
Formamide	83.3
Dimethyl formamide	68.5
Dimethyl sulfoxide	71.1
Hexamethylphosphoramide	62.8
Pyridine	64.0
Acetone	65.7
Acetic acid	79.2
Benzene	54
Sulpholane	77.5
1,2 Dimethoxyethane	62.1
Acetone/H_2O–99/1	68.1
Acetone/H_2O–90/1	76.6
Acetone/H_2O–80/1	80.7
Acetone/H_2O–70/1	83.2
Acetone/H_2O–60/1	85.5

[a] For a more complete list of Z-values, see reference 16, p. 301.

Studies of the charge-transfer spectra of pyridinium iodides may reveal their degree of association. In media of high dielectric constants the salts are expected to dissociate into the free ions, a process resulting in a decrease of the intensity of the charge transfer band. This indeed was observed by Kosower [23]. A plot of the Z-value of the solvent versus the apparent molar extinction coefficient, calculated from the absorption maximum at 2×10^{-3} M salt concentration, shows a fair linear relationship. However, this relation is not valid for solvents such as dimethylformamide, dimethylsulfoxide, acetone, acetonitrile, and a mixture of 90% pyridine and 10% H_2O. In these media the extinction coefficients are lower than expected. Apparently these reagents complex with iodide ions and therefore may enhance the dissociation of the pair into free ions. It is not known whether any separated ion pairs are encountered in these systems and to what extent, if any, this could affect the transition energy. If the separated pairs do not absorb in the charge transfer band region, we would observe a decrease of the apparent molar extinction coefficient as the fraction of separated ion pairs increases. Such a decrease should not be concentration dependent.

The pyridinium iodide system provides a useful tool in the study of the effect of ion pairing on nucleophilic reactions. Mackay and Poziomek [24] determined the rate constant of the reaction of 1-ethyl-4-cyanopyridinium iodide with methyltosylate for ion pairs and free ions in various solvents. The ion-pair dissociation constant for the equilibrium $Py \cdot R^+, I^- \rightleftarrows Py \cdot R^+ + I^-$ was determined spectrophotometrically. Solvent-separated ion pairs, if present, were treated as ion pairs and not as free ions, although the visible charge transfer band was assumed to arise only from contact ion pairs. The considerable variation of the extinction coefficient with the polarity of the solvent was interpreted as possible evidence for the presence of separated ion pairs.

Another solvent parameter, $E_T(30)$, was introduced by Dimroth et al. [25], who used the electronic transition of the pyridinium phenol betaine as an indicator of solvent polarity. The transition results in an intramolecular charge transfer, and changes in the band intensity due to ion pair dissociation are not observed. Values for $E_T(30)$ range from 63.1 kcal/mole for water to 30.9 kcal/mole for hexane; a plot of E_T values versus Kosower's Z-values is fairly linear over the entire range of solvent polarities from pure benzene to water.

3. ION PAIRING IN FLUORENYL SALTS

Fluorenyl and substituted fluorenyl carbanion salts provide convenient systems for spectrophotometric studies of structures and properties of ion pairs and their solvates. The fluorenyl carbanion is stable in amines (e.g., pyridine, ethylene diamine), ethereal solvents, and other aprotic media such as dimethylsulfoxide and hexamethylphosphoramide, but, it is rapidly protonated in most of the common protic solvents such as water and alcohols [26]. The optical spectrum of the fluorenyl carbanion reveals a sharp absorption band in the 345–375 nm, region ($\varepsilon \approx 10,000$), the exact position of the maximum depending on the size of the cation. The sharpness of the band and the considerable variation of the position of its maximum with the anion-cation distance makes the fluorenyl carbanion a sensitive probe to study the structure and solvation of ion pairs.

Changes in the absorption spectra of the fluorenyl salts were observed first on cooling their tetrahydrofuran solution [2]. The spectrum of the sodium salt in THF at three different temperatures is depicted in Fig. 2. The 356 nm absorption band decreases on lowering the temperature and a new band emerges at 373 nm. Similarly, spectral changes are observed in the

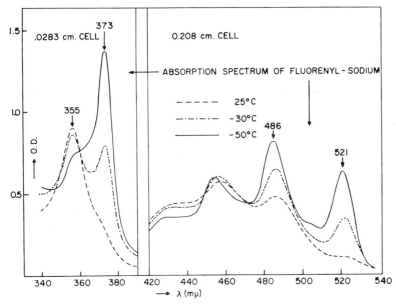

Figure 2. Optical absorption spectrum of fluorenylsodium in THF at 25, −30, and −50°C.

visible region, but their analysis is more difficult due to an extensive overlap of the relevant peaks. The reversibility of the process leading to the spectral changes and the existence of two distinct absorption bands indicate that two species coexist in rapid equilibrium, one being more abundant at low temperature, the other more stable at room temperature. The ratio of the peak heights is independent of the carbanion concentration and is not affected by the addition of sodium tetraphenylboron, a salt that is substantially ionized in THF ($K_d \approx 8 \times 10^{-5}$ M at 25°C) [19]. Furthermore, conductance studies of the fluorenyl salts [2, 59] showed that only a small percentage of free ions are present at salt concentrations of 10^{-3}–10^{-2} M. Hence, the foregoing spectral changes cannot be explained by equilibria of the type

$$(F^-, M^+)_2 \rightleftharpoons 2F^-, M^+$$
$$2F^-, M^+ \rightleftharpoons F^-, M^+, F^- + M^+$$

or

$$F^-, M^+ \rightleftharpoons F^- + M^+$$

but a rapid equilibrium between two kinds of ion pair differing in their solvation states provides a satisfactory explanation of the observed phenomena. The respective species are assumed to be contact or tight ion pairs and solvent-separated or loose ion pairs, the latter being favored at low temperatures:

$$F^-, M^+ + nS \rightleftarrows F^-//M^+$$

The interaction between the molecules of solvent and the cation constitutes the driving force for the formation of the solvent-separated ion pairs. Hence, the equilibrium between contact and solvent-separated pairs may be utilized in studying the factors affecting the interactions of cations with solvent molecules.

3.1. The Role of the Cation

The absorption maxima of a variety of anionic species show hypsochromic shifts when they become paired with cations, and the cause of this phenomenon was already discussed. It is, therefore, not surprising that the absorption bands of the fluorenyl contact ion pair shift to higher wavelength as the radius of the cation increases. This is shown in Table 3. In fact, as seen in Fig. 3, the

Table 3   Dependence of λ_{max} on the Radius of the Cation for 9-Fluorenyl Salts in THF at 25°

Cation	r_c (Å)	λ_{max} (nm)
Li$^+$	0.60	349
Na$^+$	0.96	356
K$^+$	1.33	362
Cs$^+$	1.66	364
N$^+$Bu$_4$	3.5	368
‖M$^+$	~4.5	373
Free ion		374

wavenumber v_m of the transition is linearly related to the inverse of the cation radius r_c. A similar relation was noted by Zaugg and Schaefer, who investigated the spectra of alkali salts of phenols and enols [3]. Extrapolation of the plot to $1/r_c = 0$ yields $\lambda_m = 374$ nm for the free fluorenyl anion, a value almost identical to that observed for the solvent-separated ion pairs. Further confirmation of this value is provided by the spectra obtained in THF at extremely low dilution ($\sim 10^{-5}$ M). At these very low salt concentrations the ratio of the peak heights at 373 and 356 nm for the sodium salt in THF becomes concentration dependent, an observation accounted for by the formation of a substantial fraction of free ions.

Cations affect to the same extent the different electronic transitions of the fluorenyl carbanion. The presence of a cation in the contact ion pair increases the energy of the transitions over that of the free anion, the increase being 4.0 kcal/mole for the Na$^+$ ion and 6.2 kcal/mole for the Li$^+$ ion.

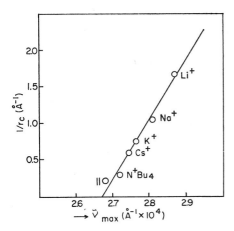

Figure 3. Correlation between wave number and the inverse of the cationic radius for contact fluorenyl ion pairs in THF at 25°C.

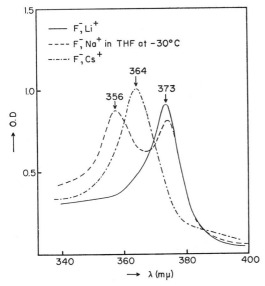

Figure 4. Absorption spectra of fluorenyl salts in THF at −30°C for various counterions.

At constant temperature and in a fixed solvent, the equilibrium between these ion pairs is greatly affected by the nature of the cation, as demonstrated by the spectra shown in Fig. 4. The interaction of THF with the large Cs^+ ions is weak [19]; consequently, F^-, Cs^+ is a contact ion pair even at $-70°C$. The same is found for the K^+ and the tetrabutyl ammonium salt. On the other hand, the fluorenyl sodium in THF yields equal fractions of the two kinds of ion pair at $-30°$, and the lithium salt virtually exists as a separated ion pair below $0°C$. In pure solvents, the lithium salt usually yields the highest fraction of separated ion pairs; their proportions decrease along the series sodium, potassium, and cesium. This order may change when the solvating agent is polydentate, for example, when the separated ion pairs are formed through complexation with macrocyclic ethers. The behavior of fluorenyl salts involving divalent cations will be discussed later.

3.2. Effect of Solvent Structure and Polarity

The formation of solvent-separated ion pairs is greatly facilitated by increased solvent polarity. Often, small changes in the solvent structure can drastically affect the equilibrium between the contact and separated ion pairs.

Typical examples of the effects exerted by solvents on the absorption spectrum of fluorenyl lithium are depicted in Figs. 5 and 6. The lithium salt is singled out for these studies because the strong specific interactions of the small Li^+ ion with solvent molecules permits studies of low polarity media in which the Na salt exists only as a contact ion pair. Moreover, the overlap of absorption peaks corresponding to the two types of ion pair is the smallest for the Li salt, and this facilitates the calculation of equilibrium constants.

The absorption maximum of the separated ion pair is usually independent of the cation radius and of the solvent even when a low-polarity solvent like tetrahydropyran ($D = 5.6$) is replaced by a highly polar medium, such as hexamethylphosphoramide ($D = 30$) or dimethylsulfoxide ($D = 45$). This indicates that, contrary to previous belief [27], solvent polarity per se does not affect the spectrum of the ion pair if its structure is retained. In most cases the spectral shifts result from the decreased influence of the cation on the carbanion, caused by specific cation-solvent interactions which lead to a partial separation of the ions. However, in some amines considerable shifts in the absorption maximum of the separated ion pairs were observed. Apparently, in such solvents solvation of the carbanion becomes important [28, 29].

The absorption maxima of the contact ion pairs often show small bathochromic shifts when the solvent polarity is increased. This effect is particularly pronounced in solvents containing small amounts of reagents capable of

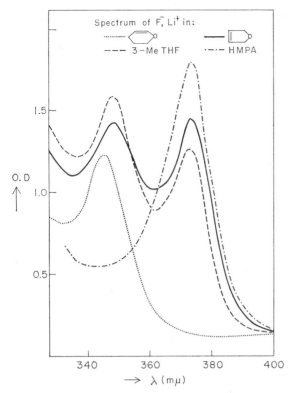

Figure 5. Optical spectrum of contact and separated ion pairs of fluorenyllithium at 25°C in 3,4 dihydropyran, ...; 3-methyltetrahydrofuran, – – – –; 2,5 dihydrofuran, ———; and hexamethylphosphoramide, – · – · – ·.

coordinating strongly with the cation, for example, dimethylsulfoxide or hexamethylphosphoramide [2, 28]. It seems that the *external* solvation of the contact ion pairs by such a reagent is responsible for this phenomenon. Complexes of this type have been observed for many organo-alkali compounds in hydrocarbon solvents mixed with ethers or amines [30–33]. Their formation greatly affects the state of aggregation and the reactivity of these species. In the case of fluorenyl contact ion pairs, the external solvation tends to disperse the cationic charge and weakens the coulombic interaction between the two ions. The tightness of the contact ion pair therefore is expected to be solvent dependent, although the resulting shifts in its absorption maximum amounts only to a few nm for the fluorenyl salt. Recent studies by Burley and Young [112] of alkali metal salts of 1,3-diphenylbut-1-ene have shown that the absorption maximum of the tight ion pair of this carbanion salt is very sensitive to the nature of the solvent. The λ_m for the

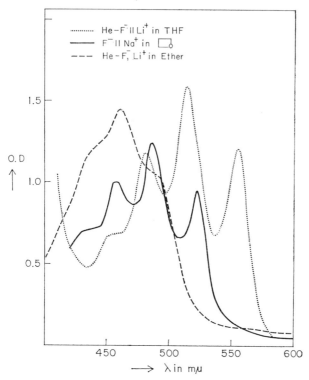

Figure 6. Visible spectrum of 9-(2-hexyl)fluorenyllithium in diethylether, – – –; and in THF, ...; also of fluorenylsodium in oxetane, ———; all at 25°C.

lithium salt changes from 467 nm in di-*s*-butyl ether to 530 nm in tetra-hydropyran. The loose ion pair absorbs at 565 nm. The changes in the contact ion pair spectrum were attributed to formation of solvated contact ion pairs. Other evidence for a variable ion pair distance in the contact ion pairs may be deduced from electron spin resonance studies of salts of radical anions, as in the study of sodium naphthalene in ethereal solvents [34, 35]. The alkali coupling constants of the contact ion pairs are sensitive to the nature of the solvent and usually decrease with increasing polarity.

The average interionic distance in the *solvent-separated* ion pairs also depends on the solvent or the nature of the entity complexed with the ion pair. However, the optical spectra of loose ion pairs are not very sensitive to the variation of the interionic distance; as mentioned earlier, they are virtually indistinguishable from the spectra of the respective free ions. Consequently, the spectral studies do not provide much information about subtle variations in the structure of different loose ion pairs.

The ratio of the concentrations of the two kinds of fluorenyl ion pair coexisting in a solvent in equilibrium with each other provides a measure of the effectiveness of the solvent in coordinating with alkali ions [2, 28]. This quantity may be calculated from the relevant absorption spectra, provided that the absorption spectra of the individual ion pairs are known. The ratios of the two ion-pair concentrations for both fluorenyllithium and its 9-substituted 2-hexyl derivative in various unsubstituted cyclic ethers at 25°C are given in Table 4. The effectiveness of these cyclic ethers in forming

Table 4   Solvent-Separated Ion-Pair Formation for Fluorenyllithium (F^-, Li^+) and 9-(2-Hexyl)fluorenyllithium (He-F^-, Li^+) in Unsubstituted Cyclic Ethers at 25°

Solvent	$\dfrac{[F^- \parallel Li^+]}{[F^-, Li^+]}$	$\dfrac{[He\text{-}F^- \parallel Li^+]}{[He\text{-}F^-, Li^+]}$
Oxetane[a]	>50	—
Tetrahydrofuran	4.6	>50
Tetrahydropyran	0.45	20
Hexamethylene oxide	0.24	2.3
2,5-Dihydrofuran	1.1	50
3,4-Dihydropyran	0.01	0.14
Furan	<0.01	0.02
Dioxolane	0.08	10
Dioxane	0.01	0.3

[a] The lithium salts of fluorenyl carbanions were not stable in oxetane. The value listed for oxetane represents that of fluorenyl-sodium, which is stable in this solvent.

separated ion pairs closely follows the change in the basicity of the respective ring oxygen atom, which may be measured by a variety of methods [36–39]. Thus the proportion of separated ion pairs is the largest in the 4-membered oxetane, the most basic of the studied cyclic ethers, and becomes progressively smaller for tetrahydrofuran (THF), tetrahydropyran (THP), and ethylene oxide. The reported basicities of these ethers decrease in the same order [38]. The 7-membered ring ether, hexamethyleneoxide, appears to be an exception, since it is less effective in separating the ion pairs than THF, although Arnet and Wu [37] found it to be more basic than the latter.

The data for unsaturated cyclic ethers also appear to reflect changes in the basicity of the respective oxygen atoms. For example, delocalization of the lone electron pair renders the 3,4-dihydropyran considerably less basic than tetrahydropyran, and furan is still less basic. Their solvating capacity follows this order. On the other hand, delocalization of the oxygen lone electron pair is negligible in 2,5-dihydrofuran, and indeed its solvating ability is not much different from that of THF.

Steric factors are also of great importance in determining the cation solvating power of ethers. In the solvent-separated ion pairs the alkali ion is coordinated with the oxygen atoms of several solvent molecules and their packing in the solvation shell may become a critical factor (see Chapter 2). Bulky substituents adjacent to the coordination site undoubtedly increase the average distance between alkali ion and the coordination site or they may decrease the number of solvating molecules in the solvation shell. This effect may account for the lower solvating power of hexamethylene oxide as compared to the less basic but smaller THF molecule. The effect of steric hindrance is clearly demonstrated by the behavior of the substituted tetrahydrofurans. As seen from the data listed in Table 5, a methyl or methoxy

Table 5   Solvent-Separated Ion-Pair Formation for Fluorenyllithium (F^-, Li^+) and 9-(2-Hexyl)fluorenyllithium (He-F^-, Li^+) in Substituted Tetrahydrofurans and Some Other Solvents at 25°

Solvent	$\dfrac{[F^- \parallel Li^+]}{[F^-, Li^+]}$	$\dfrac{[He\text{-}F^- \parallel Li^+]}{[He\text{-}F^-, Li^+]}$
Tetrahydrofuran	4.6	>50
3-Methyltetrahydrofuran	0.85	>50
2-Methyltetrahydrofuran	0.33	1.50
2,5-Dimethyltetrahydrofuran	0.02	0.07
2,5-Dimethoxytetrahydrofuran	0.04	
2-(Methoxymethyl)tetrahydrofuran	>50	
3,3-Dimethyloxetane	1.2	>50
o-Dimethoxybenzene	>50	
m-Dimethoxybenzene	0.01	
Hexamethylphosphoramide	>50	

group placed in the α position of THF makes the respective derivative a much poorer alkali ion solvating agent than the unsubstituted ether, although the basicity of the oxygen atom increases due to the substitution of hydrogen by an electron donating group. It should be stressed that the basicity, measured by the proton affinity of the ether, is only slightly affected by the steric conditions, because the proton is so small. The situation is different when the solvation of an alkali ion is considered. Substitution at the β position, as in 3Me-THF, has a much smaller effect because the substituent is situated near the periphery of the solvation shell.

The steric problems are modified when the solvating molecule possesses more than one coordination site. Such a chelating compound becomes a more powerful solvating agent, as exemplified by the behavior of 2-CH_3OCH_2-THF, and it appears to be most effective when the oxygen atoms are separated by two carbon atoms [40], the linear polyglycolethers being an outstanding example. The behavior of o-dimethoxybenzene, which, in spite of its low basicity, forms separated ion pairs with fluorenyllithium at 25°C, calls for additional comment. In this molecule the oxygen atoms are rigidly placed in a position favorable for chelation, and hence the internal rotational entropy is not affected by solvation of the cation. In contradistinction with o-dimethoxy-benzene, solvation by the more basic dimethoxyethane reduces the freedom of rotation of that ether and thus decreases the entropy of the system by forcing the DME molecule to acquire a unique conformation, which is energetically the least stable. Chelation with m-dimethoxybenzene is not expected because of the wrong orientation of its methoxy groups and therefore only contact ion pairs are formed in this solvent.

Steric hindrance in the solvation shell is more important for a small cation than for a large one. This is illustrated by the behavior of 3,3-dimethyl-oxetane. Oxetane is an excellent solvating agent and sodiumfluorenyl ion pairs exist entirely in the solvent-separated form when dissolved in this ether. However, the ratio of solvent-separated to contact pairs falls to 0.6 when the sodium salt is dissolved in 3,3-dimethyloxetane. The two methyl groups considerably increase the volume of the substituted molecules and therefore reduce their solvating ability by hindering tight packing in the solvation shell. The resulting crowding in the solvation shell appears to be even greater for the smaller lithium ion. The relevant ratio of concentrations of separated and contact pairs for lithium fluorenyl in this solvent is 1.2, only twice as high as for the sodium salt, although in other solvents the degree of solvation of lithium salts is much greater than that of the sodium salt. For example, the ratio of solvent-separated to contact pairs in THF is 4.6 for F^-, Li^+ and 0.05 for the sodium salt.

Extensive studies by Shatenshtein et al. [41–43], discussed later in this chapter, show that the yield of radical ions formed by the reduction of a hydrocarbon by metallic sodium in a series of ethers decreases in the order DME $\approx$ THF > THP > dioxolane > dioxane. This aggrees with results obtained for the fluorenyl salts, although here DME is found to be a better solvating agent than THF. For sodium fluorenyl the observed order is DME > THF > dixolane > 2MeTHF $\approx$ THP > dioxane.

In amines the higher basicity of the nitrogen atom facilitates ion-pair solvation, and in solvents like pyridine and ethylene diamine both the sodium and lithium fluorenyl exist entirely as separated ion pairs at room temperature [44]. In cyclohexylamine the lithium salt is also predominantly a loose ion pair [45], but surprisingly tetramethylethylenediamine appears to be inefficient in separating the ion pairs. This compound is known to form strong complexes with organo-lithium reagents such as butyllithium in hydrocarbon solvents [33] and dramatically increases their reactivity. The increased reactivity is attributed to an enhancement of the ionic character of the carbon-lithium bond and to a decrease in the state of aggregation of the organo-lithium compounds. However, fluorenyllithium exists in pure TMEDA at room temperature predominantly as a contact ion pair. Although this amine does increase the solubility of the fluorenyllithium in hydrocarbon solvents (apparently by external complexation with the contact ion pair), the four methyl groups apparently hinder the ion-pair separation. A close approach to the nitrogen atoms is prevented and consequently the interaction energy is substantially smaller for this amine than for ethylene diamine. In some amines considerable shifts in the absorption maxima of the separated ion pairs are observed. For example, the addition of small quantities of ethylenediamine to a dioxane solution of fluorenyllithium converts this salt to a solvent separated ion pair, but its absorption maximum is shifted by 6 nm toward shorter wavelength as compared with the λ_{max} observed in other solvents [44]. The small size of the diamine may permit the interionic distance in the separated ion pair to be relatively short. On the other hand, as suggested by Pascault [29], who investigated fluorenyl salts in ammonia, amines of this type may solvate the carbanion and thus affect its spectrum.

In highly polar solvents such as dimethylsulfoxide, hexamethylphosphoramide, or polyglycoldimethylethers, most of the fluorenyl salts exist as separated ion pairs only. The behavior of these coordinating agents will be discussed later.

3.3. Temperature and Pressure Dependence of the Ion-Pair Equilibrium

The enthalpy change ΔH_i governing the equilibrium between the two kinds of ion pair is a significant thermodynamic quantity, knowledge of which permits the discussion of the energetics of ion-pair separation. It also plays an important role in determining the temperature dependence of reactions in which the two kinds of ion pair simultaneously participate, each with its own characteristic rate constant [46, 47].

The first evidence for the existence of tight and loose ion pairs in solutions of carbanion salts came from the study of the temperature-dependent spectrum of fluorenyl sodium in THF (Fig. 2) and even more dramatic spectral changes are depicted in Fig. 7. It shows that 9(2-hexyl) fluorenyllithium in 2,5-dimethyltetrahydrofuran is transformed from a predominantly contact ion pair state at $-20°$ to virtually 100% separated ion pairs at $-40°$. The ratio of the concentrations of the two ion pairs is calculated from the spectra, but in such calculations we must take into account the changes in the

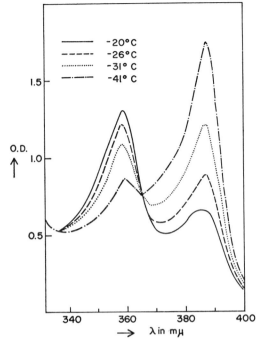

Figure 7. Temperature dependence of the contact–solvent-separated ion-pair equilibrium of 9-(2-hexyl)fluorenyllithium in 2,5-dimethyltetrahydrofuran.

spectra of the individual ion pairs arising from the variation of temperature [2]. The final results pertaining to some solvents are plotted in Fig. 8 as log K_i (K_i being the ratio of the ion-pair concentrations) versus $1/T$, and the ΔH_i values calculated from the slopes of the respective lines are listed in Table 6.

The data indicate that in ethereal solvents the formation of separated ion pairs is usually exothermic. Denison and Ramsey [48] showed that in the absence of specific ion-solvent interactions the enthalpy change for the complete separation of two ions is given by $\Delta H_a = Ne^2(1 + d \ln D/d \ln T)/aD$, where D denotes the dielectric constant of the medium and a the original distance between the charges. The separation is therefore exothermic when the factor $1 + d \ln D/d \ln T$ is negative. This is known to be the case for 2-methyl THF, THF, and DME, the respective $d \ln D/d \ln T$ values being -1.12, -1.16, and -1.28. However, the calculated exothermicity, even for a short interionic distance, would amount to only 1–2 kcal/mole, and for THP solutions the ion-pair equilibrium would be endothermic because for this solvent $d \ln D/d \ln T = -0.97$. The data of Table 6 indicate therefore

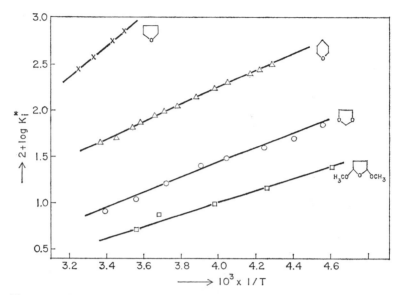

Figure 8. Plots of log K_i versus $1/T$ for fluorenyllithium in tetrahydrofuran, tetrahydropyran, 1,3-dioxolane, and 2,5-dimethoxytetrahydrofuran.

Table 6 Enthalpy and Entropy Changes for Solvent-Separated Ion-Pair Formation from the Contact Ion Pairs

Solvent	F^-, Li^+		HeF^-, Li^+	
	$-\Delta H_i$ (kcal/mole)	$-\Delta S_i$ (e.u.)	$-\Delta H_i$ (kcal/mole)	$-\Delta S_i$ (e.u.)
2,5-Dimethoxy-THF	2.9	16	—	—
2,5-Dimethyl-THF	2.0	14	10.0	50
Dioxolane	3.5	17	—	—
Δ^2-Dihydropyran	3.0	16	8.2	33
Tetrahydropyran	6.6	28		
THF	7.5	22		
2-Methyl-THF	7.5	27	9.8	32
Dioxane-THF (1:1)	3.6	14		
Dioxane-THF (2:1)	—	—	11.0	36
Hexamethylene oxide	4.3	16		

that the specific ion-solvent interaction provides a major contribution to the exothermicity of the ion-pair separation in ethereal solutions.

It is difficult to relate quantitatively the ΔH_i values with the structure of the solvent. The total enthalpy is determined by the coulombic interaction between anion and cation in the tight and loose ion pair and by the physical and specific solvation energy of the two ion pairs. The specific solvation energy depends on the structure of the solvent molecule as this affects their accommodation in the solvation shell. The role of dielectric constant is complex because at the very short distances over which the separation takes place the dielectric saturation must be taken into account. Generally, it appears that the exothermicities increase in more polar solvents, although for solvents like THP, 2-methyl THF, and THF differences in ΔH_i are almost within the experimental uncertainties. Tighter contact ion pairs are expected to be formed in the less polar solvents, and consequently smaller $-\Delta H_i$ values are anticipated for these systems.

In many solvents the fraction of solvent-separated ion pairs is much higher for the 9-substituted fluorenyllithium salt than for the unsubstituted fluorenyllithium (see Tables 4 and 5). This is caused by the higher $-\Delta H_i$ values for the former salts. For example, even in a poor solvent like 2,5-dimethyltetrahydrofuran ΔH_i is -10 kcal/mole for the 9(2-hexyl) derivative and only -3 kcal/mole for the unsubstituted salt. The exothermicity found for the substituted sodium salt is the same or even slightly smaller than that observed for fluorenyl sodium [44]. This point needs explanation and will be considered in the next section.

The effect of pressure upon the equilibrium between contact and separated ion pairs of fluorenyllithium and sodium in THF has recently been investigated by Szwarc et al. [49] and by Le Noble and Das [108]. Spectroscopic observations, extended to pressures as high as 5000 atm, show that the equilibrium shifts to the loose ion pairs as the pressure increases (see Fig. 9). This result was explained in terms of electrostriction, the observed volume contraction resulting from a tight binding of solvent molecules around the cations of the separated ion pairs. The ΔV's for the sodium and lithium salts were found to be -24.2 ml/mole and -16.4 ml/mole, respectively [49]. The smaller value for the lithium salt was interpreted to indicate a higher degree of solvation for the tight lithium ion pair than for the sodium pair. The entropy changes for the two salts are -33 e.u. and -22 e.u., respectively, the ratio of these two values being remarkably similar to that observed for the corresponding ΔV values. The binding of one mole of THF to the separated ion pairs is calculated to cause a volume contraction of about 8 ml, approximately 10% of the solvent molar volume. Such a contraction of the volume of the pure solvent requires pressures of about 2000 atm. It therefore seems that the forces contracting the solvent around the ion pairs are equivalent to those produced by a pressure of 2000 atm.

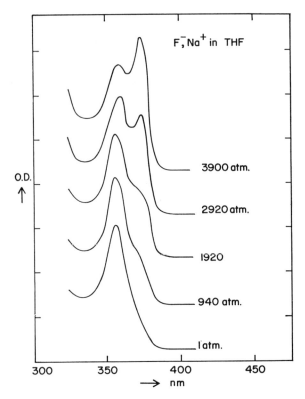

Figure 9. Pressure dependence of the contact–solvent-separated ion-pair equilibrium of fluorenylsodium in THF at 25°C.

3.4. The Effect of Anion Structure

Simultaneous existence of contact and solvent-separated ion pairs is not restricted to solutions of the fluorenyl salts. Coexistence of two species has been observed spectrophotometrically for many substituted fluorenyl salts, for carbanions other than fluorenyl, and for radical ions. Sometimes the spectra reveal two sets of distinct absorption maxima corresponding to the two kinds of ion pair, as illustrated in Fig. 10 for 2,3-benzofluorenyl sodium. Often, however, the relevant peaks overlap, preventing a clear recognition of the separate absorption spectra in mixtures of the two ion pairs, and then only a spectral shift of the anion absorption peak may be observed. Such a shift, even when caused entirely by a change in interionic distance between the two ions, does not necessarily indicate the presence of two thermodynamically distinct ion pairs. The vibrational motion of the ions in the ion pair is affected by the fluctuating environment of the neighboring

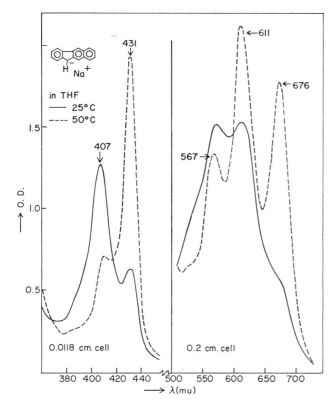

Figure 10. Optical absorption spectrum of 2,3 benzofluorenyl sodium in THF at 25°C, ———, and −50°C, − − −.

solvent molecules, and, as pointed out by Szwarc et al. [50], the potential energy curve changes with temperature. Its gradually changing shape continually modifies the interionic distance converting the tight pair into a loose pair. In such instances, those physical properties of the ion pairs that depend on the mutual interaction between cation and anion do not result from a superposition of the properties of two chemically distinct species, but are expected to undergo a gradual change with temperature. Spectroscopically, an absorption characteristic for a contact ion pair is then observed at higher temperatures, with the maximum gradually shifting to longer wavelength on lowering the temperature until an absorption maximum, characteristic for a loose ion pair, is reached at sufficiently low temperatures. This type of ion-pair behavior may be encountered in systems involving large cations like Cs⁺, or in solvents of low polarity. The potential energy barrier separating the two kinds of ion pair is then expected to be low.

An example of such a behavior was reported by Nichols and Szwarc for the sodium salt of 9,10-dihydroanthracene [51]. They pointed out that, in contrast to the planar fluorenyl salts, the alkali ion in this carbanion salt probably vibrates with a rather large amplitude as the anion flips through its planar configuration. The extent of this vibration is then gradually modified as the temperature changes. The lithium salt yields a glyme-separated ion pair on addition of glyme-3, but in the absence of the glyme the salt appears to be a contact ion pair in THF even at $-80°C$. This behavior contrasts with that of the 10-alkyl derivative of the 9-lithium salt which predominantly forms separated ion pairs in THF even at 0°C.

In general, we would expect the formation of separated ion pairs to be facilitated by increased charge delocalization of the anion. Comparison of the spectrum of fluorenyl sodium with that of 2,3-benzofluorenylsodium (see Figs. 2 and 10) in THF shows a higher fraction of separated ion pairs for the latter salt. The absorption maxima of the two ion pairs are at 401 and 431 nm, while peaks of lower intensity are found at higher wavelength, with a sharp maximum for the separated ion pair at 676 nm. The spectra of alkali salts of 1,2- and 3,4-benzofluorenyl show considerably broader bands [52, 53] and the two ion-pair bands largely overlap. However, considerable spectral shifts can be observed on changing temperature and solvent. The shifts follow the same general pattern as found for the fluorenyl salts. This is also the case for another fluorenyl derivative, the carbanion salt of 4,5-methylene-phenanthrene. The effects of temperature, counterion, and solvent on the absorption spectrum of this salt were recently reported by Casson and Tabner [54]. The absorption bands are very broad, but the data clearly show the presence of two kinds of ion pair. Spectral changes due to ion-pair equilibria were also observed for carbanions of aromatic hydrocarbons, for example, protonation products of the dinegative ions of anthracene, perylene, etc., [96, 97], and for the rather strongly delocalized carbanion of 1,3-diphenylbut-1-ene [112]. Other examples of charge delocalization effects are discussed in the section on radical ions.

Alkoxides are expected to yield tight ion pairs, even in polar solvents such as dimethoxyethane. In higher dielectric constant media the alkoxides may dissociate into free ions rather than form stable separated ion pairs, since the latter process is less dependent on the macroscopic dielectric constant. Ion-pair separation could possibly be accomplished by addition of reagents such as dimethylsulfoxide or macrocyclic ethers, particularly with the charge-delocalized phenoxides. The strong tendency of alkoxides to aggregate [55] complicates the study of these systems.

The presence of large substituents on anions favors the solvent-separated ion pair. A close approach of cation and anion is then hindered, and the larger interionic distance lowers the stability of the contact ion pair. This effect is exemplified by the different behaviour of the monoradical anion ($T^{\cdot-}$) and the dianion (T^{2-}) of tetraphenylethylene, a system studied by Roberts and Szwarc [56] and by Garst et al. [57, 58]. It appears that the geometry of the T^{-} radical anion is such that the four phenyl groups are twisted out of the plane of the C=C bond. This prevents a close approach of the Na$^+$ ion to the center of the negative charge, and the salt forms a separated ion pair in THF in the temperature range $-70°$ up to $25°$, as evidenced from its optical spectrum and its high dissociation constant. In the dianion sodium salt, the two cations can approach the respective carbanions much closer, since the free rotation around the C—C bond now allows a coplanar configuration for each Ph$_2$C-group, the two planes being mutually perpendicular. The low dissociation constant and the large exothermicity of the dianion dissociation in THF at about 25°C confirm its contact ion-pair structure. The spectral changes are also in agreement with this conclusion, as the 485 nm peak observed at 25°C changes to 510 nm at $-70°$C. In this respect the dianion of tetraphenylethylene resembles that of 1,1,4,4-tetraphenylbutane, $\overline{C}(Ph)_2CH_2CH_2(Ph)_2\overline{C}$. A THF solution of this carbanion shows bathochromic changes in its optical spectrum on lowering the temperature and on the addition of polar solvents [27]. Separate ion-pair bands are not observed due to the broadness of the transition, but the spectrum clearly results from a superposition of the absorption bands of the two types of ion pairs, with the maximum of the loose ion pair at 501 nm and those of the contact ion pairs at 460 (Li$^+$), 472 (Na$^+$), and 485 (Cs$^+$). The cesium band remains at 485 nm in THF even at $-70°$C, whereas the absorption of Na$^+$ and Li$^+$ salts shift to 501 nm.

Similar changes were observed by Waack et al. [31] for diphenylhexyllithium in mixtures of benzene and THF. When small quantities of THF are present, the tight ion pairs associate with two THF molecules. Addition of more THF results in an absorption shift and a broadening of the band. In pure THF the ion pairs are of the loose type, associated with four THF molecules. The enthalpy change was found to be -4.6 kcal/mole, while the entropy change at 22°C is -15.6 e.u.

Charge delocalization and steric hindrance in the contact ion pairs of triphenylmethyl carbanion salts are more pronounced than for diphenyl methyl carbanions. As a result, the fraction of solvent-separated ion pairs is higher for the former salts. The separation of the polystyryl salts or benzyl carbanion pairs are more difficult. Calculations based on kinetic data indicate that the fraction of separated sodium ion pairs is about 0.2 at $-60°$ in THF and 0.3 at 0°C in DME [46], while it is about 1.0 for the sodium diphenyl carbanion salt under these conditions.

The behavior of 9-alkyl substituted fluorenyl salts is puzzling. The formation of separated ion pairs is much more pronounced for 9-alkyl substituted fluorenyllithium than for fluorenyllithium itself (see Tables 4 and 5).

Steric hindrance does not seem to be an important factor here because the sodium salts do not show this difference. Moreover, a large interionic distance in the contact ion pairs would result in a smaller difference between the transition energies of the two ion pairs. But experimentally we find the same value as for fluorenyllithium, 6.2 kcal/mole [28]. Ion-pair aggregation may be partially responsible for the different behavior of the lithium salts. Reactivity studies appear to indicate that fluorenyllithium is aggregated in low-polarity media [60]. Such an aggregation may hinder the formation of separated ion pairs. The substituted salts are expected to be less aggregated, and hence ion-pair separation may be easier.

Recent work by Exner, Waack, and Steiner has indeed shown that fluorenyllithium salts are aggregated in ethereal solvents [109]. Ebulliometry shows that in THF at 25°C an appreciable fraction of 9-(2-hexylfluorenyllithium) exists as dimeric ion pairs in the concentration range 0.002–0.04M. In cyclohexane only dimers are found in the concentration range 0.01 to 0.1M. It is likely that the unsubstituted fluorenyllithium salt is even more strongly aggregated. It is also most interesting that the contact ion pair of 9-(2-hexylfluorenyllithium) absorbs at 358 nm in ether solvents [28, 109] but shows a bathochromic shift to 368 nm when present in cyclohexane [109]. Exner et al. found that addition of small quantities of diethyl-ether or THF to 9-(2-hexyl)F$^-$, Li$^+$ in cyclohexane produces a hypsochromic shift to 358 nm, with simultaneous formation of monoetherates and di-etherates of the lithium ion pair. The spectra clearly show two distinct ion-pair bands with an isosbestic point. The authors propose the existence of structurally different dimers of the lithium ion pairs in cyclohexane and ether solvents as one possible explanation of the observed spectral shifts.

4. ION-PAIR SOLVATION IN SOLVENTS CONTAINING COORDINATING AGENTS

In powerful cation solvating media such as ethylenediamine, dimethyl-sulfoxide, or polyglycoldimethyl ethers, most of the fluorenyl salts form only separated ion pairs. Even in 1,2-dimethoxyethane, fluorenyllithium exists only as a solvent-separated ion pair below 40°C. The same holds for fluorenyl-sodium in oxetane. Studies of ion-pair solvation by such reagents may be carried out in a mixture of the solvating agent with a relatively low-polarity solvent in which the fluorenyl salt predominantly exists as a contact ion pair. For example, small quantities of dimethylsulfoxide added to a fluorenyllithium solution in dioxane converts the contact ion pairs to DMSO-separated ion pairs [2].

The polyglycoldimethylethers with the general formula

$$CH_3O[CH_2CH_2O]_xCH_3,$$

referred to as glymes, are very effective alkali-solvating reagents and have often been used to enhance the reactivity of organo-alkali compounds. Their ability to coordinate with alkali ions depends on the number of available coordination sites. We discuss these systems in detail because their complexation with fluorenyl salts shows a number of interesting features which facilitate the understanding of ion-pair solvation. Even more effective than the linear glymes are the macrocyclic ethers discussed in Section 4.2.

4.1. Spectrophotometric Studies of the Coordination of Lithium and Sodium Carbanion Pairs with Glymes

The solvation of fluorenyl ion pairs by glymes was investigated by Chan, Wong, and Smid [61, 62]. A fluorenyl solution containing only contact ion pairs, say fluorenyl sodium in THF at 25°C, is titrated in vacuum with a solution of glyme containing a small amount of the fluorenyl salt to ascertain the absence of any damaging impurities. The optical spectra are recorded at appropriate intervals during the titration.

The variation of spectra which is observed when small quantities of glymes are added to a dioxane solution of fluorenyllithium or to a THF solution of fluorenyl sodium or potassium is shown in Fig. 11. The glyme-separated ion pair and the solvent-separated ion pairs have identical absorption maxima. An isosbestic point, seen in the figure, proves the stoichiometry of the solvation process. Assuming that the glyme complexation to the fluorenyl salts can be represented by the equilibrium

$$F^-, M^+ + nG \rightleftarrows F^-, G_n, M^+ \cdots K_i$$

we can obtain the number of glyme molecules, n, involved in the reaction by plotting $\log K_i^*$ versus $\log G$, where K_i^* represents the ratio $[F^-, G_n, M^+]/[F^-, M^+]$. Such plots are shown in Figs. 12 and 13. For the lithium and sodium salts of fluorenyl, the respective slopes of lines obtained for various glymes are all equal or close to unity (only for the diglyme-fluorenyl sodium system the slope deviates substantially from unity, its value being 0.85). The complexation constants for the glymes are collected in Table 7 (the number following each glyme refers to the number of oxygen atoms in the glyme, e.g., glyme-4 is triethyleneglycoldimethylether or triglyme, and x denotes the number of C_2H_4O groups).

The ability of glymes to effectively solvate cations exemplifies the importance of cooperative effects in coordination phenomena. This is already

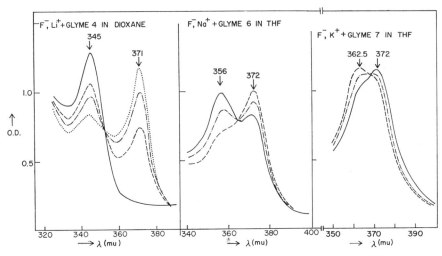

Figure 11. Spectral changes on addition of glyme 4 to F⁻, Li⁺ ($\approx 1.5 \times 10^{-3}\,M$ in dioxane), of glyme 6 to F⁻, Na⁺ ($\approx 2.10^{-3}\,M$ in THF), and of glyme 7 to F⁻, K⁺ ($5.10^{-4}\,M$ in THF; [glyme 4]. = 0.0 (———), 6.46×10^{-3} (– – –), 11.3×10^{-3} (–·–·–), and $24.3 \times 10^{-3}\,M$ (·····); [glyme 6]. = 1.34×10^{-3}(———), 2.64×10^{-3} (–·–·–) and $4.32 \times 10^{-3}\,M$ (– – –); [glyme 7]. = 1.02×10^{-3} (– – –), 2.78×10^{-3} (–·–·–), and $4.8 \times 10^{-3}\,M$ (———).

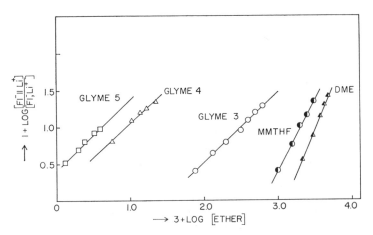

Figure 12. Complexation of 1,2-dimethoxyethane, 2-methoxymethyltetrahydrofuran, and glyme-3, 4, and 5 to fluorenyllithium in dioxane at 25°C.

116

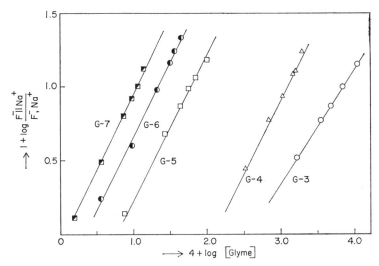

Figure 13. Glyme-separated ion-pair formation of fluorenylsodium in THF at 25° with glyme-3, 4, 5, 6, and 7.

Table 7   Equilibrium Constants[a] for Glyme-Separated Ion Pair Formation of Fluorenyl Alkali Salts at 25°C

Glyme	F^-, Li^{+b}	F^-, Na^{+b}
DME ($x = 1$)	0.055 ($n = 2.4$)	—[c]
2-(CH$_3$OCH$_2$)THF	0.25 ($n = 2$)	—[d]
Glyme-3 ($x = 2$)	3.1	1.2
Glyme-4 ($x = 3$)	130	9.0
Glyme-5 ($x = 4$)	240	170
Glyme-6 ($x = 5$)		450
Glyme-7 ($x = 6$)		800

[a] All K_i values were calculated by taking $n = 1$, except for DME ($n = 2.4$) and 2-(methoxymethyl)tetrahydrofuran ($n = 2$).

[b] Dioxane was used as solvent for the lithium salt, THF for the sodium salt.

[c] No titration carried out; fraction of solvent separated ion pairs at 25°C is about 0.9 in pure DME.

[d] At room temperature only solvent-separated ion pairs are present in the pure solvent.

apparent in solvents like 1,2-dimethoxyethane and 2-methoxymethyl THF, although at least two solvent molecules participate in the formation of the separated ion pair. For glymes with three or more oxygen atoms the formation of the glyme-separated ion pair involves only one glyme molecule, that is, $n = 1$. It should be realized, however, that in these experiments we measure the difference in solvation state between glyme-separated and contact ion pair. It is conceivable that the contact ion pair is already coordinated externally to a glyme, and that the separated ion pair actually contains two glyme molecules:

$$F^-, M^+, G + G \rightleftharpoons F^-, G, M^+, G$$

We shall deal with this problem later, when the results for fluorenyl potassium are discussed, but it can be shown from the spectrophotometric data that the lithium salt forms a 1:1 complex with glyme-4 and glyme-5 and the sodium salt a similar complex with glyme-5, 6, and 7. For the systems F^-, Li^+-glyme-3 and F^-, Na^+-glyme-3 and 4, the total glyme concentration is much higher than the carbanion concentration, and the optical data only reveal that the equilibrium between the two kinds of ion pair involves one additional glyme molecule. They do not provide information about the total number of glyme molecules associated with the contact pair.

The data of Table 7 show that in the lithium system the complexation constant K_i strongly increases with the number of 0 atoms in glyme up to glyme-4. However, a further increase in chain length leads to only a small increase in the K_i. This probably indicates that not more than four oxygens may coordinate with Li^+. A similar trend is observed for the sodium salt; K_i sharply increases and then appears to level at glyme-5. The interaction of a Na^+ ion with an oxygen atom is weaker than with the smaller Li^+ ion, and in spite of a smaller Coulombic interaction in the contact ion pair, apparently a glyme with at least five oxygen atoms is needed to obtain a complexation constant with F^-, Na^+ comparable to that found for the system F^-, Li^+ glyme-4. Because of the larger diameter of Na^+, apparently all of the five oxygen atoms can be simultaneously coordinated with the Na^+ ion forming the primary solvation shell.

The coordination of glymes or other ethers with alkali ions is not of a purely donor-acceptor type, and there is no compelling reason to assume a tetrahydral arrangement of oxygen atoms around the Li^+ or Na^+ ions [43]. For example, macrocyclic ethers with a nearly planar polyether ring of six oxygen atoms are one of the strongest coordinating agents for Na^+ ions [63].

The increased complexation ability of the higher glymes (e.g., glymes with more than five oxygen atoms for F^-, Na^+) is to be expected for statistical reasons. As pointed out by Ugelstad and Rokstad [64], the total number of possible ways W to coordinate the glyme with the cation is given by $W = q!/[p!(q - p)!]$ where q represents the number of oxygens per glyme molecule and p the average number of oxygens atoms coordinated to the cation. However, not all the combinations of five oxygen atoms are equally probable. The solvating power of ethers possessing two oxygen atoms in the chain decreases sharply when they are separated by more than two carbon atoms [41]. Hence only those combinations involving consecutive coordination sites are of importance. The number of these is $W = q - p + 1$, i.e., $W = 3$ for the system F$^-$, Na$^+$ — glyme-7 provided that $p = 5$. The observed increase in the K_i by a factor 4.5 when glyme-5 is replaced by glyme-7 is therefore reasonable. A perfect agreement is not to be expected since conformational effects may also become important when the chain length of the glyme increases.

The observed complexation constants depend on the dielectric constant of the medium, although the effects are not large. For example, on replacing THF by tetrahydropyran, the K_i values for fluorenylsodium decrease by about a factor 4. This change is small if we recall that in these two solvents the dissociation constants of F$^-$, Na$^+$ into the free ions differ by several powers of ten. The complexation of triisopropanolamineborate with fluorenylsodium shows even smaller variation; the respective constants in THF, THP, and dioxane were found to be 102, 104, and 72 [44].

Apparently dielectric saturation makes the effective dielectric constant much smaller than the macroscopic one, but another factor should also be considered. A less polar solvent yields a tighter contact ion pair, making separation more difficult. However, external solvation of a contact ion pair is more effective in polar solvents. On coordination with a glyme, at least part of these solvent molecules must be removed, and this process requires more energy for THF than for dioxane or THP. The two effects may compensate each other.

In solvents where the fluorenyl salts exist wholly or partially as separated ion pairs, the values for the glyme complexation constants may be entirely different from those reported in Table 7. Competitive solvation of the alkali ion by solvent molecules and by glyme determines the stability of the glyme complex. A particular glyme may form a stable complex with the sodium salt, whereas in the same solvent the complex with the lithium salt may be unstable. Such a case was encountered in a study of the complexation of glymes with the coronene radical anion and is discussed in a later section.

The temperature dependence of formation of glyme-separated ion pairs of fluorenylsodium is shown in Fig. 14. The values of ΔH_i and ΔS_i derived from these plots are listed in Table 8. With the exception of glyme-3, the enthalpy values do not greatly differ from the ΔH_i of -7.6 kcal/mole found for the formation of solvent-separated ion pairs in pure THF [2]. Therefore, the effectiveness of glymes as coordinating agents, as compared to solvents like THF, is mainly due to a smaller loss of entropy. Actually, separation of

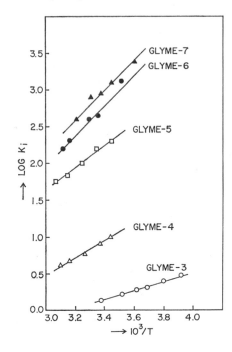

Figure 14. Temperature dependence of the glyme-separated ion-pair equilibrium F^-, $Na^+ + G \rightleftarrows F^-$, G, $Na^+(K_i)$ in THF for different glymes.

two oppositely charged ions from a contact ion-pair distance to that of a separated ion pair involves a considerable loss of entropy, probably as much as 20–30 e.u. Hence the observed value for ΔS_i of approximately -14 to -18 e.u. for glyme-separated ion-pair formation indicates release of one or two THF molecules when the glyme becomes coordinated to the ion pair. This is not unlikely as the contact ion pairs are probably externally solvated by a few THF molecules. The formation of solvent-separated ion

Table 8   Enthalpies and Entropies of Formation of Glyme-Separated Ion-Pair from the Fluorenyl Sodium Contact Ion Pair in THF

Glyme	$-\Delta H_i$ (kcal/mole)	$-\Delta S_i$ (e.u.)
3	2.8	9
4	5.4	14.5
5	7.1	14
6	9.2	18.5
7	9.0	17

pairs in pure THF involves a considerably larger loss of entropy, 30 e.u., because several THF molecules become bound to the Na$^+$ ion [2] in the course of this process.

4.2. Complexation of Glymes with Fluorenyl Potassium. External Solvation of Contact Ion Pairs

The complexation behavior of glymes with fluorenyl potassium appears to differ from that of the lithium and sodium salts. The plots of log K_i^* versus log [G] are appreciably curved, as shown in Fig. 15. For glyme-6 and 7 the ratio [F$^-$, G, K$^+$]/[F$^-$, K$^+$] appears to reach a constant value at higher concentrations of glyme. On the other hand, the plots for glyme-4 and 5, although initially curved, eventually become linear with a slope of one.

The behavior of glyme-6 and 7 can be rationalized in the following way. Contact ion pairs can be specifically solvated without formation of separated ion pairs. For example, ethers form solvation complexes with the contact ion pairs of polystyryllithium [65], polyisoprenyllithium [66], and diphenyl-hexyllithium [31] in hydrocarbon solvents. It is therefore not unreasonable to

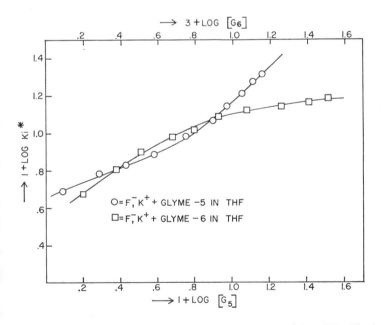

Figure 15. Plots of log K_i^* versus log [glyme] for F$^-$, K$^+$ in THF with glyme-5 (◯) and glyme-6 (☐).

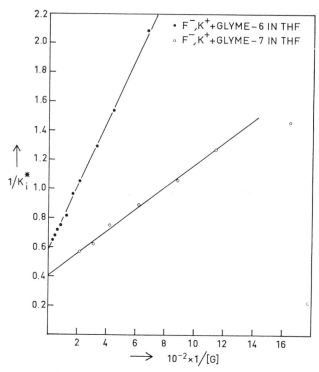

Figure 16. Plots of $1/K_i^*$ versus $1/$[glyme] for F$^-$, K$^+$ in THF with glyme-6 (●) and glyme-7 (○).

expect a competition between formation of glymated contact ion pairs and glyme-separated ion pairs when glymes are added to fluorenyl salts. Such a situation, in fact, was encountered in the work of Slates and Szwarc [67] when glymes were added to a mixture of sodium and biphenyl in tetrahydropyran. The results (to be discussed in the section on radical ions) were explained by postulating the formation of both glyme-separated ion pairs and the isomeric glymated contact ion pairs of biphenylsodium, and on this basis the equilibrium constant of the transformation, glymated contact pair $\rightleftarrows$ glyme-separated pair, was calculated.

Let us assume that both glyme-containing ion pairs are formed in the F$^-$, K$^+$-glyme system:

$$\text{F}^-, \text{K}^+ + \text{G} \rightleftarrows \text{F}^-, \text{K}^+, \text{G} \qquad (K_1)$$

$$\text{F}^-, \text{K}^+, \text{G} \rightleftarrows \text{F}^-, \text{G}, \text{K}^+ \qquad (K_2)$$

The experimental quantity K_i^*, measured from the optical spectra, is then given by

$$K_i^* = \frac{\text{F}^-, \text{G}, \text{K}^+}{\text{F}^-, \text{K}^+ + \text{F}^-, \text{K}^+, \text{G}}$$

or

$$\frac{1}{K_i^*} = \frac{1}{K_2} + \frac{1}{K_1 K_2 G}.$$

Indeed, for the systems F$^-$, K$^+$-glyme-6 and F$^-$, K$^+$-glyme-7 plots of $1/K_i^*$ versus $1/G$ are linear, as shown in Fig. 16, their slopes and intercepts giving K_1 and K_2. The results were shown to be independent of the concentration of F$^-$, K$^+$.

The preceding treatment predicts that the ratio of concentrations of separated and contact ion pairs should not exceed K_2 (see also Fig. 15). However, for a sufficiently high concentration of glymes all the contact ion pairs are eventually converted to the glyme-separated ion pairs. This is particularly noticeable for the systems F$^-$, K$^+$-glyme-4 and F$^-$, K$^+$-glyme-5, where plots of $1/K_i^*$ versus $1/G$ are curved and extrapolate to the origin, that is, $K_i^* \to \infty$ for sufficiently large [G].

The behavior of the last two systems can be explained by including a third solvation step in which the glymated contact ion pairs are converted to separated ion pairs on addition of more glyme. The system is now described by the following three equilibria:

$$\text{F}^-, \text{K}^+ + \text{G} \rightleftarrows \text{F}^-, \text{K}^+, \text{G} \qquad (K_1)$$
$$\text{F}^-, \text{K}^+, \text{G} \rightleftarrows \text{F}^-, \text{G}, \text{K}^+ \qquad (K_2)$$
$$\text{F}^-, \text{K}^+, \text{G} + \text{G} \rightleftarrows \text{F}^-, \text{G}, \text{K}^+, \text{G} \qquad (K_3)$$

and

$$K_i^* = \frac{\text{F}^-, \text{G}, \text{K}^+ + \text{F}^-, \text{G}, \text{K}^+, \text{G}}{\text{F}^-, \text{K}^+ + \text{F}^-, \text{K}^+\text{G}} = \frac{K_2 + K_3 G}{1 + 1/K_1 G}$$

For high glyme concentrations, as in the system involving glyme-4 or glyme-5, it is reasonable to assume $K_1 G \gg 1$, that is, all contact ion pairs are externally glymated. This leads to

$$K_i^* = K_2 + K_3 G.$$

Indeed, for the F$^-$, K$^+$-glyme-4 and F$^-$, K$^+$-glyme-5 systems in THF or THP the plots of K_i^* versus G are linear, as exemplified by Fig. 17. Apparently, in the concentration range of glyme needed to form glyme-separated ion pairs all fluorenylpotassium ion pairs are already externally coordinated with glyme-4 or glyme-5.

Values for the equilibrium constants K_1, K_2, and K_3 are given in Table 9. The K_1 values, which describe the external coordination with the contact ion pairs, show only a relatively small increase with increasing chain length of

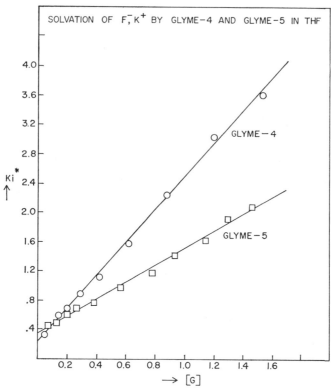

Figure 17. Plots of K_i^* versus [glyme] for F^-, K^+ in THF at $25°$ with glyme-4 and 5, respectively.

Table 9   Equilibrium Constants for Glyme-Coordinated Ion-Pair Formation of Fluorenyl Potassium[a] in THF and THP at 25°C

Glyme	Medium	$K_1(M^{-1})$	K_2	$K_3(M^{-1})$
Glyme-3	THF	—	0	0.21
Glyme-4	THF	>50	0.24	2.27
	THF[b]	>50	0.30	2.2
	THP	>50	0.18	1.2
Glyme-5	THF	>50	0.35	1.07
	THF[b]	>50	0.28	0.92
	THP	>50	0.18	0.69
Glyme-6	THF	250	1.75	—
Glyme-7	THF	530	2.5	—
	THP	1,700	2.7	—

[a] Concentration of F^-, K^+ between $1 \times 10^{-3} M$ and $5 \times 10^{-4} M$.
[b] Concentration of F^-, $K^+ \approx 6 \times 10^{-3} M$.

the glyme. The limited availability of space on the periphery of the K^+ ion in the F^-, K^+ contact ion pair does not allow coordination with more than four or five oxygen atoms, and changing to glyme-6 or glyme-7 increases K_1 by a statistical factor only. The K_1 value does depend on the solvent medium. The glyme displaces one or more of the externally bound solvent molecules more easily in THP than in THF. A higher K_1 value in THP is therefore justified.

The K_2 is large for glyme-6 or 7. Isomerization of the glymated contact ion pair to a separated species involves a loss in Coulombic interaction energy, and the weak oxygen $-K^+$ interaction apparently requires the coordination of at least six oxygen atoms to force the separation. The effect of dielectric constant appears to be small again, the K_2 values being somewhat lower in THP than in THF.

The ion-pair separation induced by a second glyme molecule is most pronounced for glyme-4 and appears to decrease for the larger glymes. This may indicate that the K^+ ion can coordinate with a maximum of 7 or possibly 8 oxygen atoms, and larger glymes may become less effective because of steric hindrance. However, ion pairs coordinated with two glymes are apparently formed at high glyme concentrations even with glyme-6 or 7 since K_i^*, after reaching a constant value (see Fig. 15), eventually increases again, and only separated ion pairs are found in pure glyme-6 or 7.

The external glymation of contact ion pairs is not limited to potassium ion pairs. Although it was not observed on addition of glyme-5 to F^-, Na^+ in THF (i.e., only separation occurs), addition of glyme-4 is likely to yield glymated contact ion pairs. However, the experimental conditions were most probably such that $K_1G \gg 1$. Evidence for species such as F^-, G, Na^+, G is provided by nuclear magnetic resonance data on the system F^-, Na^+-glyme-4 [62]. The work of Slates and Szwarc [67] directly demonstrates external solvation of a sodium ion pair by both glyme-3 and glyme-4.

Other evidence for the solvation of alkali ions by two glyme molecules comes from kinetic studies of the anionic polymerization of polystyrylsodium in mixtures of THP and glyme-4. Shinohara et al. [68, 69] found that the free Na^+ ion is coordinated with two glyme-4, although with only one glyme-5 molecule. Chemical analysis of various crystalline glyme complexes may be desirable and could provide further information regarding the stoichiometry of the glyme complexes.

Although the glyme is complexed only to the cation, the structure of the anion may strongly affect the complexation constant, particularly when the change in anion structure results in a change in the solvation state of the ion pair. As long as the interionic distances in the respective ion pairs are comparable, the glyme complexation constants K_i should not be greatly affected. For example, the K_i value of glyme-5 at 25°C for sodium naphthalene in THP (calculated from electron spin resonance data [70]) and that for poly-styrylsodium in THP (kinetic studies [69]) are 200–300 M^{-1} and 90 M^{-1}, respectively, compared to 170 for F^-, Na^+ in THF at 25°C. In the absence of glyme all these salts form contact ion pairs under these conditions. On the other hand, high complexation constants of the order of $10^4 M^{-1}$ [71, 72] are found for the system glyme-5-triphenylenesodium or coronenesodium in THF, where solvent-separated pairs exist even in the absence of glyme. The reverse is found for coronenelithium, the complexation with glyme-4 being less effective than that of glyme-4 with fluorenyllithium in dioxane.

The strong solvating power of glymes with respect to alkali ions has been observed by other investigators; Down et al. [73], for example, studied the solubility of sodium and its alloy with potassium in different glymes by measuring the intensities of the resulting blue metal solutions. Although no quantitative data were obtained, the intensities of the blue solutions were shown to increase with the number of oxygen atoms in the glyme. Ugelstad and his co-workers [64, 74] studied the effect of glyme structure on the rate of isomerization of 3-butenylbenzene and of allylbenzene induced by potassium t-butoxide, and on the rate of the reaction of alkaliphenoxides with butyl-halides. The experiments (carried out in the pure glymes) show a large enhancement of rates with an increase in the chain length of the glyme. For example, glyme-5 increases the rate of reaction of sodium phenoxide with butylbromide by a factor of 180 compared to the rate of the same reaction in DME. With potassium phenoxide, the respective rates in glyme-3, 4, 5, and 7 are 8, 51, 72, and 200, all relative to DME. The increase in the rate is largely due to an increase in the concentration of reactive free alkoxide ions as a result of the glyme interaction with the free alkali ion. Whether glyme-separated ion pairs also play a role is not known. The possibility of ion-pair association may complicate the interpretation of the results, particularly for reactions involving planar phenoxides. The degree of association may change on the addition of glymes.

Recent studies by Shinohara et al. [68, 69] have shown that addition of small quantities of glymes dramatically increases the rate of anionic polymerization of styrene in tetrahydropyran. For example, at a polystyrylsodium concentration of $5 \times 10^{-5} M$, the polymerization rate increases by nearly a factor of 200 when glyme-5 is present at a concentration of $10^{-3} M$. The formation of reactive glyme-separated ion pairs and an increase in the fraction of reactive free ions was found to be responsible for this effect.

4.3. Interactions of Macrocyclic Polyethers with Fluorenyl Ion Pairs

Pedersen [63, 75] has recently developed a class of very strong cation-binding complexing agents referred to as macrocyclic polyethers or crown ethers. Some of these compounds (depicted below) are dibenzo-18-crown-6 (I), dicyclohexyl-14-crown-4 (II) and monobenzo-15-crown-5 (III). The first number refers to the total number of atoms in the ring, the second number corresponds to the number of ring oxygen atoms.

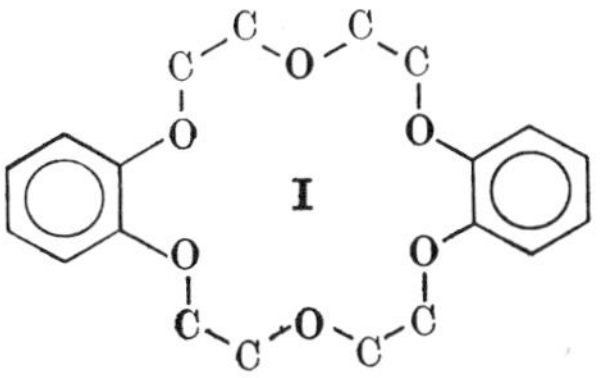

I 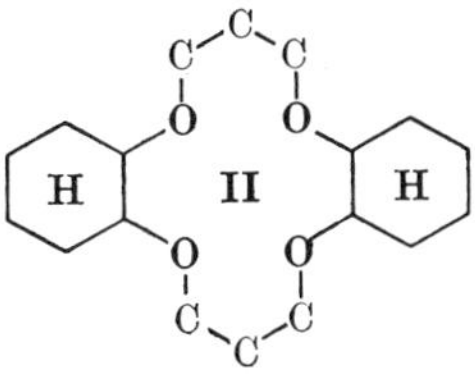   II 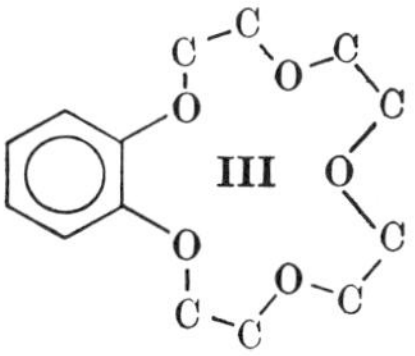   III

The crown ethers considerably increase the solubility of inorganic salts in nonpolar media [63] and form crystalline complexes with many salts [76, 77, 80]. Potentiometric measurements prove the existence of stable crown complexes of free alkali ions in water and methanol [81]. The crown compounds also exert specific effects on the cation transport across biological membranes [83–85].

The stability of these complexes depends on the size of the cation relative to that of the hole of the macrocyclic polyether, the charge of the ion, the number of ring oxygen atoms, their basicity, coplanarity, and symmetrical placement, steric hindrance in the polyether ring, and the extent of ion association with the solvent [63]. For example, the small Li^+ ion fits the hole of a 14-crown-4 polyether, but the larger Na^+ and K^+ ions do not. However, the Na^+ and K^+ ions can be accommodated by the cavity of an 18-crown-6 ether. The stability of the complex is increased as more oxygen atoms are available for coordination, provided they are favorably located in the polyether ring. In this respect, the dibenzo-18-crown-6 (I) is one of the best complexing agents for Na^+ and K^+ ions, although the dicyclohexyl derivative is claimed to complex even better on account of its more basic oxygen atoms [63].

The complexation of macrocyclic polyethers with fluorenyl ion pairs resembles the behavior of polyglycoldimethylethers. Addition of dibenzo- or dicyclohexyl-18-crown-6 to fluorenylsodium in THF or THP yields a 1:1 complex absorbing at 372 nm, the optical spectrum being identical to that of the solvent-separated ion pair of F^-, Na^+ [78]. The same is observed with monobenzo-18-crown-6 [79]. The respective complexation constants were found to be larger than 10^7 M^{-1}. This compares with a value of only 450 M^{-1} for the linear pentaglyme [62]. Complexation of the linear polyethers to alkali ions is of course accompanied by a much more unfavorable entropy change than when crown ethers are involved.

Addition of monobenzo-15-crown-5 (C5) to F^-, Na^+ (both concentrations being about 5.10^{-4} M, with C5 being in a slight excess) in THP as solvent causes a pronounced shift in the contact ion pair maximum from 354 to 359 nm. A shoulder around 370 nm reveals the presence of separated ion pairs, but their fraction increases only slightly on addition of excess C5. A careful analysis of this system [79] shows that crown complexed contact ion pairs (λ_m 359 nm) are produced in addition to crown separated ion pairs (λ_m 373 nm), similar to what is observed with glyme-6 or 7 and F^-, K^+ [62]. The two modes of complexation compete with one another, and at high C5 concentration a constant ratio of F^-, C5, Na^+/F^-, Na^+, C5 = K_2 is obtained.

A similar complexation is observed in other crown-fluorenyl salt systems, such as fluorenyl potassium with monobenzo-18-crown-6 and the two isomers of dicyclohexyl-18-crown-6 [79]. Plots of K_i^* versus [crown] are shown in Fig. 18 for a few of these systems (see the previous section for the pertinent equations). The system C5 − F^-, K^+ appears to deviate significantly from all the other systems, with the ratio K_i^* of separated over contact ion pairs showing a sharp increase at higher C5 concentrations. Apparently, a second C5 molecule is complexed to the F^-, K^+, C5 ion pair, similar to that observed for glyme-4 and glyme-5 with F^-, K^+ [62]. A linear correlation is expected between K_i^* and [C5] under conditions where all contact ion pairs are present as F^-, K^+, C5 species, i.e., $K_i^* = K_2 + K_3[C5]$. For this system, K_2 is close to zero, that is, very few F^-, C5, K^+ species are present. Complexation constants for a number of crown-fluorenyl salt systems are collected in Table 10.

The dependence of K_2 on solvent can be attributed largely to the fact that the crown compound must remove a few solvent molecules before it can complex externally to a contact ion pair. On the other hand, complexation leading to separated ion pairs probably does not necessitate removal of the entire solvation shell due to the planar conformation of the complexed crown ethers. The equilibrium between the two kinds of ion pair is therefore more accurately described by F^-, M^+, C + $nS \rightleftarrows F^-$, C, M^+, S_n, where S denotes a solvent molecule. Less polar solvents would therefore favor the crown-complexed contact ion pairs.

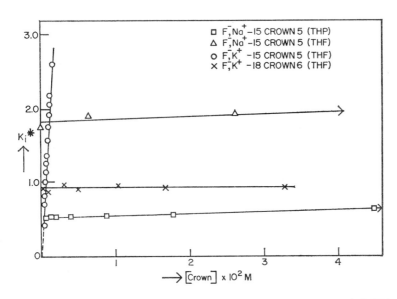

Figure 18. Plots of K_i^* versus [crown] for the systems C5-F$^-$, Na$^+$ (THF and THP), C5-F$^-$, K$^+$ (THF), and C6-F$^-$, K$^+$ (THF); [fluorenyl salt] $\approx 6.10^{-4}$ M.

Table 10 Complexation Constants^a of Macrocyclic Polyethers with Fluorenyl Ion Pairs in Ethereal Solvents at 25°C

Ion Pair	Crown	Solvent	$K_1 \times 10^{-3}$ M^{-1}	K_2	$K_3 M^{-1}$	$K_i \times 10^{-3}$ M^{-1}
F$^-$, Na$^+$	C5	THF	9.2	1.8	3.5	16.5
		THP	>20	0.52	2.8	> 10
	C6	THF				>20,000
		THP				>40,000
	DC6	THP				>40,000
F$^-$, K$^+$	C5	THF	~ 5	~0.2	1,840	~ 1
	C6	THF	>10	0.93		> 9
		THP	>10	0.55		> 5.5
	DC6A	THF	>30	1.80		>54
	DC6B	THF	>40	0.80		>32

^{a} Complexation constants refer to the following equilibria:

$$F^-, M^+ + C \rightleftarrows F^-, M^+, C \qquad (K_1)$$
$$F^-, M^+ + C \rightleftarrows F^-, C, M^+ \qquad (K_i)$$
$$F^-, M^+, C \rightleftarrows F^-, C, M^+ \qquad (K_2)$$
$$F^-, M^+, C + C \rightleftarrows F^-, C, M^+, C \qquad (K_3)$$

129

A smaller polyether ring appears to favor the formation of externally complexed contact ion pairs, especially when the cation does not easily fit into the hole of the crown. The hole of C5 is barely large enough to accommodate a Na^+ ion [80, 81], and the cation may slightly protrude from the plane of the ring, with the oxygen atoms most probably in a nearly planar conformation such that an electron-rich environment exists just above the ring. This situation resembles that found by X-ray crystallography for solid complexes of dicyclohexyl-18-crown-6 and RbCNS [76, 77].

The observation that F^-, K^+ complexes with two C5 molecules is in agreement with potentiometric measurements of Frensdorff [81], who found that cyclohexyl-16-crown-5 forms 1:1 and 2:1 complexes with free K^+ ions in methanol. Other evidence comes from X-ray data on C5-KI crystals, which show that a K^+ ion is sandwiched between two C5 ethers [110]. Moreover, recent studies [82] on the cation-binding properties of polymers containing C5 crown ethers as pendent groups have demonstrated a pronounced enhancement in the efficiency of the C5 group in complexing large cations such as K^+ and Cs^+ in comparison to that of monomeric monobenzo-15-crown-5. This can be attributed to cooperative effects involving neighboring C5 moieties, which will be particularly important when a cation such as K^+ or Cs^+ forms stable 2:1 complexes with the monomeric crown ethers. The maximum number of F^-, K^+ ion pairs that can be bound to this polymer was shown to be approximately half of the total number of crown units present, whereas with F^-, Na^+ each crown unit can bind one F^-, Na^+ ion pair [111].

A 2:1 complex of C5 with F^-, Na^+ appears to be unstable, although there is an increase in K_i^* at higher C5 concentration. The low K_3 value (see Table 10) is probably caused by increased repulsion between the oxygen atoms of the two C5 molecules when small cations are sandwiched between the two crown ethers. The same is true for the system C6 — F^-, K^+. The cation in both systems can be located close to the center of the polyether hole, which also makes complexation of a second crown molecule unfavorable.

5. COMPETITIVE COMPLEXATION IN MIXTURES OF ION-PAIR SALTS. SELECTIVITY OF MACROCYCLIC ETHERS

For some of the macrocyclic ethers only a lower limit could be obtained for the value of the complexation constant with alkali ion pairs, and therefore no conclusions could be drawn about the order of selectivity with respect to complexation to the various alkali ions. However, this information may be acquired by observing the optical spectrum of a mixture of two salts to which the complexing agent is added. For example, when the dibenzo-18-crown-6 complex of fluorenylpotassium (λ_m 372 nm) in THF is mixed with an equimolar quantity of fluorenylsodium (λ_m 356 nm), the crown compound,

E, is almost completely transferred from the potassium to the sodium salt:

$$F^-, E, K^+ + F^-, Na^+ \rightleftarrows F^-, K^+ + F^-, E, Na$$
$$\lambda_m = 372nm, \quad 356nm, \quad 362nm, \quad 372nm$$

The reaction is instantaneous; the maximum of the F^-, Na^+ contact ion pair disappears and a new maximum appearing at 362 nm is identified with that of the F^-, K^+ contact ion pair.

The results obtained with dimethyldibenzo-18-crown-6 were confirmed by nuclear magnetic resonance measurements [78]. When the crown ether is complexed to F^-, Na^+, the nmr line of the 16 aliphatic polyether ring protons splits up into two peaks of equal intensity which are shifted upfield by 0.75 ppm and 1.1 ppm, respectively, due to the diamagnetic anisotropy of the aromatic fluorenyl ring. Below 0°C, a slow exchange spectrum is observed when the crown ether is in excess. The exchange reaction F^-, E, $Na^+ + E^* \rightarrow$ F^-, E^*, $Na^+ + E$ (E and E^* being the dimethyldibenzo-18-crown-6) proceeds with an activation energy of 12.5 kcal/mole. No slow exchange spectrum of this crown ether with F^-, K^+ is observed, even at a temperature as low as -60°C, indicating that in THF the crown is more tightly bound to Na^+ than to K^+.

Studies of other alkali fluorenyl salts show a selectivity order $Na^+ >$ $K^+ > Cs^+ > Li^+$ for their complexation with the 18-crown-6 ether [78]. The comparatively low value for Li^+ is partially due to its small ionic diameter (1.20 Å) as compared to the diameter of the hole of the 18-crown-6 compound which is approximately 3 Å [80]. However, the solvation state of the ion pairs, and therefore the solvent medium itself also affects the order of selectivity. In THF, fluorenyllithium at 25°C is predominantly a solvent-separated ion pair, and complexation with a crown compound requires at least partial removal of the THF solvation shell around the Li^+ ion. The other salts are all contact ion pairs under these conditions. It is therefore not surprising that a different complexation order with respect to alkali ions is found when water or methanol is used as solvent [63, 81, 83, 84], that is, $K^+ > Cs^+ > Na^+ > Li^+$. Both Na^+ and Li^+ are strongly solvated by H_2O or CH_3OH, and these solvents can more effectively compete with the macrocyclic ether than THF. Although THF is also more strongly bound to Na^+ than to K^+, the difference is expected to be less than with more polar solvents. Moreover, the sodium and potassium fluorenyl salts are contact ion pairs in THF, whereas in water and methanol the alkali ions are probably present as fully solvated free ions.

Even a *small* change in solvent basicity can affect the order of complexation. For example, in an equimolar mixture of F^-, Na^+, dibenzo-18-crown-6, and F^-, K^+, the crown compound is preferentially complexed to the potassium salt when oxetane is the solvent. In this solvent, F^-, Na^+ forms a separated ion pair while F^-, K^+ is a contact ion pair, and complexation to the sodium salt would therefore require at least partial removal of the oxetane solvation shell. Intrinsically, however, the dibenzo and dicyclohexyl-18-crown-6 are more strongly bound to Na^+ than to K^+.

Temperature may also affect the order of complexation. For example, F^-, Na^+ in THF forms a contact ion pair at 25°C but a solvent-separated ion pair at -70°C, whereas the potassium salt remains a tight ion pair within this temperature range. Hence the observed selectivity of the dibenzo-18-crown-6 in THF at 25°C, $Na^+ \gg K^+$, may reverse at low temperature.

Complexation does not necessarily lead to ion-pair separation. Some of the macrocyclic ethers and glymes may complex externally to a contact ion pair. Hence we may visualize a reaction of the type

$$C^-, E, M_1^+ + C^-, M_2^+ \rightleftharpoons C^- \parallel M_1^+ + C^-, M_2^+, E$$

Mixing equimolar quantities of C^-, M_2^+ and C^-, E, M_1^+ may not show any change in spectrum, yet the complexing agent, E, could have transferred to C^-, M_2^+ while the ion pair C^-, M_1^+ remains separated by acquiring a solvation shell of solvent molecules. In such a case, optical studies would not yield the correct information regarding the complexing abilities of E with respect to the two ion-pair salts.

Competition experiments can also be useful in cases where ion-pair solvation does not lead to changes in the optical spectra (e.g., tight and loose ion pairs could have a similar spectrum) or where the nuclear magnetic resonance spectrum of the complexing agent is essentially not affected by its association with the ion or ion pair. For example, no change in optical spectrum is observed when glymes are complexed to $NaBPh_4$, and the NMR proton shifts of the complexed glyme are also very small. Nevertheless, information on the glyme complexation process can be obtained by adding to the $NaBPh_4$-glyme system a salt which, on complexation with glyme, does show a change in optical spectrum. For example, when $NaBPh_4$ is added to the dibenzo-18-crown-6 complex of F^-, Na^+ in THF, the optical spectrum of the latter changes and the F^-, Na^+ contact ion-pair absorption band appears [78]. This indicates that a partial transfer of the macrocyclic ether to $NaBPh_4$ takes place:

$$F^-, E, Na^+ + Na \parallel BPh_4 \rightleftharpoons F^-, Na^+ + Na, E, BPh_4$$

The boron salt is shown here as a separated ion pair in THF [19]. It is probable that a few molecules of THF are released when the THF solvation shell is replaced by the macrocyclic ether. The equilibrium constant of this reaction was found to be 2. No value could be obtained for the complexation constant of this macrocyclic ether with $NaBPh_4$, since the equilibrium constant F^-, $Na^+ + E \rightleftharpoons F^-$, E, Na^+ is larger than 10^7 M^{-1}. Other macrocyclic ethers or glymes, however, have lower complexation constants with F$^-$, Na$^+$, and competition experiments with NaBPh$_4$ may yield the complexation constant for this salt.

Other salts behave similarly. For example, the foregoing equilibrium with NaClO$_4$ instead of NaBPh$_4$ yields a value of 0.04 for the equilibrium constant. This salt is a tight ion pair in THF (its dissociation constant in THF is very low [18]), and an efficient binding with the crown compound probably can be accomplished only at the expense of a considerable amount of coulombic interaction if the interionic distance in the Na$^+$, ClO$_4^-$ ion pair is enlarged in the process.

The competition experiments may also be carried out by using nuclear magnetic resonance. In this case, a salt is chosen which will induce strong shifts in the complexing agent when it becomes associated with the salt. For this purpose, salts like triphenylene sodium, coronene salts, or fluorenyl salts can be used, the first two salts giving strong paramagnetic downfield shifts and the third a diamagnetic upfield shift of the glyme protons. The method may be illustrated by the following example: When NaBPh$_4$ is added to the glyme-5 complex of triphenylene sodium (Tr$^-$, Na$^+$) in THF, the glyme is partially transferred to NaBPh$_4$:

$$Tr^-, G, Na^+ + Na^+ \parallel BPh_4 \rightleftarrows Tr^- \parallel Na^+ + Na, G, BPh_4$$

The resulting NMR spectrum of the glyme protons is the average of that shown by the two types of glyme complexes [71]. The equilibrium constant was found to be about 0.6; however, it is slightly concentration dependent, possibly indicating some degree of aggregation of ion pairs at the comparatively high salt concentrations ($\sim$0.1 M). In this system no change in optical spectrum is observed, since the radical anion salt remains a separated ion pair even in the absence of glyme. This would be different in MeTHF, where Tr$^-$, Na$^+$ is a contact ion pair at room temperature [10]. Both the optical and NMR spectra may then be used to determine the equilibrium constant. Moreover, the equilibrium constant for the reaction Tr$^-$, Na$^+$ + G $\rightleftarrows$ Tr$^-$, G, Na$^+$ can also be determined in 2MeTHF ($K \approx 10^4 \ M^{-1}$), and therefore the complexation constant for the reaction NaBPh$_4$ + G $\rightleftarrows$ Na, G, BPh$_4$ can be calculated.

6. FLUORENYL SALTS OF DIVALENT CATIONS

The structure and properties of ion pairs containing divalent cations are less known than those of the alkali salts. We are again faced with the problem of finding a system that can serve as a probe to study ion-pair solvation. On the whole, it is more difficult to prepare and purify divalent salts of carbanions and radical anions than the corresponding alkali salts. The solubilities of the former in ethereal solvents are often low. Nevertheless, a number of divalent salts have recently been prepared and studied and their behavior shows some interesting features.

The barium salt of the fluorenyl carbanion was prepared by Hogen Esch and Smid [86] and by Pascault [29]. It can be obtained in a pure form by stirring in THF a mixture of 1,1-diphenylethylene and fluorene (a slight excess of fluorene is sufficient) on a barium mirror for a few days at 25°C. On cooling, orange crystals of barium difluorenyl can be isolated from the solution. A THF solution of the salt shows a sharp absorption maximum at 347 nm (see Fig. 19). The fraction of separated ion pairs is low even at

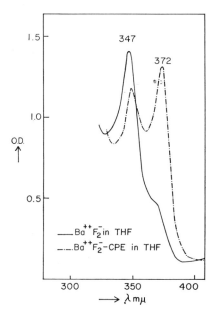

Figure 19. Optical absorption spectrum of barium fluorenyl and its 1 : 1 complex with dimethyldibenzo-18-crown-6 between 320 and 400 nm in THF at 25°C.

−70°C, whereas for the sodium salt it exceeds 0.9 under these conditions. In DME at 25°C the barium salt is also essentially a contact ion pair, and its behavior resembles more that of F^-, K^+ than that of F^-, Na^+. The barium salt probably has a sandwich-type structure, and the solvation of the barium, particularly the external solvation of the contact ion pair, is therefore hindered.

Not surprisingly, the conductance data reveal a low dissociation constant, K_d, for the equilibrium $Ba^{2+}, F_2^{2-} \rightleftarrows (Ba, F)^+ + F^-$, the value in THF being $3 \cdot 10^{-9}$ M^{-1} at 25°C, compared to 6.2×10^{-7} and 1.6×10^{-7} M^{-1}, respectively, for the sodium and potassium salts [59]. Besides the possibility that Ba^{2+}, F_2^{2-} may be a tighter ion pair than the alkali ion pairs, it should also be realized that the alkali ion pairs on dissociation yield fully solvated free alkali ions, whereas the barium salt produces a tight $[Ba, F]^+$ ion pair and solvation occurs only on one side of the Ba^{2+} ion.

The strontium salt yields higher fractions of separated ion pairs [87]. The spectrum of this salt in THF at $-25°C$ reveals two absorption maxima of approximately equal height at 347 and 371 nm, indicating an appreciable proportion of separated ion pairs. It would be interesting to see whether, under suitable conditions, a complete conversion to separated ion pairs may occur by a stepwise process:

$$F^-, M^{2+}, F^- \rightleftarrows F^-, M^{2+} \parallel F^-$$

$$F^-, M^{2+} \parallel F^- \rightleftarrows F^- \parallel M^{2+} \parallel F^-$$

Information about such processes could also be deduced from studies of the ion-pair dissociation over a sufficiently large temperature range. At 25°C, the K_d is $1.3 \cdot 10^{-7}\ M^{-1}$ for Sr^{2+}, F_2^{2-} in THF, much higher than that for the barium salt.

Ion-pair separation can also be accomplished by adding complexing agents such as glymes or macrocyclic ethers. A mixture of equimolar quantities of dibenzo-18-crown-6 and barium difluorenyl (or the corresponding strontium salt) in THF produces a 1:1 complex, the spectrum of which, shown in Fig. 19, reveals approximately equal fractions of tight and loose ion pairs. Their proportion is not changed on further addition of the macrocyclic ether. A similar behavior is observed with glyme-6. Apparently the solvation with the glyme or crown compound is asymmetric, with formation of species of the type F^-, Ba^{2+}, E, F^-. Another possible structure of the complex is shown in Fig. 20, with the Ba^{2+}-crown complex sandwiched in between the two fluorenyl moieties. It is possible that the Ba^{2+} ion rapidly vibrates through the hole of the macrocyclic ether ring, and the ring conformation changes accordingly during this motion.

An optical spectrum showing only one absorption maximum at 372 nm is obtained in THF on mixing equimolar quantities of barium or strontium difluorenyl with cryptates [87], a class of very powerful cation binding reagents recently discovered by Lehn et al. [88, 89]. The cryptate

$$N \underset{(CH_2CH_2O)_2 - CH_2CH_2}{\overset{(CH_2CH_2O)_2 - CH_2CH_2}{\left\langle (CH_2CH_2O)_2 - CH_2CH_2 \right\rangle}} N$$

forms a very stable and poorly soluble 1:1 complex with the two salts. Models of the cryptate show that the Ba^{2+} cation is tightly held in the cryptate cage, and its displacement is difficult even in water. Since the Ba^{2+} cation is symmetrically surrounded by the cryptate molecule, it is not surprising that

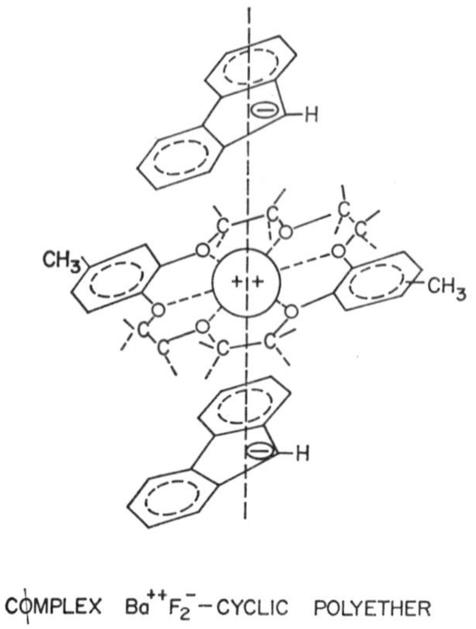

COMPLEX Ba^{++}F$_2^-$—CYCLIC POLYETHER

Figure 20.   Possible structure of the barium fluorenyl-dimethyldibenzo-18-crown-6 complex in THF or pyridine.

the fluorenyl complex shows only one absorption peak, that of the separated ion pair. A similar one-peak spectrum is obtained by adding ethylenediamine or hexamethylphosphoramide to a THF solution of bariumdifluorenyl, but it is not known how many molecules are complexed to the barium salt or whether the ion-pair separation involves a multistep mechanism.

The proton NMR spectrum of the dibenzo-18-crown-6 complex of barium-difluorenyl in pyridine (its solubility in THF is only $\sim 10^{-2}$ M) shows upfield shifts of 0.5 and 1.4 ppm for the aliphatic polyether ring protons of the crown compound. The NMR pattern resembles that of the complex with F$^-$, Na$^+$ in pyridine [78], but the shifts with the barium salt are larger, probably because the diamagnetic anisotropy at the polyether ring protons is enhanced due to the presence of the second fluorenyl ring. The rate of exchange F$^-$, Ba^{2+}, E, F$^-$ + E* $\rightarrow$ F$^-$, Ba^{2+}, E*, F$^-$ + E in pyridine at 60° is of the order of 500 M^{-1} sec^{-1}, while for F$^-$, Na$^+$ at this temperature in THF the value is close to 10^5 M^{-1} sec^{-1} and probably even higher in pyridine. The exchange reaction with the barium complex is probably sterically hindered, and the cation is also more tightly bound to the crown compound than Na$^+$. This can also be concluded from the fact that in an equimolar mixture of the crown compound, F$^-$, Na$^+$ and Ba^{2+}, F$_2^{2-}$, the macrocyclic ether is exclusively complexed with the barium salt.

7. STRUCTURE AND SOLVATION OF ION PAIRS OF RADICAL ANIONS

Much of our knowledge of the structure and solvation of ion pairs is derived from studies involving radical anions of aromatic hydrocarbons and ketones. These species, whose existence has been known for a long time [90–93] are formed by transfer of an electron from a metal M (usually an alkali or alkaline earth metal) to an aromatic hydrocarbon or ketone. When carried out in solution, the solvation of the radical ion and its counterion contributes significantly to the exothermicity of the reaction.

The equilibrium constant for the reaction between metal and hydrocarbon (or ketone),

$$M + A \rightleftarrows M^+, A^{\overline{\cdot}}$$

depends on the ionization potential of the metal, the electron affinity of the ketone or aromatic hydrocarbon, the temperature and the nature of the solvent (see also Chapter 5). Solvent, temperature, cation, and type of anion radical determine to a large extent whether free ions or ion pairs are formed, and whether the ion pairs are separated or in contact, or associated to larger aggregates. All these factors influence the equilibrium constant of the reaction, and its value and temperature dependence therefore provide information helpful in studies of ion-pair structures and their solvates.

Much of our knowledge of ion-pair structures of radical ions has been obtained from the elegant electron spin resonance studies of Weissman, Hoijtink, de Boer, Hirota, Symons, and many other investigators, and this work is fully reviewed in Chapters 5 and 8. Studies utilizing optical absorption spectra and conductance measurements [50, 95] as research tools have also yielded important information on the behavior of radical anion salts. Various conclusions that have emerged from these investigations parallel those found for carbanions. This is to be expected, since many of the investigated phenomena are the result of cation-solvent interactions, and the structure of the anion, be it a radical or a carbanion, is often not relevant for the investigated problem.

The most extensive investigations dealing with equilibria between alkali metals and aromatic hydrocarbons in ethereal solvents are those of Shatenshtein and his co-workers [40–43]. The equilibrium constant of the electron-transfer reaction from alkali metal to an aromatic hydrocarbon is often too high to permit quantitative studies of such systems. However, when naphthalene and particularly biphenyl are the acceptors, the conversion into radical ions is only partial under proper conditions. The equilibria can then be studied spectrophotometrically at various temperatures by stirring an ethereal solution of the hydrocarbon over an alkali mirror and recording the spectrum of the resulting radical ion. Sufficient time must be allowed to reach equilibrium, especially at low temperature. Some of Shatenshtein's results for the biphenyl and naphthalene systems are shown in Tables 11 and 12. Although the basicity of the coordinating solvent is an important factor in determining the value of the equilibrium constant, the steric factors are usually dominant. This is clearly demonstrated by comparing the data for a

Table 11 Equilibrium Constant $K = [B^{-}, Na^+]/[B]$ for the Reaction
Sodium + Biphenyl (sol) $\rightleftarrows B^{-}, Na^+$ (sol)[a]

T (°C)	MME	1,2DMPr	THF	MeTHF[b]	DEE	THP	1,3DMPr
40°	0.12	0.09	0.10				
30°	0.28	0.20	0.20				
20°	0.75	0.49	0.36	0.02	0.07		
10°	2.55	1.40	0.66	0.036	0.11		
0°	7.0	5.0	1.50	0.055	0.19	0.06	
−10°			2.90	0.11	0.39	0.10	0.12
−20°				0.20	1.25	0.17	0.34
−30°				0.45	8.7	0.29	1.20
−40°				1.18		0.48	

MME = 1,2 methoxyethoxyethane; 1,2 DMPr = 1,2 dimethoxypropane; THF = tetrahydrofuran; MeTHF = 2 methyltetrahydrofuran; DEE = 1,2 diethoxyethane; THP = tetrahydropyrane; 1,3 DMPr = 1,3 dimethoxypropane.
[a] Data taken from A. I. Shatenshtein, E. S. Petrov, and M. I. Belousova, *Organic Reactivity*, 1, 191 (1964) (Tartu State University, Estonia, U.S.S.R.).
[b] Taken from reference 67.

Table 12 Effect of Solvent Structure on the Equilibria
Na + Biphenyl $\rightleftarrows Na^+, B^{-}$ and Na + Naphthalene $\rightleftarrows Na^+, N^{-}$

Solvent[a]	$\dfrac{[Na^+, B^{-}]}{[B_0]}$	$\dfrac{[Na^+, N^{-}]}{[N]_0}$
CH_3OCH_3	0.02	0.2
$C_2H_5OC_2H_5$	0.01	0.02
$CH_3OC_2H_4OCH_3$	1.0	1.0
$CH_3OC_2H_4OC_2H_5$	0.6	1.0
$CH_3OC_2H_4OC_3H_7$	0.22	0.85
$C_4H_9OC_2H_4OC_4H_9$	0.1	0.2
$CH_3OCH_2OCH_3$	0	0
$CH_3O(CH_2)_2OCH_3$	1.0	1.0
$CH_3O(CH_2)_3OCH_3$	0.06	0.5
$CH_3O(CH_2)_4OCH_3$	0.03	0.2
$CH_3O(CH_2)_5OCH_3$	—	0.05

[a] Data taken from reference 41.

series of linear glycol ethers (Table 12), which show a sharp drop in the yield of radical ion as the size of the terminal alkyl groups of the ethers increases. The length of the aliphatic chain between the oxygen atoms is also important, since their solvating power is due to a cooperative effect of the two oxygen atoms. The highest yield of radical anions is obtained for the ethylene glycol ethers, largely because a comparatively stable five-membered chelate ring is formed. Solvents which have the acetal structure O—C—O are poor solvating agents for cations, and the equilibrium constants are also found to be low when the oxygen atoms are separated by more than two CH_2 groups, because of a more unfavorable entropy change. This may depend somewhat on the size of the counterion. For example, according to Shatenshtein, the equilibrium constant for B^{-}, K^{+} is larger in 1,3-dimethoxy-propane than in 1,2-dimethoxyethane; the reverse is true for Na^{+}.

The smallest cation usually gives the highest equilibrium constant in a particular solvent ($Li^{+} > Na^{+} > K^{+} > Cs^{+}$) because of the higher heat of solvation. However, dioxane was reported to give 4% biphenyl reduction with Na, but none with Li. Several interpretations of this result are possible. In very low dielectric constant media ion-pair aggregation may occur, and this could affect the ion-pair solvation in an unfavorable way (see the results on fluorenyllithium). Also, radical anions are rather unstable in dioxane, particularly the lithium salts, and rapid destruction of radical ions could lead to formation of alkoxides. In such a case, we would not observe the blue color of the biphenyl anion. Ionic impurities, like alkoxides, affect the equilibrium constant, since they interact with the radical ion pair [67]. Hence care must be taken to avoid any destruction of radical ions, especially in low dielectric constant media.

The radical anion pair equilibria are exothermic, as already observed by Hoijtink and others [4]. The data of Table 13 (see reference 41 for other solvents) show that the extent of exothermicity depends on the solvating power of the solvent, and it often increases as the temperature is lowered (i.e., the van't Hoff plots are curved). Some of Shatenshtein's $\log K - 1/T$ plots show a rather abrupt change of the slope at a particular temperature; for example, for 1,2-methoxyethoxyethane or its mixture with tetrahydrofuran (the latter solvent shows two breaking points of increasing slopes). This was interpreted evidence for a stepwise mechanism of the cation solvation, more solvent as molecules being coordinated to the alkali ion at lower temperatures. Although this conclusion is certainly correct and has also been verified for other ion pairs [e.g., the stepwise solvation of $NaAl(Butyl)_4$ in mixtures of hexane and small quantities of THF (see Reference 94)], the sharp transition in these plots is somewhat surprising. Some of the ΔH values—those in diglyme-heptane— appear to be unusually high, and ion-pair aggregation in this low dielectric constant medium may be a contributing factor.

Table 13   Enthalpy and Entropy Changes for the Reaction
$$\text{Na} + \text{Biphenyl} \rightleftarrows \text{B}^{\overline{\cdot}}, \text{Na}^+$$

Solvent[a]	T (°C)	$-\Delta H$ (kcal/mole)	$-\Delta S$ (e.u.)
THP	−45 to 0	6.8	31
DEE	−15 to 20	9.6	38
	−30 to −15	22.0	86
MeTHF[b]	−53 to 25	9.9	43
THF	−10 to 40	11.2	40
1,3-DMPr	−35 to −10	15.5	63
1,2-DMPr	0 to 45	16.5	58
1,2-DME	0 to 45	17.4	60
THF-	10 to 30	5.5	21
heptane	−10 to 10	8.5	31
(6:1)	−20 to −10	14.0	54
Glyme-3-			
heptane	25 to 45	31	100
(1:1.4)			

[a] See Table 11 for abbreviations; 1,2-DME = 1,2-dimethoxyethane;
[b] Data from reference 67.

Numerous solvation states of ion pairs complicate the interpretation of the results. Tight ion pairs are abundant in the less polar solvents and solvent-separated ion pairs in the more polar media and at lower temperatures. The difference in ΔH values observed at lower and higher temperature ranges therefore should not differ much from the enthalpy change of the contact ion-pair–solvent-separated ion-pair equilibrium. For example, the ΔH difference in THF between the highest and lowest temperature range is 8.5 kcal/mole for $\text{B}^{\overline{\cdot}}$, Na^+ formation. The ΔH_i for the fluorenylsodium ion-pair equilibrium is −7.6 kcal/mole. For 1,2-dimethoxyethane the difference is 7–8 kcal/mole, again similar to that found for the carbanion pair equilibrium in DME [47]. On the other hand, the $-\Delta H$ for $\text{B}^{\overline{\cdot}}$, Na^+ formation in MeTHF is equal to 9.9 kcal/mole, and its value remains constant down to −60°. This indicates that above this temperature no solvent-separated ion pairs are formed in this solvent, a conclusion supported by the optical absorption spectrum of B, Na^+ in MeTHF [67].

A careful study of the equilibrium between sodium and biphenyl was carried out by Slates and Szwarc [67] in mixtures of THP (and of MeTHF) with glyme-4 (triglyme) and glyme-5 (tetraglyme). They utilized a more sophisticated technique to determine spectroscopically the temperature dependence of the equilibrium constant and the structure of the species formed on coordination of the glymes to the $B^{\bar{\cdot}}$, Na^+ ion pairs.

Addition of glyme-4 to a mixture of biphenyl and biphenyl sodium in THP or MeTHF in equilibrium with a sodium mirror led to an increase in the fraction of radical ions, indicating that an equilibrium of the type

$$B^{\bar{\cdot}}, Na^+ + nG \rightleftarrows B^{\bar{\cdot}}, Na^+(glyme)_n \qquad (K_E)$$

is established in addition to

$$B + Na \rightleftarrows B^{\bar{\cdot}}, Na^+ \qquad K$$

The second equilibrium is maintained even in the absence of glyme. From the spectroscopic data an apparent equilibrium constant

$$K_A = \frac{[B^{\bar{\cdot}}_1 + B^{\bar{\cdot}}, E]}{[B_0 - B_1^{\bar{\cdot}} - B_1^{\bar{\cdot}}E]}$$

is obtained, where B_0 represents the initial concentration of biphenyl, $B_1^{\bar{\cdot}}$ the concentration of $B^{\bar{\cdot}}$, Na^+, and $B^{\bar{\cdot}}$, E the concentration of glyme-coordinated species. It can be shown that

$$\frac{[K_A - K]}{K} = K_E E^n$$

Plots of log $(K - K)$ versus log E showed $n = 1$, indicating that only one glyme molecule is complexed to the $B^{\bar{\cdot}}$, Na^+. This is in agreement with results for fluorenylsodium in mixtures of glymes and THF, but it should be stressed that the method based on the contact ion-pair–solvent-separated ion-pair equilibrium yields the difference in solvation state between the two kinds of ion pair, whereas the biphenyl-sodium equilibrium gives the total number of glyme molecules coordinated with the ion pair.

Slates and Szwarc noted that the increase in radical ion concentration on addition of glyme-4 or 5 could not be explained by the increase in the fraction of glyme-separated ion pairs only. The fraction of the glyme-separated species in the solution can be determined spectroscopically, since their absorption maximum is at 406 nm, whereas that of the contact ion pair is at 400 nm. Although the difference is only 6 nm, two separate peaks are visible, and the spectrum was shown to result from the superposition of the spectra of two thermodynamically distinct species. The fraction of glyme-coordinated ion pairs, however, was found to be considerably larger than that of the glyme-separated ion pairs, and the authors suggested that glyme-coordinated contact ion pairs and glyme-separated ion-pairs are simultaneously formed. Two glymation equilibria should therefore be considered:

$$B^{\overline{}}, Na^+ + G \rightleftarrows B^{\overline{}}, Na^+, G \qquad (K_e)$$
$$B^{\overline{}}, Na^+ + G \rightleftarrows B^{\overline{}}, G, Na^+ \qquad (K_i)$$

While the value of K_E can be used to calculate the total fraction of glyme-coordinated ion pairs, the two absorption peaks at 400 and 406 nm allow calculation of the fraction of glyme-separated ion pairs. Hence K_e and K_i and the temperature dependence of these equilibrium constants could be determined. The relevant thermodynamic parameters are listed in Table 14. The heat and entropy of external coordination of glymes with the $B^{\overline{}}$, Na^+ contact ion pair appear to be the same for glyme-3 and glyme-4. Only glyme-4 was shown to yield glyme-separated ion pairs, but it is likely that at higher glyme-3

Table 14   Thermodynamic Parameters of Formation of Glyme-Coordinated Ion-Pair of Biphenyl$^{\overline{}}$, Na^+ in THP and 2MeTHF[a]

$$B^{\overline{}}, Na^+ + G \rightleftarrows B^{\overline{}}, Na^+, G \qquad [K_e(\Delta H_e, \Delta S_e)]$$
$$B^{\overline{}}, Na^+ + G \rightleftarrows B^{\overline{}}, G, Na^+ \qquad [K_i(\Delta H_i, \Delta S_i)]$$

Glyme	Solvent	ΔH_e (kcal/mole)	ΔS_e (e.u.)	ΔH_i (kcal/mole)	ΔS_i (e.u.)
Glyme-3	THP	−4.5	−8.6	—	—
Glyme-4	THP	−4.6	−8.4	−7.0	−17
Glyme-4	MeTHF	−3.6	−3.6	−5.4	−9.7

[a] Taken from reference 67.

concentrations separated ion pairs are formed also with that glyme. For fluorenylsodium, the K_i value for glyme-3 is only 1.2, and at concentrations of 0.1 M glyme only 10% of the ion pairs are glyme separated. In the experiments with $B^{\overline{}}$, Na^+ the highest glyme-3 concentration was 0.042 M.

Studies of Slates and Szwarc were extended to systems involving equilibria between metallic sodium or potassium and solutions of naphthalene [35]. It was shown that contact-ion pairs of sodium naphthalenide in diethyl ether may be externally coordinated with two molecules of tetrahydrofuran when the latter ether is added to the solution, whereas no coordination takes place with potassium naphthalenide under identical conditions. It was shown also that two molecules of diethyl ether are released when tetrahydrofuran is coordinated. Coordination with dimethoxyethane was investigated in tetrahydropyrane solution. The spectrophotometric data were augmented by ESR studies.

Changes in the optical spectra of lithium, sodium, and potassium salts of the radical anions of polyacenes and polyphenyls resulting from a variation of solvent or temperature were reported by Hoijtink and his co-workers [4]. The band shifts are in the order of $100-1100$ cm^{-1}, as compared to 2200 cm^{-1} for the fluorenyllithium salt. The spectra at higher wavelength were initially ascribed to free ions. It seems, however, that they are due to the formation of separated ion pairs, the spectra of such pairs being identical with those of the free ions.

The difference in the absorption maxima for the contact and separated ion pairs are often small. For example, the respective maxima for sodium naphthalene are at 323 and 326 nm [2], and those of sodiumbiphenyl are at 400 and 406 nm [67]. Nevertheless, two distinct maxima can be seen simultaneously under proper conditions, indicating the presence of two chemically distinct ion pairs. For terphenylsodium in THF two separated bands are observed at 833 and 909 nm, the latter being more pronounced at lower temperature. This difference in absorption maxima was used by Biloen [98] to determine spectrophotometrically the dissociation constant of the salt of this radical ion in THF.

A pronounced difference can be detected in the spectra of the two kinds of ion pair of triphenylenesodium [10]. A contact ion-pair spectrum is found in 2MeTHF at room temperature, whereas in THF at 25°C the spectrum shows only separated ion pairs. An equimolar mixture of triphenylenesodium and tetraglyme (10^{-2} M) in 2MeTHF yields a spectrum similar to that in THF, implying that glyme-separated ion pairs are formed. The optical spectra of the two kinds of ion pair are depicted in Fig. 21.

Dilution of the 1:1 complex of triphenylenesodium and tetraglyme yields a mixture of the two ion-pair spectra, and the appearance of isosbestic points shows the stoichiometry of the reaction $Tr^{-}, Na^{+} + G \rightleftarrows Tr^{-}, G, Na^{+}$. The fraction of the two ion pairs can be calculated by computer simulation of the experimental spectra at various concentrations of salt. The equilibrium constant was found to be about 10^{4} M, slightly lower than obtained from electron spin resonance studies of the same system [10]. It is interesting that this value is much higher than that found for the complexation of tetraglyme with sodiumnaphthalene ($200-300$ M^{-1}; see reference 70) and that with fluorenylsodium (170 M^{-1}, see reference 62). Obviously, the structure of the anion affects the complexation constant, and apparently charge delocalization in the triphenylene anion facilitates separation of the two ions by the tetraglyme. A similar behavior was also reported for the salts of the coronene radical anion [72]. Of course, when the glyme complexation involves a separated ion pair, the values of the glyme complexation constants could be

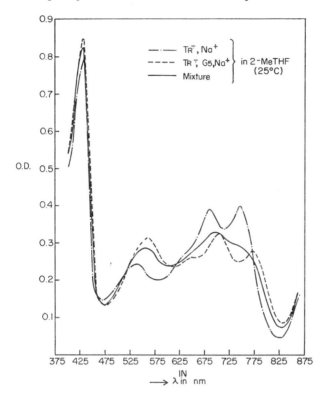

Figure 21. Optical absorption spectrum of the triphenylene sodium contact ion pair (–·–·–), its glyme-5 separated ion pair (– – – –), and a mixture of the two ion pairs (————) in 2MeTHF at 25°C.

very different. The complexation constant of triglyme with coronene lithium in THF is much smaller than that found for fluorenyllithium, because the coronene ion pair is already separated by THF molecules.

The system triphenylenesodium in 2MeTHF was also studied spectroscopically by Van Broekhoven (10) as a function of temperature. The results were interpreted as an indication of a change in ion-pair structure from contact ($T > -70°C$) to solvent separated ($T < -90°C$), the ΔH_i being -10 kcal/mole. On the other hand, the sodium coupling constant measured by ESR changes from 0.9 gauss at 80°C to 0 gauss at $-50°C$. At $-50°C$ the optical spectrum still resembles that of a contact ion pair. The free ion fraction in both cases was shown to be small. Hence, by ESR standards the ion pair may be termed a separated ion pair at $-50°C$, although the optical spectrum apparently identifies the species under the same conditions as a contact ion pair. It is therefore clear that application of different techniques for studies of ion pairs is highly desirable.

The importance of utilizing different techniques in the study of ion-pairs structure cannot be overstressed. While certain variations in the solvation state or structure of an ion pair may be revealed by one method of investigation, the same changes may remain undetected when a different technique is applied. A slight change in the relative position of the alkali ion with respect to the plane of the radical anion, for example, arising from temperature change, could considerably affect its ESR spectrum but may leave the optical spectrum unaffected. One method may be more sensitive to differences in the structure of tight ion-pair species, whereas another experimental approach may detect only a change from a tight to a loose ion-pair structure, or large variations in the structure of loose ion pairs. It is interesting therefore to note that the extensive ESR studies of Hirota on the sodium salts of naphthalene and anthracene have indeed shown evidence for the existence of two different kinds of tight ion pair [34].

The optical spectra of solutions of the monoradical anions and dianions of fluorenone, benzophenone, benzil, xanthone, etc., have also yielded evidence for the existence of dynamic equilibria between various types of ionic species. Warhurst et al. [1] observed red shifts in the optical spectra of ketyls with increasing radius of the counterion or increasing polarity of solvents. For example, the absorption maxima for the Li^+, Na^+, K^+, and Cs^+ salts of fluorenone in dioxane are 448, 452, 463, and 465 nm, respectively, and a similar gradation was found for salts of benzophenone, of the radical ion and dianion of dibiphenyl ketone, and of other ketones. The cation effects were explained in terms of a perturbation of the molecular energy levels of the ketyl by the cationic field, a perturbation which decreases as the cation radius increases. The frequency maxima of the absorption band were found to be roughly proportional to the inverse of the interionic distance, which was taken as $r_c + 2$, r_c being the cation radius in Å. The use of a more polar solvent, say THF or DME, shifts the spectrum of the benzophenone radical ion salt of a given cation to higher wavelength. This was interpreted by Warhurst [1] and by Garst [99] as evidence for strong cation-solvent interactions, which would decrease the cationic field strength at the anion and at the same time increase the interionic distance.

Subsequent studies by Hirota and Weissman [100] and Garst et al. [55, 101] have shown that much of the solvent effect appears to have a rather different origin. In the ketyl radical ion the charge is localized largely on the oxygen atom. This favors formation of tight ion pairs, even in such solvents as DME. Although external solvation of the cation of the ion pair very likely exists, it may not affect the spectrum very much, as was shown to be the case for fluorenyl carbanion salts. On the other hand, ion-pair association of alkoxides in ethereal solutions is a common phenomenon, and formation of ionic aggregates, such as ion quadrupoles, in ketyl solutions is to be expected. Indeed, the experiments of Hirota show that the fluorenone ketyl spectra in solvents such as THF or MeTHF are concentration dependent. For example, two distinctly different absorption bands are found for the fluorenone⁻, Na^+ in 2MeTHF; a 450 nm band at high concentration changing to a band at 525 nm on dilution. In solvents of low polarity like toluene and cyclohexane, the intensities of the bands decrease. This is accompanied by a decrease in the ESR signal of the monoradical anion.

The data were interpreted by assuming the following equilibria:

$$Ar_2-\underset{\underset{Na^+Na^+}{\overset{|}{O^-}\ \overset{|}{O^-}}}{\overset{I}{C-C}}-Ar_2 \rightleftarrows Ar_2CO^{\bar{\cdot}} \underset{Na^+}{\overset{Na^+}{\diamond}} OCAr_2 \overset{II}{\rightleftarrows} 2Ar_2CO^-, Na^+$$

$$III\downarrow\uparrow$$

$$2Ar_2CO^{\bar{\cdot}} + 2Na^+$$

Equilibrium I represents the change from a diamagnetic dimer to a paramagnetic dimer, the first species representing a colorless picolate, which is favored in nonpolar solvents such as cyclohexane. Slightly polar solvents like dioxane or ether, or small quantities of an ether like DME added to cyclohexane or toluene, already facilitate formation of the paramagnetic dimer. The validity of Beer's law for a number of ketyls in dioxane, as shown by Warhurst et al. [1], does not provide real evidence for monomeric species. More likely, the dissociation is so small that over a wide concentration range no noticeable change in aggregate concentration is observed, and Beer's law is still obeyed. Furthermore, the disappearance of the ESR signal does not necessarily indicate formation of a picolate-type structure. Some very interesting color changes have been observed for ketyls of phenanthrenequinone and naphthoquinone [103,115]. The first ketyl shows a change in color from a red paramagnetic monomer to a green diamagnetic dimer, with the red colored species favored at lower concentrations and in more polar solvents. It is believed that in the green diamagnetic dimer no ordinary covalent bond is formed (as in the picolates) but that the close proximity of the two radical anions and their resulting strong interaction causes the species to exist in a singlet state. Interesting information concerning the structure of such diamagnetic dimers was deduced by Staples and Szwarc [115] from studies of kinetics of their formation. Similar phenomena have been observed with phenazine radicals [104].

From the optical spectra, Hirota and Weissmann were able to deduce the thermodynamic parameters for the dissociation of the sodiumfluorenone dimer [100]. The respective dissociation constants at 23° in DME, THF, and 2MeTHF are $2 \times 10^{-3}\,M$, $5 \times 10^{-4}\,M$, and $2 \times 10^{-5}\,M$. The $\Delta H°$ values are -2.5 kcal/mole, -1.3 kcal/mole, and "small"; the $\Delta S°$ values are -29, -31, and -37 e.u. The low ΔH values suggest that the monoradical anion pair is probably tight. This is also indicated by the cation effect on the dimer dissociation constant, which decreases in the order $Cs^+ > Rb^+ > K^+ > Na^+ > Li^+$. Specific solvation, at least in 2MeTHF, is apparently not strong enough to induce ion-pair separation even for the lithium salt, and it is unlikely that separated ion pairs are formed to any great extent even in solvents like DME.

No detailed studies were carried out with glyme-type solvents, but it is likely that predominantly glymated contact ion pairs are formed (see the section on fluorenyl carbanions). It is interesting that addition of macrocyclic ethers to fluorenone$^-$, Na^+ in THF yields a spectrum identical to that of the monomeric species in DME, $\lambda_m = 530$ nm [107]. This suggests that even in the presence of these powerful complexing agents we may still be dealing essentially with contact ion pairs, although the NMR spectrum of the solution shows that the cyclic polyether is complexed to the ketyl salt.

There are indications from ESR data that also in the case of ketyls a variety of tight ion pairs, solvated and nonsolvated dimeric ion pairs, etc., may exist in ethereal solutions [105, 106], and the last word has certainly not been said about these interesting systems. However, a thorough discussion of these ESR data is outside the scope of this chapter.

Investigations by Zaugg and Schaefer [3] and by Garst et al. [55] regarding the effects of solvents and counterion on the optical spectra of alkali phenoxides and enolates show a parallel behavior with the ketyl systems. Bathochromic shifts are observed in DME on increasing the size of the counterion, indicating a tight ion-pair structure for these species. The effect is absent in dimethylformamide, where free ions are the predominant species [3]. It is probable that in the ether solvents ion-pair aggregation is at least partially responsible for the observed spectral changes of phenoxides and enolates.

Changes in ketyl spectra were used by Garst et al. [101] to measure solvent polarity. Although from the viewpoint of cation-solvent interactions many similarities exist between this system and the contact–solvent-separated ion-pair equilibrium of fluorenyl carbanion salts, we should not expect scales of solvent polarity based on the two systems to be identical. In the ketyl systems, the dielectric constant of the medium may play an important role, as the ion-pair aggregation depends largely on this parameter. This is not the case for the contact–solvent-separated ion-pair equilibrium of fluorenyl carbanions, which was found to be largely independent of dielectric constant in mixtures of solvents containing a strongly solvating entity. Nevertheless, cation solvation as determined by the basicity and geometrical structure of the solvent will also be important in the ketyl system, as the ketyl-solvent complexes are expected to weaken the dimer structure and favor the monomeric species.

REFERENCES

1. H. V. Carter, B. J. McClelland, and E. Warhurst, *Trans. Faraday Soc.*, **56**, 455 (1960).
2. T. E. Hogen Esch and J. Smid, *J. Am. Chem. Soc.*, **87**, 669 (1965); **88**, 307, 318 (1966).
3. H. E. Zaugg and A. D. Schaefer, *J. Am. Chem. Soc.*, **87**, 1857 (1965).
4. K. H. J. Buschov, J. Dieleman, and G. J. Hoijtink, *J. Chem. Phys.*, **42**, 1993 (1965).
5. N. S. Bayliss and E. G. McRae, *J. Phys. Chem.*, **58**, 1002 (1954).
6. (a) E. M. Kosower and P. E. Klinedinst, Jr., *J. Am. Chem. Soc.*, **78**, 3493 (1956); (b) E. M. Kosower and J. C. Burbach, *J. Am. Chem. Soc.*, **78**, 5838 (1956).
7. T. R. Griffiths and M. C. R. Symons, *Mol. Phys.*, **3**, 90 (1960).
8. C. S. Irving, G. W. Byers, and P. A. Leermakers, *J. Am. Chem. Soc.*, **91**, 2141 (1969).
9. F. Feichtmayr and J. Schlag, *Ber. Bunsenges. physik. Chem.*, **68**, 95 (1964).
10. J. A. M. van Broekhoven, Ph.D. Thesis, University of Nijmegen, The Netherlands, 1970, p. 68.
11. M. Smith and M. C. R. Symons, *Trans. Faraday Soc.*, **54**, 338 (1957).
12. M. Smith and M. C. R. Symons, *Trans. Faraday Soc.*, **54**, 346 (1957).
13. M. Smith and M. C. R. Symons, *Disc. Faraday Soc.*, **24**, 207 (1957).
14. M. J. Blandamer, T. E. Gough, and M. C. R. Symons, *Trans. Faraday Soc.*, **59**, 1748 (1963).
15. M. J. Blandamer, T. E. Gough, and M. C. R. Symons, *Trans. Faraday Soc.*, **62**, 286 (1966).
16. E. M. Kosower, *An Introduction to Physical Organic Chemistry*, Wiley, New York, 1968.
17. M. J. Blandamer, T. E. Gough, and M. C. R. Symons, *Trans. Faraday Soc.*, **62**, 301 (1966).
18. D. N. Bhattacharyya, C. L. Lee, J. Smid, and M. Szwarc, *J. Phys. Chem.*, **69**, 612 (1965).
19. C. Carvajal, K. J. Tölle, J. Smid, and M. Szwarc, *J. Am. Chem. Soc.*, **87**, 5548 (1965).
20. D. C. Pepper, private communication.
21. M. J. Blandamer, T. E. Gough, T. R. Griffiths, and M. C. R. Symons, *J. Chem. Phys.*, **38**, 1034 (1963).
22. M. J. Blandamer, T. E. Gough, and M. C. R. Symons, *Trans. Faraday Soc.*, **60**, 488 (1964).
23. E. M. Kosower, *J. Am. Chem. Soc.*, **80**, 3253 (1958); **80**, 3261 (1958); **80**, 3267 (1958).
24. R. A. Mackay and E. J. Poziomek, *J. Am. Chem. Soc.*, **92**, 2432 (1970).
25. K. Dimroth, C. Reichardt, T. Siepmann, and F. Bohlmann, *Ann.*, **661**, 1 (1963).
26. C. D. Ritchie, *J. Am. Chem. Soc.*, **91**, 6749 (1969).
27. R. Waack and M. A. Doran, *J. Phys. Chem.*, **67**, 148 (1963).

28. L. L. Chan and J. Smid, *J. Am. Chem. Soc.*, **90**, 4654 (1968).

29. J. P. Pascault, Thesis, l'Universite de Lyon, France, 1970.

30. S. Bywater, and D. J. Worsfold *Can. J. Chem.*, **40**, 1564 (1962).

31. R. Waack, M. A. Doran and P. E. Stevenson, *J. Am. Chem. Soc.*, **88**, 2109 (1966).

32. C. G. Screttas and J. F. Eastham, *J. Am. Chem. Soc.*, **87**, 3276 (1965).

33. A. W. Langer, *Trans. N.Y. Acad. Sci.*, *Sec II*, **27**, 741 (1965).

34. N. Hirota, *J. Am. Chem. Soc.*, **90**, 3603 (1968).

35. L. Lee, R. F. Adams, J. Jagur-Grodzinski, and M. Szwarc, *J. Am. Chem. Soc.*, **93**, 4149 (1971).

36. S. Searles and M. Tamres, *J. Am. Chem. Soc.*, **73**, 3704 (1951).

37. E. M. Arnett and C. Y. Wu, *J. Am. Chem. Soc.*, **84**, 1684 (1962).

38. S. Searles, M. Tamres, and E. R. Lippincott, *J. Am. Chem. Soc.*, **75**, 2775 (1953).

39. M. Brandon, M. Tamres, and S. Searles, *J. Am. Chem. Soc.*, **82**, 2129 (1960).

40. A. I. Shatenshtein, E. S. Petrov, M. I. Belousova, K. G. Yanova, and E. A. Yakovleva, *Dokl. Akad. Nauk. SSSR*, **151**, 353 (1963).

41. A. I. Shatenshtein, E. S. Petrov, and E. A. Yakovleva, *J. Polymer Sci. C*, **16**, 1729 (1967).

42. E. A. Kovrikhnykh, F. S. Yakushin, and A. I. Shatenshtein, *Reakts. Sposobnost. Org. Soedin, Tarta, Gos. Univ.*, **3**, 209 (1966).

43. A. I. Shatenshtein and E. S. Petrov, *Usp. Khim.*, **36**, 269 (1967).

44. R. Smyk, M.S. Thesis, College of Forestry, Syracuse University, Syracuse, New York, 1968.

45. A. Streitwieser, Jr., J. H. Hammons, E. Ciuffarin, and J. I. Brauman, *J. Am. Chem. Soc.*, **89**, 59 (1967).

46. T. Shimomura, K. J. Tölle, J. Smid, and M. Szwarc, *J. Am. Chem. Soc.*, **89**, 796 (1967).

47. T. Shimomura, J. Smid, and M. Szwarc, *J. Am. Chem. Soc.*, **89**, 5743 (1967).

48. J. T. Denison and J. B. Ramsay, *J. Am. Chem. Soc.*, **77**, 2615 (1955).

49. S. Claesson, B. Lundgren, and M. Szwarc, *Trans. Faraday Soc.*, **66**, 3053 (1970).

50. P. Chang, R. V. Slates, and M. Szwarc, *J. Phys. Chem.*, **70**, 3180 (1966).

51. D. Nicholls and M. Szwarc, *Proc. Roy. Soc. A*, **301**, 223 (1967).

52. A. Streitwieser and J. I. Brauman, *J. Am. Chem. Soc.*, **85**, 2633 (1963).

53. T. E. Hogen Esch, Ph.D. Thesis, University of Leiden, The Netherlands, 1967, p. 35.

54. D. Casson and B. J. Tabner, *Chem. Soc. B*, **1969**, 572.

55. J. F. Garst, R. A. Klein, D. Walmsley, and E. R. Zabolotny, *J. Am. Chem. Soc.*, **87**, 4080 (1965).

56. R. C. Roberts and M. Szwarc, *J. Am. Chem. Soc.*, **87**, 5542 (1965).

57. J. F. Garst, E. R. Zabolotny, and R. S. Cole, *J. Am. Chem. Soc.*, **86**, 2257 (1964).

58. J. F. Garst and E. R. Zabolotny, *J. Am. Chem. Soc.*, **87**, 495 (1965).

59. T. Ellingsen and J. Smid, *J. Phys. Chem.*, **73**, 2712 (1969).

60. T. E. Hogen Esch and J. Smid, *J. Am. Chem. Soc.*, **89**, 2764 (1967).

61. L. L. Chan and J. Smid, *J. Am. Chem. Soc.*, **89**, 4547 (1967).

62. L. L. Chan, K. H. Wong, and J. Smid, *J. Am. Chem. Soc.*, **92**, 1955 (1970).

63. C. J. Pedersen, *J. Am. Chem. Soc.*, **89**, 7017 (1967).

64. J. Ugelstad and O. A. Rokstad, *Acta Chem. Scand.*, **18**, 474 (1964).

65. S. Bywater and D. J. Worsfold, *Can. J. Chem.*, **40**, 1564 (1962).

66. M. Morton, L. J. Fetters, and E. E. Bostick, *J. Polymer Sci. C*, **1**, 311 (1963).

67. R. V. Slates and M. Szwarc, *J. Am. Chem. Soc.*, **89**, 6043 (1967).

68. M. Shinohara, J. Smid, and M. Szwarc, *J. Am. Chem. Soc.*, **90**, 2175 (1968).

69. M. Shinohara, J. Smid, and M. Szwarc, *Chem. Comm.*, *London*, **1969**, 1232.

70. K. Höfelmann, J. Jagur-Grodzinski, and M. Szwarc, *J. Am. Chem. Soc.*, **91**, 4645 (1969).

71. E. de Boer, A. M. Grotens, and J. Smid, *J. Am. Chem. Soc.*, **92**, 4742 (1970).

72. E. de Boer, A. M. Grotens, and J. Smid, *Chem. Comm.*, **1970**, 1035.

73. J. L. Down, J. Lewis, B. Moore, and G. Wilkinson, *J. Chem. Soc.*, **1959**, 3767.

74. J. Ugelstad, A. Berge, and H. Listou, *Acta Chem. Scand.*, **19**, 208 (1965).

75. C. J. Pedersen, *J. Am. Chem. Soc.*, **92**, 391 (1970).

76. D. Bright and M. R. Truter, *Nature*, **225**, 176 (1970).

77. D. Bright and M. R. Truter, *J. Chem. Soc.*, B, **1970**, 1545.

78. K. H. Wong, G. Konizer, and J. Smid, *J. Am. Chem. Soc.*, **92**, 666 (1970).

79. U. Takaki, T. E. Hogen Esch, and J. Smid, *J. Am. Chem. Soc.*, in press.

80. C. J. Pedersen, *J. Am. Chem. Soc.*, **92**, 386 (1970).

81. H. K. Frensdorff, *J. Am. Chem. Soc.*, **93**, 600 (1971).

82. S. Kopolow, T. E. Hogen Esch, and J. Smid, *Macromolecules*, **4**, 359 (1971).

83. G. Eisenman, S. M. Ciani, and G. Szabo, *Fed. Proc.*, **27**, 1289 (1968).

84. G. Eisenman, S. Ciani, and G. Szabo, *J. Membrane Biol.*, **1**, 294 (1969).

85. S. G. A. McLaughlin, G. Szabo, G. Eisenman, and S. Ciani, *Biophys. Soc. Abstr.*, **10**, 96a (1970).

86. T. E. Hogen Esch and J. Smid, *J. Am. Chem. Soc.*, **91**, 4580 (1969).

87. T. E. Hogen Esch and J. Smid, unpublished results.

88. B. Dietrich, J. M. Lehn, and J. P. Sauvage, *Tetrahydron Letters*, **34**, 2885 (1969).

89. J. M. Lehn, J. P. Sauvage, and B. Dietrich, *J. Am. Chem. Soc.*, **92**, 2916 (1970).

90. W. Schlenk and E. Bergmann, *Ann.*, **463**, 1 (1928); **464**, 1 (1928).

91. N. D. Scott, J. F. Walker, and V. L. Hansley, *J. Am. Chem. Soc.*, **58**, 2442 (1936).

92. D. Lipkin, D. E. Paul, J. Townsend, and S. I. Weissman, *Science*, **117**, 534 (1953).

93. S. I. Weissman, J. Townsend, D. E. Paul, and G. E. Pake, *J. Chem. Phys.*, **21**, 2227 (1953).

94. E. Schaschel and M. C. Day, *J. Am. Chem. Soc.*, **90**, 503 (1968).

95. R. V. Slates and M. Szwarc, *J. Phys. Chem.*, **69**, 4124 (1965).

96. N. H. Velthorst and G. J. Hoijtink, *J. Am. Chem. Soc.*, **87**, 4529 (1965).

97. N. H. Velthorst and G. J. Hoijtink, *J. Am. Chem. Soc.*, **89**, 209 (1967).

98. P. Biloen, Ph.D. Thesis, University of Amsterdam, The Netherlands, 1968, p. 35.

99. J. F. Garst, C. Hewitt, D. Walmsley, and W. Richards, *J. Am. Chem. Soc.*, **83**, 5034 (1961).

100. N. Hirota and S. I. Weissman, *J. Am. Chem. Soc.*, **86**, 2538 (1964).

101. J. F. Garst, D. Walmsley, C. Hewitt, W. R. Richards, and E. R. Zabolotny, *J. Am. Chem. Soc.*, **86**, 412 (1964).

102. K. Maruyama, *Bull. Chem. Soc. (Japan)*, **37**, 553 (1964).
103. N. Hirota, in *Radical Ions*, E. T. Kaiser and L. Kevan, Ed., Interscience, New York, 1968.
104. K. H. Hausser and J. N. Murrell, *J. Chem. Phys.*, **27**, 500 (1957).
105. N. Hirota, *J. Am. Chem. Soc.*, **89**, 32 (1967).
106. N. Hirota, *J. Phys. Chem.*, **71**, 127 (1967).
107. U. Takaki and J. Smid, unpublished results.
108. W. J. le Noble and A. R. Das, *J. Phys. Chem.*, **74**, 3429 (1970).
109. M. M. Exner, R. Waack, and E. C. Steiner, *Abstracts 161st Nat. Meeting Am. Chem. Soc.*, Los Angeles, 1971, Orgn. no. 184.
110. M. Truter, private communication.
111. Z. Machacek, T. E. Hogen Esch, and J. Smid, forthcoming.
112. J. W. Burley and R. N. Young, *J. Chem. Soc.* (B), 1018 (1971).
113. J. F. Coetzee and C. D. Ritchie, eds., *Solute-Solvent Interactions*, Marcel Dekker, New York, 1969.
114. Y. Karasawa, G. Levin, and M. Szwarc, *J. Am. Chem. Soc.* **93**, 4614 (1971). *Proc. Roy. Soc.*, in press.
115. T. L. Staples and M. Szwarc; *J. Am. Chem. Soc.*, **92**, 5022 (1970).

4

Infrared and Raman Studies of Ions and Ion Pairs

WALTER F. EDGELL

Department of Chemistry, Purdue University, Lafayette, Indiana

1. INTRODUCTION

The discovery [1] that alkali ions vibrate in solution offers a new funda-mental way of obtaining information about electrolytic solutions. The infrared bands at long wavelength associated with this phenomenon arise from the excitation of quantum states of the solution connected with motion

153

of solution elements adjacent to the alkali ion. We hope therefore to obtain information about the forces acting on the ions and the structure of solutions at ions from the spectroscopic study of ion vibration.

In this chapter, we examine the ion-vibration phenomena and their implications for solution structure. This reveals the nature of the ion sites and the presence of ion pairs. Some insight into just what constitutes an ion pair is obtained by this approach. Thermodynamics of intersite conversion are considered for one case. Finally, the net force acting on the alkali ions is investigated and considerations are given to the origin of this force.

Infrared bands arising from the vibration of alkali ions appear as broad bands of medium intensity in the upper region of the far infrared range or in the lower portion of the mid-infrared range. Figure 1 shows a typical set of

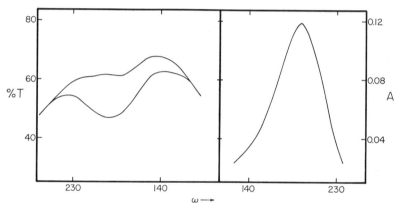

Figure 1. Infrared absorption band due to the vibration of the solution elements at the Na^+ for $NaCo(CO)_4$ in THF. Left panel, upper curve: pure THF; left panel, lower curve: $0.2M$ $NaCo(CO)_4$ in THF; right-hand panel: computed absorbance of the vibrating sites.

absorption curves. On the left, the upper curve shows the transmission of the solvent in the cell, while the lower curve is that of the solution. A constant amount of zero suppression was used for both runs. The absorbance $A = \log(T_0/T)$ computed from these curves, is shown in the right-hand panel of this figure. Table 1 lists the frequency of the absorption maximum for a number of alkali metal salts in several solvents. THF is a typical solvent [2] of low dielectric constant (7.6), whereas DMSO is one whose constant is rather high (45.0). DMSO has a relatively strong absorption band near $400 \ cm^{-1}$, which interferes with the measurement of the Li^+ ion bands. Consequently it was replaced by DMSO-d_6, to move this solvent band out of the way, when Li salts were investigated.

Table 1 Frequencies of Infrared Bands From Alkali Ion Vibrations

Salt	Solvent	Frequency
$Li^+, Co(CO)_4^-$	THF	$413\ cm^{-1}$
$Na^+, Co(CO)_4^-$	THF	192
$K^+, Co(CO)_4^-$	THF	142
Li^+, BPh_4^-	THF	412
Li^+, NO_3^-	THF	407
Li^+, Cl^-	THF	387
Li^+, Br^-	THF	378
Li^+, I^-	THF	373
Na^+, BPh_4^-	THF	198
Na^+, I^-	THF	184
Li^+, NO_3^-	DMSO-d_6	425
Li^+, Cl^-	DMSO-d_6	425
Li^+, Br^-	DMSO-d_6	424
Li^+, I^-	DMSO-d_6	424
Na^+, BPh_4^-	DMSO	203
$Na^+, HCr_2(CO)_{10}^-$	DMSO	200
$Na_2^+, Cr_2(CO)_{10}^{2-}$	DMSO	200
Na^+, NO_3^-	DMSO	200
$Na^+, Co(CO)_4^-$	DMSO	199
Na^+, I^-	DMSO	194
$Na^+, Co(CO)_4^-$	Piperidine	183 (274)
$Na^+, Co(CO)_4^-$	Pyridine	180

2. THE EFFECT OF CATION ON FREQUENCY

This effect, clearly demonstrated in Table 1 for the Li^+, Na^+, and K^+ salts of $Co(CO)_4^-$ in the THF, is evident in all of the results which show that the vibration is primarily that of the alkali ion. These frequencies do not vary as expected for the motion of a rigid, solvated alkali ion, that is, when one or more solvent molecules are moving as a unit with the cation. Nor are the frequencies exactly inversely proportional to the square root of the alkali ion mass. This may be caused by the variation of the force on the alkali ion going from Li^+ to K^+. On the other hand, it is also consistent with several near-neighbor solution entities—such as solvent molecules or anions—having small displacements in the vibration in addition to the displacement of the alkali ion.

3. THE EFFECT OF ANION ON ALKALI ION FREQUENCY

The near-neighbor environment of the alkali ion in solution may be explored by examining the effect of varying the anion on the cation frequency. The bands for the four Li^+ ion salts in DMSO-d_6 are all superimposable with an absorption maximum at 425 ± 3 cm^{-1}. Essentially the same results are obtained for the sodium salts in DMSO, with the possible exception of NaI, since the measurements tabulated have an uncertainty of ± 4 cm^{-1}. We may conclude that the alkali ion vibrational frequency in DMSO is independent of the anion.

Contrast, however, the behavior of the salts in THF. The Li^+ ion frequency occurs at 413 cm^{-1} when the anion is $Co(CO)_4^-$ and it shifts progressively for six anions to a value of 373 cm^{-1} for the I^- ion. This shift is both real and pronounced since the tabulated values are uncertain by ± 3 cm^{-1}. This same uncertainty applies to the tabulated values of the Na^+ ion frequency for Na^+, $Co(CO)_4^-$ and Na^+, BPh_4^- in THF, whereas that for NaI is perhaps twice as large. Thus the frequency of these Li^+ and Na^+ salts in THF vary with anion.

4. THE EFFECT OF SOLVENT ON ALKALI ION FREQUENCY

Solvent molecules must be part, if not all, of the near-neighbor shell around the alkali ion in solution and therefore they might exert a significant force on the alkali ion in its vibration. Two kinds of comparison may be made between data in Table 1 which bear upon this point. First, the frequency of the Li^+ shows a pronounced difference when the same salt is dissolved in THF and in DMSO. Differences also occur for the Na^+ ion salts but are smaller.

Second, compare the Na^+ ion vibration for Na^+, $Co(CO)_4^-$ dissolved in THF, DMSO, piperidine, and pyridine. The frequency shifts from 180 cm^{-1} in pyridine to 199 cm^{-1} in DMSO. It is seen that the solvent has indeed a role in determining the net force acting on the alkali in its vibration.

5. EXISTENCE OF SOLUTION STRUCTURE AT THE CATION

Perhaps the most important fact about these infrared bands is their existence. For this means that structure exists in solution at the alkali ion whose lifetime is substantially longer than the period of vibration. Such structure must then exist for a time substantially longer than 10^{-13} sec. An upper value for the lifetime of this structure of $\sim 10^{-7}$ sec may be obtained by considering the rate of ligand attachment to alkali ions [2]. This estimate is in agreement with ion-pair lifetime estimates near the larger value obtained for sodium napthalenide and similar aromatic radical ions from electron spin resonance measurements [3].

The experimental results cited here lead to some conclusions about the solution structure. The fact that the alkali ion frequency varies with anion for the salts in THF places the anion in the near-neighbor environment of the cation. Solvent molecules must also be near neighbors of the alkali ion on physical grounds and make a significant contribution to the force causing the alkali ion vibration in both THF and DMSO solutions. This suggests a model for this vibration in THF in which the alkali ion vibrates in a cage formed by solvent molecules and anion; this means both solvent molecules and anion(s) are near neighbors of the cation. Cage elements, as well as the alkali ion, move in the vibration as suggested by the frequency variation with both cation change and cation isotope substitution as well as by momentum conservation requirements.

The alkali ion also may be thought of as vibrating in a cage when these salts are dissolved in DMSO. However, the near neighbors of the alkali ion are then solvent molecules as shown by the independence of the frequency from anion variation.

At least two kinds of solution structure can give a cage as just described for DMSO solutions. In one, the cation and anion together with solvent molecules are coupled in a single structural unit of some stability with one or more solvent molecules between the two ions. This is a model for a solvent-separated ion pair. In the other, the anion occupies less stable structural positions in the solution, most likely at greater distances from the cation and its solvent near neighbors. These are "free" ions. These two cases are expected to give rise to very similar spectra. This kind of a cage is called a solvent-surrounded cation whatever solution structure is involved.

Other types of solution structure may exist at ions. Two that may be expected to be found are cluster ions, of which the triple ion is the most likely example, and ions involving more than one kind of solvent molecule in the structure at the ion when mixed solvents are involved.

6. CONCENTRATION AND THE ALKALI ION BAND INTENSITY

In THF the alkali ion band shifts with anion and hence this band is believed to arise from the vibration of a contact ion pair as described previously. But is there only one kind of ion site for salts in THF? How the population of a given ion site varies with gross salt concentration depends upon the equilibria between the several kinds of site* present. For this reason, an examination of the variation of alkali ion band intensity with salt concentration is of interest. The results for Na^+, $Co(CO)_4^-$ in THF are shown in Fig. 2.

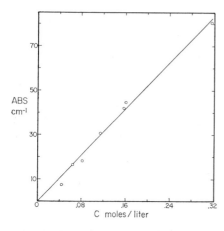

Figure 2.   The concentration dependence of the integrated absorbance of the vibration at the Na^+ for $NaCo(CO)_4$ in THF.

The simplest interpretation of these results is that there is only one kind of ion environment present, the contact pair. However, this result may also arise in other ways. Suppose that both contact and solvent-separated ion pairs coexisted in the THF solution. Then the equilibrium would be

$$S\ Na^+, A^- + S = S\ Na^+, S, A^-$$

Here A^- is the anion and S is a solvent molecule. The solvent molecule to the left of the cation in the preceding formula is written to emphasize the cage nature of the model; all the near-neighbor solvent molecules are not shown (for simplicity). The ratio of the populations of the contact and solvent-separated ion pairs is independent of the salt concentration in the case being considered. Hence the band intensity arising from either or both solution components would vary linearly with salt concentration.

The same intensity result would obtain in the case when several cation environments contribute to the band in substantial amounts, even when the concentration ratio changes, providing the molar absorbance of these environments were essentially the same. The molar absorbance of a localized vibration in solution is given by

$$\text{ABS} = K\left(\frac{\partial \mu}{\partial Q}\right)^2; \qquad \frac{\partial \mu}{\partial Q} = \sum_i \left(\frac{\partial \mu}{\partial q_i}\right)\left(\frac{\partial q_i}{\partial Q}\right) \tag{1}$$

* The term "site" used in this chapter is equivalent to the term "type of ion" or ion pair used in other chapters of this book.

where $\partial\mu/\partial Q$ is the change of electric moment with the normal coordinate for the motion and q_i is the displacement of the ith charge center, for example, the alkali ion.

To investigate this question, compare absorbances for a contact ion pair and a solvent-surrounded cation (free ion or solvent-separated ion pair). It is suitable for present purposes to use the simple model of Fig. 8 (see page 168) for the vibrational motions (see the following discussion). In terms of these models,

Contact ion pair:

$$\text{ABS} = Ke^2 \frac{M + A + S}{M(A + S)}$$

Solvent-surrounded cation: (2)

$$\text{ABS} = Ke^2 \frac{2S}{M(M + 2S)}$$

where M, A, and S are the mass of the cation, anion, and solvent molecule, respectively, and e is the electron charge. This leads to an expected ratio of about 1.3 for the absorbance of the Na^+ vibration in a contact ion pair to that of the Na^+ in a solvent-surrounded cation for $NaCo(CO)_4$ in THF. Since the assumptions seem more likely to fail in ways which make the ratio more nearly one, we must consider it possible that this band arises from the vibration of the Na^+ in more than one environment, all of which occur at about the same frequency.

We may note that $A + S \approx M + A + S$ and $2S \approx M + 2S$. Hence Eq. 2 leads us to expect that the intensity of the alkali ion bands would be approximately inversely proportional to alkali ion mass. Qualitative observations on these bands are in agreement with this expectation.

7. MULTIPLICITY OF ION SITES IN SOLUTION

The question about the number of ion sites in these solutions was raised in Section 6. Although one infrared band associated with alkali ion motion is found in each solution, it would be premature to infer the presence of only one kind of alkali ion site in each investigated system. We need a more sensitive test of ion environment than the observation of the shift of alkali band frequency with anion change.

The intramolecular vibrations of a polyatomic ion are sensitive to its surroundings. Consequently, the internal vibrations of a polyatomic anion may serve as a probe of its solution environment and thus that of the cation. The CO stretching vibrations of the $Co(CO)_4^-$ are sensitive to ion environment and salts of this anion are soluble in a variety of solvents. Hence the $Co(CO)_4^-$ ion has been used in a series of anion-probe studies.

The $Co(CO)_4^-$ ion is tetrahedral. This symmetry gives rise to a triply degenerate CO stretching frequency of type F_2, which is active in both infrared and Raman spectra, and a nondegenerate CO stretching mode, type A_1, which is active only in the Raman spectrum. The F_2 vibrations have been observed as a strong infrared and Raman band near 1900 cm^{-1}; the A_1 mode appears as a weak Raman band near 2000 cm^{-1} [4]. When the $Co(CO)_4^-$ ion is found in a solution environment where the *significant* forces which act on it during its vibration are tetrahedral, we can expect a single strong infrared band in the 1900 cm^{-1} region. However, when this anion is in an environment where the significant forces acting on it during its vibration do not have this high symmetry, the threefold degeneracy of the F_2 vibrations is lifted and two or three bands can be expected to appear near 1900 cm^{-1}, depending upon the symmetry of the forces exerted on it by its environment.

A test of the ability of the anion-probe method to reveal the multiplicity of ion sites in solution was made by examining the 1900 cm^{-1} region of the infrared spectra of Na$^+$, $Co(CO)_4^-$ dissolved in a series of solvents (DMF, DMSO, DME, pyridine, THF, and piperidine). Figure 3 shows the result

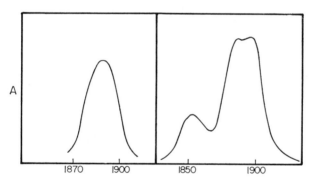

Figures 3 and 4. Left panel: infrared band from CO stretching modes of anion for NaCo(CO)$_4$ in DMSO. Right panel: infrared band from CO stretching modes of anion for NaCo(CO)$_4$ in THF.

obtained in DMSO. Only a single strong band appears at 1888 cm^{-1} implying that each CO group of the $Co(CO)_4^-$ ion sees the same kind of environment. Two kinds of solution structure can produce this result. The first are the "free" ions in which both Na$^+$ and $Co(CO)_4^-$ ions see only solvent molecules as their near neighbors and move independently of each other. The second is the solvent-separated ion pair in which the cation and anion share a solution structure of some (relative) stability where there is one, and possibly more, solvent molecules between the cation and anion. We assume in this case that the solvent molecule(s) separating the cation and anion mutes the force of the cation on the anion. These constitute the solution structure referred to earlier as the solvent-surrounded cation. The results for Na^+, $Co(CO)_4^-$ in DMF and DME are similar to those for DMSO.

The 1900 cm^{-1} band system is more complex for the salt dissolved in pyridine, THF, or piperidine. Typical of these, and of intermediate complexity, is the results for THF shown in Fig. 4. Besides the band at 1887 cm^{-1}, we find prominent bands at 1895 and 1855 cm^{-1}. When traces of water are added to this solution, both the bands at 1895 and 1855 cm^{-1} lose intensity, whereas that at 1887 gains intensity. The same thing occurs when the temperature is dropped. On the basis of this evidence, the band at 1887 cm^{-1} may be assigned to the anion vibrating in a different environment from that which generates the 1895 cm^{-1} band and, further, the band at 1855 cm^{-1} arises from the anion in the same environment that generates the 1895 cm^{-1} band. Thus two anion sites giving rise to prominent but different bands exist for Na^+, $Co(CO)_4^-$ in THF.

The evidence from the 1900 cm^{-1} infrared band also suggest two anion sites in pyridine and at least two and more probably three sites in piperidine. We may conclude that multiple ion sites are common with Na^+, $Co(CO)_4^-$ solutions and that they occur even in solutions that show only one alkali ion vibration band in the infrared spectrum.

8. IDENTIFICATION OF ION SITES FOR Na⁺, Co(CO)₄⁻ IN THF

The infrared spectrum of Na^+, $Co(CO)_4^-$ in THF shows two bands in the CO stretching region. A very weak band appears at 2003 cm^{-1} in addition to the band of complex envelope previously described at 1890 cm^{-1}. The band at 1890 cm^{-1} can be divided into three components on the bases of the band areas which change with temperature or with the addition of traces of water. This was done with the aid of a computer with the results shown in Fig. 5 for the $0.025M$ solution. The frequency of the band components are found in Table 2. The laser Raman spectrum was recorded photoelectrically in the same region and treated similarly with the results found in this table.

To interpret the data of Table 2, we need to know what might be expected for the $Co(CO)_4^-$ ion in various possible solution environments. The pattern of the CO stretching frequencies shown by the anion depends upon the symmetry of the effective force field at the specific site. The major possibilities are shown schematically in Fig. 6.

Various symmetries may arise in several ways. As discussed previously, the

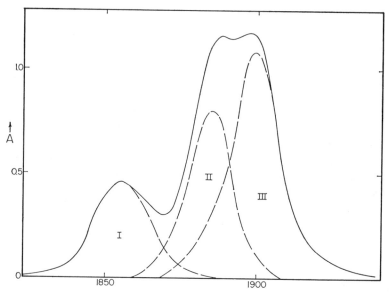

Figure 5.   Partition of the anion band into band components for NaCo(CO)$_4$ in THF.

T_d field may arise from a "free" ion in which each of the CO groups are subjected to the same significant solution forces. At the same time, this field could be expected for the anion in a solvent-separated ion pair. Recall, however, the effect of CCl$_3$H, as well as water, upon the NO$_3^-$ band in solution [5, 6] and therefore the possibility of strong interaction of solvent with anion should not be neglected. When this interaction is strong enough to make a noticeable contribution to the force felt by a vibrating CO group, the effective symmetry at the Co(CO)$_4^-$ ion is no longer T_d if the anion is not symmetrically solvated. If one CO group is different from the other three,

Table 2   Anion Carbonyl Frequencies and Assignments for the Infrared and Raman Band Components. NaCo(CO)$_4$ in THF

Environment	Infrared	Raman	Assignment
I	1886 cm^{-1}	1889 $(0.76)^a$	T_d, F_2
		2005 (~ 0.1)	T_d, A_1
II	1855	1857 (0.74)	C_{3v}, A_1
	1899	1905 $(0.76)^a$	C_{3v}, E
	2003	2005 (~ 0.1)	C_{3v}, A_1

^{a} Single depolarization measurement made at band maximum near 1890 cm^{-1}.

the C_{3v} symmetry expectations will apply; if two are different from the others, C_{2v} selection rules will apply.

In a contact ion pair, the Na+ will have a pronounced effect upon the Co(CO)$_4^-$ ion. When this interaction is monodentate or tridentate, a cylindrical axial field is introduced and the symmetry is C_{3v}. As a result, the degeneracy of the F_2 frequency is partially lifted to produce in its place a doubly degenerate E frequency and a nondegenerate A_1 frequency. The E frequency can be expected to produce a strong infrared band and a strong, depolarized Raman band. The A_1 can be anticipated in the infrared spectrum as a band of medium intensity and in the Raman spectrum as a medium-intense band which is polarized (see, however, the following discussion). In addition to these, we will have another A_1 frequency which is closely related (similar atom displacements, etc.) to the A_1 frequency of the T_d case. It will appear as a weak, polarized Raman band at or near the position of the corresponding A_1 frequency for the T_d case. Any intensity this frequency shows in the infrared spectrum arises from the difference between the unique CO group and the other three, which is induced by the Na+. As a consequence it will yield a weak infrared band. It is also possible that a contact ion pair might be formed as a result of a bidentate interaction of the Na+ ion with the anion. In this case, the symmetry is C_{2v} with the expectations of Fig. 6.

Triple ions are also relevant. We consider here only the case in which two

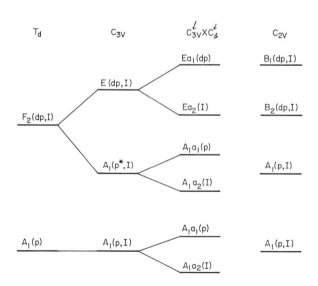

Figure 6. Schematic representation of the CO stretching frequencies expected from the Co(CO)$_4^-$ anion in various types of solution sites (see text).

anions are in intimate contact with one Na^+ and where the two anions are equivalent, on opposite sides of the cation, and are cylindrically distorted from a tetrahedral field. The expectations for such a case are conveniently treated by the methods proposed for larger molecules [7]. The effective point group is

$$G = C_{3v}{}^l \times C_s{}^i$$

and the selection rules are given in Fig. 6. The capital letters (E and A_1) describe the symmetry under the operations of the local group $C_{3v}{}^l$; the lower-case letters describe the symmetry under the interchange group $C_s{}^i$ (a_1 modes are symmetric to the interchange, a_2 modes are antisymmetric to the interchange). The Ea modes are twofold degenerate, whereas the Aa modes are nondegenerate. It can be seen that the same number of infrared bands and the same number of Raman bands are expected for the C_{3v} case and for the $G = C_{3v}{}^l \times C_s{}^i$. The difference between the two cases lies in the fact that the frequency of each infrared band coincides with that of a Raman band for the contact ion pair, whereas no coincidences occur between infrared and Raman bands for this triple ion. In the event, however, that the anions as described are not on opposite sides of the Na^+, all six frequencies become active in both infrared and Raman spectra.

We are now in a position to compare the data of Table 2 with the expectations discussed above and in Fig. 6. It is apparent that the data for the $Co(CO)_4^-$ environment of type I corresponds to the anion in a T_d situation. The infrared data for the type II environment satisfies the expectations for both a C_{3v} and a $G = C_{3v}{}^l \times C_s{}^i$ condition. Since the differences between the frequencies of the corresponding infrared and Raman band components are small and within the experimental uncertainties, these data best satisfy the criteria for a C_{3v} condition. The assignment of the observed band components to the allowed frequencies on this basis is made in Table 2. One feature of the assignment deserves comment. As can be seen, the expectation that depolarized bands would yield an intensity ratio in the polarization measurement of $\rho = 0.75$ and that for polarized bands $\rho < 0.75$ is satisfied for all but the A_1 band at 1855 cm^{-1}. It has been shown, however, that a value of ρ near 0.75 is expected for this mode in the rather special case here. It results from the form of the mode of vibration and the fact that both the structural units off the C_3 axis and that on the C_3 axis are CO groups.

It follows from this discussion that the $Co(CO)_4^-$ ion in the type I environment may exist as a free ion or as a solvent-separated ion-pair. The type II environment is consistent with the anion existing as an unsymmetrically solvated free ion, or forming a contact ion-pair, or a triple ion where the anion-anion coupling is small. Both I and II types of NaCo(CO$_4$) are in equilibrium. Possibilities for this equilibrium include the following:

$$A^- + S = SA^- \tag{3}$$

$$Na^+A^- = Na^+ + A^- \tag{4}$$

$$Na^+A^- + S = Na^+SA^- \tag{5}$$

$$Na^+SA^- = Na^+ + A^- + S \tag{6}$$

$$A^-Na^+A^- = Na^+ + 2A^- \tag{7}$$

$$Na^+ + A^-Na^+A^- = 2Na^+SA^- \tag{8}$$

The symbol A$^-$ is used for the symmetrically solvated anion, S for a solvent molecule, and SA$^-$ for the unsymmetrically solvated anion. Na$^+$, Na$^+$A$^-$, and Na$^+$SA$^-$ refer to the sodium ion, the contact ion pair, and the solvent-separated ion pair—all solvated.

The integrated absorbance E of an infrared band component arising from a given ion site divided by the cell thickness t is proportional to the population (concentration) of that site. By the usual considerations applying to equilibrium, the population of the sites will vary with gross salt concentration in a manner determined by the equilibria involved. Consequently, a study of the variation of E/t with salt concentration can throw light on the equilibria involved between the two kinds of ion environments existing in this solution. Such a study was made for NaCo(CO)$_4$ in THF in the concentration range from 0.001 to 0.03M. The absorbance versus frequency (cm^{-1}) curves were resolved into the three band components shown in Fig. 5 with the aid of a computer; the results are seen in Fig. 7. The population of environment I (component II of infrared band) and of environment II (component I and component II of infrared band) vary linearly with salt concentration. This eliminates equilibria like 4, 6, 7, and 8 from consideration as dominant factors. Therefore triple ion sites are not major solution components in the concentration range of this study. The results are consistent with *either* equilibrium 3 or equilibrium 5 but are not consistent with the presence at the same time of substantial amounts of both groups of ions (A$^-$, SA$^-$) and (Na$^+$A$^-$, Na$^+$SA$^-$).

The equivalent conductance of NaCo(CO)$_4$ in THF solutions have been measured [8]. The values of Λ vary from 9 to 16 ohm^{-1} in the foregoing concentration range while the value of Λ^0 may be estimated roughly as 110 ohm^{-1}. These values are not consistent with A$^-$ and SA$^-$ as the major anion sites in this solution.

We conclude from these studies that the salt NaCo(CO)$_4$ exists in THF solution primarily as solvent separated ion pairs (type I environment) and contact ion pairs (type II environment). This does not preclude the presence

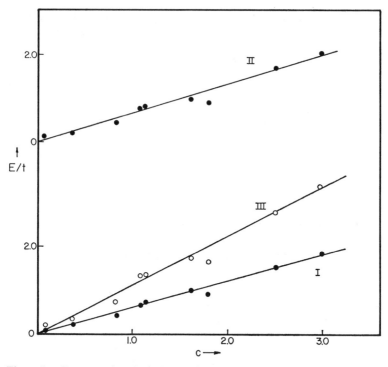

Figure 7. Concentration dependence of the integrated absorbance per unit cell thickness (E/t) for the three components of the 1890 cm^{-1} band for $NaCo(CO)_4$ in THF. The units are E/t in 10^2 cm^{-2}; c in 10^{-3} moles/liter.

of other anion sites in smaller amounts. For example, the equivalent conductance values require the presence of some free ions. And the forms of Λ versus $\sqrt{c}$ curve implies the presence of some triple ions in the higher concentration range of this study.

9. THE CHARACTER OF THE FORCE FIELD DEDUCED FROM THE CATION-LASER RAMAN SPECTRA

We turn now to the question of the nature of the forces acting on the alkali ion during its vibration. The intensity of a vibration appearing in a Raman spectrum is proportional to the square of the change in the polarizability of the charge clouds of the system with the vibrational deformation. It is well known that stretching vibrations of covalent bonds give strong Raman bands. In contrast, the vibration of two nondeformable charge clouds would not appear in the Raman spectrum. It follows that some estimate of the covalent character of the forces acting on the alkali ions may be obtained by examining the strength of the alkali band in the Raman spectra. To this end, the laser Raman spectra of $NaCo(CO)_4$, $Na_2Cr_2(CO)_{10}$, and $NaHCr_2(CO)_{10}$ in THF and DMSO solutions and of $NaCo(CO)_4$ in DMF were examined. A Raman line which could be assigned to the Na^+ vibration was not observed. In addition, long photographic exposures were made on LiCl dissolved in THF without finding a Li^+ band. To evaluate the LiCl results, the Raman band due to CCl stretching at 459 cm^{-1} in CCl_4 was taken as a standard for comparison. The CCl_4 was dissolved in THF and the intensity of the 459 cm^{-1} band was measured as a function of the CCl_4 concentration. This line could be still observed when the CCl_4 concentration was greater than $1/150$ of the salt concentration used in the LiCl runs. We may conclude that the polarizability change in any Li^+ vibration is at least an order of magnitude $(\sqrt{150})$ smaller than that for the CCl_4 vibration. It follows that covalent forces, if they occur at all, are small and that these alkali ion vibrations take place primarily under the influence of ionic forces.

10. NET FORCE ON THE ALKALI ION

The alkali ion is subjected during its vibration to forces from a number of solution elements. These forces cannot be determined individually from a single frequency of vibration. However, the net force which resists the vibrational motion (per unit displacement of the alkali ion) can be determined. This net force constant provides a valid parameter to characterize the vibration since it represents the potential energy of the system. Moreover, it is directly related to the short-range forces in electrolytic solutions which are so difficult to obtain from thermodynamic and conductivity measurements.

A knowledge of the displacements of each solution element in the alkali ion vibration is required to compute the net force constant. At first glance this appears to be a major obstacle. On closer examination, however, one finds a satisfactory way out of this dilemma. The motion is highly localized, as shown by isotope studies, and reasonable models consistent with this situation give similar results. We estimate these forces here using the simplest of models. The model taken for the displacement in a contact ion pair is shown in Fig. 8a. Here $a = M/(A + S)$, which is determined by the condition of conservation of momentum M, A, and S are the alkali ion, anion, and solvent masses, respectively. Figure 8b shows the model for the displacements in a solvent-surrounded cation environment. In this case, $a = M/2S$. The net force constant for the vibration is given by the equation

$$K_{net} = 4\pi^2 c^2 \omega^2 T_{net} \tag{9}$$

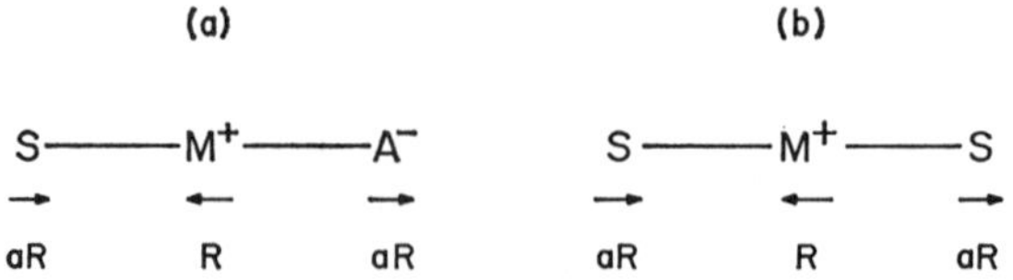

(a)   (b)

S————M$^+$————A$^-$ S————M$^+$————S

→   ←   →   →   ←   →

∂R   R   ∂R   ∂R   R   ∂R

Figure 8. Simple-model displacements for the solution vibration at the alkali ion site: (a) contact ion pair; (b) solvent surrounded cation.

where ω is the frequency of vibration in cm^{-1} and T_{net} is the kinetic energy constant defined through the relation

$$\text{kinetic energy} = T_{net}\frac{\dot{R}^2}{2}$$

R is the alkali ion displacement in terms of which all other displacements are expressed as illustrated in Fig. 8. For the models of Fig. 8,

Contact ion pair:

$$T_{net} = \frac{M(A + S + M)}{A + S}$$

Solvent-surrounded cation:

$$T_{net} = \frac{M(2S + M)}{2S}$$

(10)

The net force constants contained in Table 3 were calculated from the data of Table 1 by Eqs. 9 and 10.

Table 3   The Net Force Constant for Alkali Ion Vibrations

Salt	Solvent	K_{net}^a
LiCo(CO)$_4$	THF	0.72
LiNO$_3$	THF	0.71
LiCl	THF	0.65
LiBr	THF	0.61
LiI	THF	0.59
LiBPh$_4$	THF	0.71
NaCo(CO)$_4$	THF	0.55
NaI	THF	0.51
NaBPh$_4$	THF	0.56
KCo(CO)$_4$	THF	0.54
Li$^+$ salts	DMSO	0.77
Na$^+$ salts	DMSO	0.62

[a] Units are mdynes/A of alkali ion displacement.

11. THE ORIGIN OF K_{net}

In this section, we show that net force constants of the magnitude found in Table 3 can be understood in terms of the several kinds of force known to exist in electrolytic solutions. Consider the ion pair that was first examined theoretically by Bjerrum [9]. Recent treatments include those by Fuoss and co-workers [10], Dennison and Ramsey [11], and Pettit and Bruckenstein [12]. These authors used an expression for the potential energy of an ion pair in solution which contains the dielectric constant in the denominator. Similar treatments for a free ion in solution also depend in the same way upon the dielectric constant [13]. Since the dielectric constant for DMSO is six times that for THF, we would expect a much smaller force constant in DMSO than in THF if expressions of this type could be applied to ion vibration. The results of Table 3 show that this is clearly not the case. Improved models of ion solvation have been formed by summing the specific electrostatic interaction of the ion with the solvent molecules in the near-neighbor shell while considering the remaining solvent molecules as a dielectric continuum [14–16]. Although these models give a better representation of the electrostatic interactions, they are all hard shell models and cannot describe the forces on the alkali ion near its equilibrium position without substantial modifications.

We have modified such a model by representing each solvent molecule or polyatomic ion in the neighborhood of the alkali ion as a polarizable charge cloud. Repulsion between ions or molecules is then calculated as the repulsion between the charge clouds of each entity. Molecules beyond this neighborhood are treated as a dielectric continuum. This leads to the potential energy of interaction of the cluster in the cavity of the continuum:

$$U = \sum_{ij} \left[A_{ij} \exp \left(- \frac{r_{ij}}{\rho} \right) + \frac{q_i q_j}{r_{ij}} + \frac{C_{ij}}{r_{ij}^6} \right] - \frac{1}{2} \sum_i \alpha_i \mathbf{F}_{ij} \cdot \mathbf{F}_{ij} + U_p$$

$$\mathbf{F}_{ij} = \sum_j \frac{q_j \mathbf{r}_{ij}}{r_{ij}^3} \tag{11}$$

The first term gives the repulsion energy expressed in the exponential form, the second the coulombic interaction, the third the London dispersion term, the fourth accounts for the interaction of the induced dipole in each cloud with the field which creates it, and the last term is the interaction energy with the continuum. The first sum includes all cloud pairs. Here q_i is the charge of the ith cloud and α_i is its polarizability; r_{ij} is the distance from the center of the ith to the center of the jth cloud.

The condition that U must be a minimum with regards to the position (and orientation) of each entity at equilibrium serves to determine the A_{ij} if the equilibrium values of the r_{ij} are supplied. Conversely, this may be used to compute the equilibrium r_{ij} when the A_{ij} are supplied. The net force constant is obtained from

$$K_{net} = \frac{\partial^2 U}{\partial R^2} = \sum_{ab} \frac{\partial^2 U}{\partial r_a \partial r_b} \frac{\partial r_a}{\partial R} \frac{\partial r_b}{\partial R} \qquad (12)$$

where a and b run over the different coordinates which appear in U and the derivatives are evaluated at the equilibrium positions.

The ion-pair vibrations of the Li^+, Na^+, and K^+ salts of the $Co(CO)_4^-$ in THF solutions were first considered. The specific model taken for the calculation retained only the interactions of the alkali ion with the close near-neighbor charge centers of the first shell entities (anion, solvent). Values for the required parameters were estimated from the known repulsion constants for gaseous alkali halides, known dipole moments, and polarizabilities, and the quantum calculations of Nieuwport [17] for $Co(CO)_4^-$. The resulting values of K_{net} depend upon assumptions about parameters and geometry, but they are of the order of magnitude of the values in Table 3.

This type of calculation will be illustrated for the case of a solvent-surrounded cation. A simplification of Equation 11 is obtained for the model of Fig. 8b by representing each solvent molecule as a dipole μ with polarizability α which has its repulsion center located a distance b from the dipole center. Neglecting the small terms which come from U_p, we obtain

$$K_{net} = \left[\frac{(2S + M)}{2S}\right]^2 \left\{\frac{2A}{\rho^2} \exp\left[-\frac{(r - b)}{\rho}\right] - \frac{12e\mu}{r^4} - \frac{20e^2\alpha}{r^6} - \frac{144\alpha_i\mu^2}{r^8}\right\}$$

$$(13)$$

For the case of Li^+ salts in DMSO, we take $A = 1300$ eV from an estimation of the oxygen atom radius in DMSO (r_i), and the Li^+ radius (r_j) together with the relation between A and $(r_i + r_j)$ found for gaseous alkali halides. For DMSO, $\mu = 4.3D$, $\alpha \doteq 7 \times 10^{-24}$ cm³ and $b \doteq 0.9$ Å; for Li^+, $\alpha_i = 0.03 \times 10^{-24}$ cm³; finally the nearly universal value of 0.32 Å is accepted for ρ. Equation 13 yields $K_{net} = 0.64$ mD/Å of Li^+ displacement in excellent agreement with the observed value of 0.72 mD/Å. In fact, it is necessary to move the dipole center only slightly closer to the repulsion center in DMSO to obtain exact agreement. Again we see that the general magnitude of K_{net} can be understood in terms of standard electrostatic and repulsion forces.

These calculations show several facts that are useful in comparing experimental and theoretical values for the energetic quantities in electrolytic solutions and which are nicely illustrated by the Li^+ (in DMSO) calculations discussed earlier. Figure 9 shows a plot of the potential energy (dashed line) as a function of the displacement R of the alkali ion from its

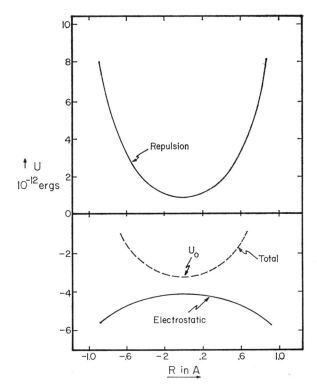

Figure 9. Contributions to the energy U for Li$^+$ in DMSO plotted vs. the alkali ion displacement in the cation mode.

equilibrium position in the mode of Fig. 8b. Its value at $R = 0$ gives the potential energy of the Li$^+$ at its equilibrium position. It can be seen that the major part of the U_0 comes from the electrostatic terms. Since U_0 leads directly to the energy of solvation of the ion, it follows that this kind of measurement will be a primary source of information about the electrostatic terms. This conclusion is well known.

In Fig. 10, we see a plot of the curvature of U (dashed line) against R. Its value at the equilibrium position is K_{net}. We note that the major contribution to K_{net} comes from the repulsion terms. It follows that these vibration studies should be a primary source of information about the repulsion terms in U_0. Since U_{rep} is not a negligible part of U_0, its inclusion in U_0 should lead to an improvement over hard shell models. A combination of the data from vibrational and energy studies may lead to a balanced determination of the parameters of U_0.

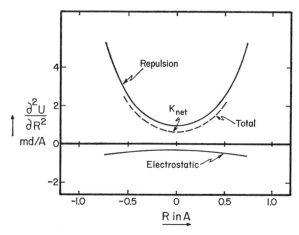

Figure 10.   Contributions to the energy resulting from the curvature $(\partial^2 U/\partial R^2)$; for Li^+ in DMSO plotted vs. the alkali ion displacement in the cation mode.

12. THE NATURE OF THE ION PAIR

The concept of an ion pair was introduced by Bjerrum to cope with the problem of solvated ions when they are close together in solution. In recent years evidence has appeared that an ion pair has greater stability than implied by the Bjerrum model. The next simplest model for a contact ion pair, consistent with such findings, is that of the "diatomic" species Na^+A^- in a solvent cavity. As a result of the spectroscopic studies of alkali ion vibration in solution, a more realistic picture seems to be slowly emerging. It is based upon the similarity of the results for the solvent-surrounded cation and the contact ion pair and upon the implications of the theoretical treatment of the dynamics of the process. It is supported by our current understanding of the effect of pressure on the alkali ion band and is consistent with the results of isotope study [1]. According to this model, the alkali ion (Li^+, Na^+) is enclosed in a disorded "solvent-lattice" of some relative stability. The order in this small region is greater than that in an ion-free region of the solvent but decreases very rapidly as we move away from the cation. In the contact ion pair, the anion replaces one (or more) solvent molecules to become one of the near neighbors of the cation. In the solvent-separated ion pair, the anion is pictured as being located somewhat further away from the cation in the structured region with one (or more?) solvent molecule in the intervening space. In this model the alkali ion is vibrating much like a free ion in a cage. Of course, the cage components also move in the vibration to conserve momentum. Whether this model will survive in the light of further research remains to be seen. Needless to say this is a main objective of the current work in our laboratory.

The present work provides evidence concerning the structure of the contact ion pair of $NaCo(CO)_4$ in THF. Since the local symmetry of the solution structure has been established as C_{3v} at the $Co(CO)_4^-$ ion, it remains to determine whether the interaction between cation and anion involves one CO group (monodentate) or three CO groups (tridentate). It can be shown that the deformation of the unique CO group in the A_1 mode of the 1890 cm^{-1} group of vibrations of the contact pair is about three times as large as that of each of the other three CO groups. On the other hand, the unique CO group is not deformed in the E modes. The Na$^+$ should have a primary effect on the CO group or groups with which it directly interacts, but it is expected to affect only slightly the other CO groups of the structure. It follows that if the Na$^+$ interacts directly with the unique CO group then the E modes will be little affected by the Na$^+$ and hence little shifted in frequency from the F_2 frequency observed for the solvent-separated ion pair at 1887 cm^{-1}. On the other hand, if the interaction is tridentate, the E modes would show a greater shift from 1887 cm^{-1} than the A_1 mode. The experimental result shows therefore that the interaction is monodentate. Further, because the A_1 mode lies below 1887 cm^{-1} we can conclude that the effect of the Na$^+$ interaction is to weaken the binding between the O and C atoms of the CO group with which it directly interacts.

13. THERMODYNAMICS OF ION INTERSITE CONVERSION

The computer resolution of the 1890 cm^{-1} band for $NaCo(CO)_4$, illustrated in Fig. 5, may be applied to data obtained over a temperature range. This makes it possible to estimate the populations of the solvent-separated ion-pair structure and the contact ion-pair structure. Thus the equilibrium constants of conversion of tight pairs into loose ones is calculated; the results are shown in Table 4. From the slope of the straight line representing the ln K versus $1/T$, we obtain $\Delta H = -3.7$ kcal and $\Delta S = -14$ e.u. The increase in order that results when a solvent molecule is transferred from the bulk solvent to its position between the cation and anion in the solvent-separated ion pair accounts for the negative value of ΔS.

The ΔH value is lower than that obtained for the same equilibrium for two other sodium salts in THF, using different techniques. Hogen-Esch and Smid [18] find -7.6 for sodium fluorenide while Grutzner, Lawlor, and Jackman [19] report -8.2 for sodium triphenylmethanide and -6.7 for sodium fluorenide. These are salts of large carbanions and the differences in the ΔH values should be sought in that fact. We may write a potential

Table 4   Thermodynamic Quantities for the Ion Intersite Conversion[a]

$$Na^+A^- + S = Na^+SA^-$$

$t(°C)$	K	$\Delta H°/T$ (cal/°K mole)	$\Delta S°$ (cal/°K mole)
29	0.45	−12	−14
2	0.82	−13	−14
−20.5	1.8	−15	−13
−42	2.6	−16	−14

[a] System: $NaCo(CO)_4$ in THF.

energy expression similar to the preceding one for each kind of ion pair. Comparing the two expressions shows that it is useful to consider the conversion process as the sum of the two steps,

$$Na^+A^- = Na^+ \,\square\, A^- \qquad (14)$$

$$Na^+ \,\square\, A^- + S = Na^+SA^- \qquad (15)$$

In the first step, the cation and anion are moved apart in the solution to their position in the solvent-separated ion pair, creating a hole between them. Solvent fills this hole in the second step. The enthalpy change involved in the second step is negative and is dominated by the interaction of the added solvent molecules* with cation. Small (negative) terms also come from the interaction of the added solvent molecules with the anion and small (positive) terms from the desolvation of the added solvent molecules. As long as the same number of solvent molecules are added in the conversion process, the enthalpy change for this step will show little variation with anion, but it will increase substantially with each additional solvent molecule added to the near-neighbor shell of the cation. The enthalpy change in the first step is positive and arises from separating the coulombic charges of the cation and its near-neighbor solvent molecules from those of the anion and its near-neighbor environment from the contact pair positions to the solvent-separated pair positions, plus that involved in pushing back the solvent to make the hole. This final contribution should be nearly the same in two conversion processes in the same solvent if the same number of solvent molecules are added to the contact pair to form the solvent-separated pair. In that case, any substantial difference in ΔH for the first step would arise from the charge-separation process and would be more positive the more the anion charge is concentrated near the cation in the contact pair.

---

* The plural is used for generality, but it is anticipated that one solvent molecule will most often be involved.

We can now see a rational explanation for the difference in ΔH for the ion-pair conversion process in THF between $NaCo(CO)_4$, on the one hand, and the sodium salts of the carbanions, on the other. The charge of the anion is more diffuse for the carbanions and this will result in making ΔH for the conversion process involving the carbanions more negative by making the enthalpy change in the first step (Eq. 14) less positive than for the process with the $Co(CO)_4^-$ ion. In light of the finding in this study that the attachment of the Na^+ to the $Co(CO)_4^-$ is monodentate, we would expect one THF molecule to be displaced from the cation near-neighbor shell on forming the contact ion pair from the solvent-separated pair. However, the foregoing carbanions are large and flat and it is just possible that they might displace two solvent molecules from the cation near-neighbor shell in the conversion to the contact pair to contribute to the observed ΔH difference.

REFERENCES

1. W. F. Edgell and A. T. Watts, *Abstracts, Symposium on Molecular Structure and Spectroscopy*, Ohio State University, June, 1965. See also W. F. Edgell, A. T. Watts, J. Lyford, IV, and W. Risen, Jr., *J. Am. Chem. Soc.*, **88**, 1815 (1966). For subsequent work described also in this chapter see W. F. Edgell, J. Lyford, IV, R. Wright, W. Risen, Jr., and A. T. Watts, *J. Am. Chem. Soc.*, **92**, 2240 (1970); W. F. Edgell, J. Lyford, IV, A. Barbetta, and C. I. Jose, *J. Am. Chem. Soc.*, **93**, 6403 (1971); W. F. Edgell and J. Lyford, IV, *J. Am. Chem. Soc.*, **93**, 6407 (1971).
2. M. Eigen, *Coordination Chemistry: Seventh International Conference*, Butterworths, London, 1963.
3. N. Atherton and S. I. Weissman, *J. Am. Chem. Soc.*, **83**, 1330 (1961); M. Szwarc, *Symposium on Interionic Forces in Condensed Phases*, Lincoln, Nebraska, October, 1970.
4. W. F. Edgell and J. Lyford, IV, *J. Chem. Phys.*, **52**, 4329 (1970).
5. D. E. Irish and A. R. Davis, *Can. J. Chem.*, **46**, 943 (1968).
6. A. R. Davis, J. W. Macklin, and R. A. Plane, *J. Chem. Phys.*, **50**, 1478 (1969).
7. W. F. Edgell, *J. Phys. Chem.*, **75**, 1343 (1971).
8. W. F. Edgell, M. T. Yang, and N. Koizumi, *J. Am. Chem. Soc.*, **87**, 2563 (1965). More recently, Mr. Jack Fisher repeated these measurements with nearly the same results.
9. N. Bjerrum, *Kgl. Danske Videnskab Selskabs*, **7**, No. 9 (1926).
10. R. M. Fuoss and F. Accascina, *Electrolytic Conductance*, Interscience, New York, 1959.
11. J. Dennison and J. Ramsey, *J. Am. Chem. Soc.*, **77**, 2615 (1955).
12. L. Pettit and S. Bruckenstein, *J. Am. Chem. Soc.*, **88**, 4783 (1966).
13. M. Born, *Z. Physik*, **1**, 45 (1920).
14. J. D. Bernal and R. H. Fowler, *J. Chem. Phys.*, **1**, 515 (1933).

15. D. D. Eley and M. G. Evans, *Trans. Faraday Soc.*, **34**, 1093 (1938).

16. A. D. Buckingham, *Disc. Faraday Soc.*, **24**, 15 (1957).

17. W. Nieuwport, Charge Distribution and Chemical Bonding in the Metal Carbonyls $Ni(CO)_4$, $Co(Co)_4^-$, and $Fe(CO)_4$, Phillips Research Reports 1965, No. 6, Eindhoven, Netherlands.

18. T. Hogen-Esch and J. Smid, *J. Am. Chem. Soc.*, **88**, 307 (1966); **88**, 318 (1966).

19. J. Grutzner, J. Lawlor, and L. Jackman, *J. Am. Chem. Soc.*, scheduled for 1972.

5

Electron Spin Resonance Studies of Ion Pairs

J. HOWARD SHARP* and MARTYN C. R. SYMONS

Department of Chemistry, University of Leicester, Leicester, England

* Present address: Department of Chemistry, The New South Wales Institute of Technology, Sydney, Australia.

1. INTRODUCTION

This chapter is concerned almost exclusively with ESR data pertaining to ion pairs. Since the ESR technique can be applied only to paramagnetic species, we are concerned with unusual ions such as benzosemiquinone, and hence these studies may seem to be of a highly specialized nature. This is not the case, however. The results are quite general, and may be extrapolated to simple ions without hesitation because the magnetic properties necessary for ESR studies are in no sense involved in the phenomena of ion pairing and ionic solvation.

Our approach is deliberately qualitative, since more quantitative aspects are discussed in this book by de Boer and Sommerdijk in Chapter 8. We have endeavored, where possible, to illustrate what may often seem to be rather complex concepts by reference to simplified models and diagrams. Attention is focused primarily upon the information that can be extracted from hyperfine coupling constants and line-widths, but some general comments about the significance of these ESR parameters are also included. The more general solvent effects, other than those influencing the behavior of ion pairs, have not been discussed. These have been briefly reviewed elsewhere [1]. We have also omitted any reference to ESR results for transition metal complexes. An extensive review of relaxation phenomena pertinent to our theme has recently appeared and it can be consulted for mathematical details [2].

1.1. Symbols and Definitions

The following symbols are used in the text:

MeCN	Acetonitrile
DMF	N,N-dimethylformamide
Diglyme	bis(2-methoxy ethyl)ether
Tetraglyme	bis(2(2'-methoxyethoxy)ethyl)ether
HMPA	Hexamethylphosphortriamide
DME	1,2-Dimethoxyethane
THF	Tetrahydrofuran
MTHF	2-Methyltetrahydrofuran
THP	Tetrahydropyran
DEE	Diethyl ether
TMED	Tetramethylethylenediamine
i-PrOH	i-Propanol
t-BuOH	t-Butyl alcohol
t-AmOH	t-Pentanol
t-HexOH	t-Hexanol
S	Any solvent molecule
H	Magnetic field
g	Average g-value ($g = \frac{1}{3}[g_{xx} + g_{yy} + g_{zz}]$)
g_e	The g-value of the free electron, $g_e = 2.0023$
a and a_{iso}	The isotropic hyperfine coupling
A	Hyperfine coupling tensor
A_0	Isotropic hyperfine coupling for one electron in the appropriate s-orbital of the atom
SD	Spin Density
ρ_M	Spin density on metal atom M $\left(\rho_M = \dfrac{a_{iso}}{A_0}\right)$ when only s-orbitals are occupied
I	Nuclear spin
M_I	Nuclear spin quantum number
ω	Exchange frequency
Δ	Separation between exchanging lines
δ	Line-width
α, β	Electrons with opposite spins
D, E	Spin Hamiltonian parameters for triplet-state species, governing the zero-field interactions
J	Electron-exchange energy in triplet-states
I.P.	Ion pair

The term ion pair is subdivided for convenience into the following categories:

Contact ion pair	cation and anion in direct contact (sometimes referred to as "tight" ion pair)
Solvent-shared ion pair	Cation and anion bridged by a specific solvent molecule
Solvent-separated ion pair	Cation and anion held electrostatically but with an unspecified number of solvent molecules separating them (sometimes referred to as "loose" ion pair)
Cationic, neutral, and anionic ion triplets	Clusters of three ions bearing positive, zero, and negative charges: $M^+A^-M^+$, $A^-M^{2+}A^-$, $A^-M^+A^-$

All of the preceding categories are distinct from "free" ions, which are perturbed only by solvent molecules.

Negative spin density	If, at a given instant, the major positive spin density associated with an unpaired electron is α, then this may cause other nominally paired electrons to favor either α or β spin, depending upon their location in the molecule. Those favoring β spin give rise to negative spin density

1.2. Plan of the Review

In recent years, more light has been shed on the subject of ion-pair formation by the application of electron spin resonance than by that of any other physical or chemical technique. Few of those who applied this technique to chemical problems as early as 1955 had any concept of the wealth of detailed and unambiguous information that it was soon to reveal. The detection of high-resolution spectra for radical anions and cations [3–5] was the first step, but the discovery by Weissman and co-workers that, in ethereal solution, each individual feature of the ESR spectra of benzophenone ketyl [6] and of naphthalene anions [7] were split into four by coupling to ^{23}Na was undoubtedly the most significant step forward. This result, as we will explain in detail later, proved quite conclusively, in a manner which required no involved reasoning or calculation, that the anions were in close association with cations in a 1:1 ratio, the individual ion pairs having lifetimes long compared with the inverse of the line-widths—that is, long compared with $\sim 10^{-6}$ sec.

Since that time many similar observations have been made, and in Section 2 we summarize the results and attempt to explain some of the trends and variations observed. Other, more subtle factors have since come to light. For example, Reddoch and others [8–10] have shown that ion-pair formation modifies the spectra of radical anions even though direct coupling to the cations may not be detected. This arises because the electron-nuclear hyperfine coupling is proportional to the local spin density in the anion, and if this is perturbed by the field of the cation, it results in a change in the magnitude of the hyperfine coupling. This distribution is, of course, solvent dependent as well, so that care has to be exercised in assigning the cause of any observed shifts, and hence this indication of ion pairing is less compelling. Indeed, one of the more interesting results in this area is that local hydrogen bonding to solvent molecules generally causes a greater perturbation than that caused by cations in the absence of such solvent molecules [11]. These changes are summarized and discussed in Section 3.

Another outstanding result revealed by ESR studies is that certain cations and anions in ion-pair units may have preferred relative orientations which have lifetimes sufficiently long to cause specific line-broadening effects, or even, in some cases, to give ESR spectra quite different from those of the normal, symmetrically solvated anions. A good example is the semiquinone anion. This is symmetrical normally, but small cations prefer to reside close to only one of the oxygen atoms at a time, thus making the anion strongly asymmetric. If this structure exists long enough, the distortion shows up clearly in the ESR spectrum, and indeed, the extent of the distortion is clearly "written" in the spectrum. If, however, the lifetime of a particular structure is in the region of 10^{-6} sec, certain features of the spectrum will be far broader than others, and from this extra broadening the lifetime of the alternative structures can be calculated. These processes are explained and exemplified in Section 4.

These are the main topics of the chapter. However, ESR can also give information about different types of ion pair and the extent of solvent participation. The tendency to form ion clusters may also be detected in the form of cationic, neutral, or anionic triple ions (i.e., $M^+A^-M^+$, $M^+A^{2-}M^+$, $A^-M^{2+}A^-$, or $A^-M^+A^-$). These aspects are touched on in both Sections 4 and 5.

Finally, in Section 6, certain neglected aspects are outlined. The complimentary role played by NMR is not given much space because this is dwelt upon far more fully by de Boer and Sommerdijk in Chapter 7. And the very relevant topic of disproportionation is given no more than a mention because this subject, as well as other electron-transfer processes, is fully discussed by Jagur-Grodzinski and Szware in the second volume of this book.

1.3. ESR Parameters and Spin Densities

The ESR spectrum of a radical can be used purely as a fingerprint, or it can be used to deduce information regarding the form of the molecular orbital containing the unpaired electron from the values of the observed hyperfine coupling constants.

The way in which a spectrum is analyzed to give information such as the types and number of coupled nuclei and the coupling constants is discussed in detail in various ESR texts [12–14] and will not be elaborated here. Suffice it to say that with the species discussed here the spectral interpretations are all unambiguous and serve to identify, with considerable confidence, the species responsible for the spectra. Thus the reader who is inexperienced in the art of ESR spectral interpretation can, we believe, rely heavily upon the identifications given in the text.

The extraction of information about the wavefunction of the unpaired electron is a more intricate problem. Here we outline the way in which hyperfine coupling parameters can be roughly translated into spin densities at the nuclei concerned. Again, the reader who is not too concerned with these details can safely omit this section, provided he is prepared to accept that approximate spin densities can indeed be derived from the data.

Nearly all the work reviewed here is concerned with liquid-phase spectra, which give, directly, only the magnitudes of the traceless parts of the g-tensors and hyperfine tensors. The isotropic hyperfine coupling, a_{iso}, is a measure of the spin density in the atomic s-orbital of the nucleus concerned. This may be positive or negative, but the sign is not given directly by the spectra. The sign can, however, be inferred from such factors as line-widths, or, under favorable conditions, it can be obtained directly from NMR spectra (see Section 6). The magnitude of the coupling can be converted into a measure of the spin density by dividing by the A_0 values given in Table 1. These A_0 values are either experimental (H·, and the alkali-metal atoms) or calculated from the best available wavefunctions. They are a measure of the coupling that would be expected for unit population of the appropriate s-orbital of the neutral atom. Apart from possible errors in these calculations, there are at least two other sources of uncertainty in the spin-density values. One is the neglect of overlap that is implied here, and the other is the neglect of orbital expansion or contraction that can arise if the effective nuclear charge varies. The first does not, in practice, appear to be very serious for the systems discussed here; the second probably is serious only for the estimates of spin density on the alkali-metal cations. These spin densities are usually very small, so that we could argue that the cation should be used for comparison, not the neutral atom. This means that the extension of the outer s-orbital is reduced relative to that of the atom. The effect of this

Table 1   Properties of the Relevant Atoms

Atom	Mass Number	Nuclear Spin	Natural Abundance (%)	A_0 (gauss)	Electron Affinity[b] of the Cation (eV)	Cation Radius (Å)		
						c	d	e
Li	6	1	7.43	54.29^a	5.390	0.60	1.17	0.94
	7	$\frac{3}{2}$	92.57	143.37^a				
Na	23	$\frac{3}{2}$	100	316.11^a	5.138	0.95	1.35	1.17
K	39	$\frac{3}{2}$	93.08	82.38^a	4.339	1.33	1.69	1.49
	41	$\frac{3}{2}$	6.91	45.32^a				
Rb	85	$\frac{5}{2}$	72.8	361.07^a	4.176	1.48	1.80	1.63
	87	$\frac{3}{2}$	27.2	$1,219.25^a$				
Cs	133	$\frac{7}{2}$	100	819.84^a	3.893	1.69	2.00	1.86
H	1	$\frac{1}{2}$	99.98	506.86^a				
C	13	$\frac{1}{2}$	1.108	$1,130^f$				
N	14	1	99.635	552^f				
O	17	$\frac{5}{2}$	0.0037	$1,660^f$				
F	19	$\frac{1}{2}$	100	$17,200^f$				

[a] Calculated from experimentally determined zero field splittings. J. P. Goldsborough and T. P. Kohler, *Phys. Rev.*, 133A, 135 (1964).
[b] Electron affinity of the cation taken to be equivalent to the ionization potential of the metal atom. Ionization potentials from *Bond Energies, Ionization Potentials, and Electron Affinities*, V. I. Vedeneyev, L. V. Gurvich, V. N. Kondrat'yev, V. A. Medvedev, and Ye. L. Frankevich, Arnold, London, 1966.
[c] Pauling cation radii. L. Pauling, *Nature of the Chemical Bond and the Structure of Molecules and Crystals*, 3rd ed., Cornell University Press, Ithaca, New York, 1960.
[d] Cation radii in vacuo (calculated) [27].
[e] Crystallographic radii (experimental) [26].
[f] Calculated from the SCF wave functions [13].

neglect, which would be difficult to allow for correctly because of the uncertain role of the solvent, is to reduce the actual spin densities. In other words, the spin densities given in the tables are upper limits.

This *s*-character may be direct—that is, the molecular orbital of the unpaired electron may include a contribution from the *s*-orbital—or it may be indirect, stemming from the presence of other electrons in the molecule. Perhaps the simplest way of expressing this is to invoke the concept of spin polarization, whereby the spin of the unpaired electron may influence other "paired" electrons to favor like, or unlike, spin on the nucleus concerned. This is described, for the Ċ—H fragment, by McConnell [15] and is shown diagrammatically in Scheme I. It is generally found, for π-radicals, that the maximum apparent *s*-character attainable is about 4% of the atomic values [16].

When considering line-widths we need to distinguish between "slow-exchange" and "fast-exchange" situations. These are subjective in the sense that they relate specifically to ESR data. Two frequency factors are relevant; the widths of the lines and their separation. Consider the equilibrium $A \rightleftharpoons B$. In the slow-exchange region, separate ESR spectra would be obtained for A and B. The controlling factor is now the line-width. As the rate increases, this width will eventually become comparable with the separation, at which time a new, weighted average line will take the place of the separate lines. We are now in the fast-exchange region, and further rate increases will cause line-narrowing. This is elaborated in Section 4, and an idea of times and widths can be derived from Fig. 11 on page 226. Generally, lifetimes in the region of 10^{-6}–10^{-8} sec are derived from the line-width measurements.

1.4. Equilibria between Distinct Species versus Continuous Change

Situations often arise when these two possibilities are confused. We define two limiting cases. In one there are two species, A and B, in an equilibrium; which can be moved to favor A or B under constraints that do not alter the physical properties of A or B. The optical spectra of such mixtures should show separate features for A and B, which would give rise to isosbestic points if they overlapped. The ESR spectra would show separate features for A and B if the equilibration rates were low, but averaged spectra would be obtained if fast equilibration were involved. It is this averaging that gives rise to the problems that are outlined in Section 4.

In the other extreme, we envisage no clear-cut equilibria, but rather a species, A, some of whose properties are strongly dependent upon the nature of the environment. If, then, the same constraints were applied, these properties of A would change continuously, from one limit of the constraint to the other. The optical spectra would then be characterized by a steady shift or change in width, and, in general, no isosbestic points would be generated [17].

The ESR parameters would also show steady shifts during the course of the constraint. This situation will have many similarities to the fast-equilibrium situation outlined earlier, and these alternatives may well be difficult to distinguish. Often, concurrent optical measurements would effect the required distinctions.*

Unfortunately, real situations are often less obvious, and the parameters of both A and B in an equilibrium process may be dependent upon the environment. In that case a very complete analysis over wide ranges of

---

* An alternative method that allows the distinction between these two extreme situations is discussed in Chapter 7, page 304. [Editor].

different constraints would be necessary; ESR studies have not normally been broad enough to warrant attempts at unravelling such a situation. The distinction between these equilibria situations has been thoroughly discussed by Szwarc in two excellent reviews [18, 19].

2. HYPERFINE COUPLING TO DIAMAGNETIC GEGENIONS

In this section we are concerned with the detection of ion pairs by the appearance of hyperfine features in the ESR spectra of radical anions which are caused by interaction between the unpaired electrons and the nuclei of the diamagnetic cations. For example, in the spectrum of the ion pair A^-Na^+, every feature in the ESR spectrum of A^- will be split into a quartet of lines resulting from coupling to ^{23}Na nuclei, which have $I = \frac{3}{2}$. (The four lines then come from nuclei which have $M_I = +\frac{3}{2}, +\frac{1}{2}, -\frac{1}{2},$ and $-\frac{3}{2}$, each orientation being equally probable to a very good approximation.)

Our concern in this section is with the magnitude of this interaction rather than with the conditions required for its detection. We start by outlining the factors that are most likely to contribute to this magnitude, which, of course, is related to the spin density on the cation in question.

We then pass in Section 2.2 to a detailed outline of the trends detected as various experimental factors are systematically varied; many of these trends are conveniently summarized in Figs. 1 and 3 and Tables 2 and 4.

We then turn, in Section 2.3 to a brief consideration of the rare but highly significant information available from solid-state studies, since this can answer, directly, some of the problems posed in Section 2.1.

After discussing the facts, we consider, in Section 2.4, attempts at interpretation of some of these trends. Readers may prefer to study this section first, making brief excursions to the figures or tables as required.

2.1. Mechanisms of Spin-Transfer

The simplest model for describing the acquisition of positive spin density by the cations in the ion pairs (A^-M^+) is that of electron transfer from A^- to M^+. This will be favored by low ionization potentials for the anions and high electron affinities for the cations. A trend with electron affinities of the cations is certainly observed in most cases, but trends with anion ionization potentials are less clear (see Section 2.2). For such a transfer, suitable overlap of the appropriate molecular orbital of the anion and the outer s-orbital of the cation must occur, but, as has been stressed by many authors, this need not be precise—that is, the cation can wander over the surface of the anion and, provided this is rapid relative to the resulting fluctuations in $a(M^+)$, a weighted average value for $a(M^+)$ would result.

Atherton and Weissman [7] used this approach to explain the temperature dependence of the sodium hyperfine coupling in sodium naphthalenide, pointing out that at low temperatures the position of the sodium ion would approach the centre of the anion, where the molecular orbital of the unpaired electron has a node. Some approximate calculations that utilized this model have been presented by Aono and Oohashi [20].

An alternative mechanism for transfer of spin density to the *s*-orbital of the cation is that of inner-shell spin polarization, or configuration interaction [21]. This mechanism was invoked by de Boer to explain his observation that negative spin density is acquired by certain cations in pyracene anion ion pairs at low temperatures [21]. Again, the "vibrations" of the cation relative to the anion diminish on cooling and more time is spent close to the nodal plane of the unpaired electron. At the node the normal mechanism of spin transfer, discussed earlier, cannot operate for the unpaired electron. To explain the acquisition of negative spin density, mixing of various excited states is considered, the most important being the case where inner *s*-electrons of the cation are moved into the outer *s*-orbital. This can certainly, in principle, give rise to negative spin density. This mechanism has also been invoked by Hirota [22] to explain his results for metal ketyls and by Dodson and Reddoch for naphthalene salts [23].

Another mechanism [24] involves the direct passage of negative spin density from the anion to the outer *s*-orbital of the metal. This model can be understood by reference to Schemes I and II. In I we see a simple picture of the way negative spin density is acquired by protons in π-radicals [15]. In II this same mechanism is applied to an ion pair. The negative spin density favored in the outer reaches of the lone-pair orbital will be acquired

I

II

by the adjacent cation provided there is some weak covalent bonding. In the position shown in II, the normal mechanism cannot operate and hence the coupling constant will be negative, but as the cation oscillates out of the nodal plane toward the π-system so positive spin density will be acquired. It is possible that a related mechanism operates in such ion pairs as those of naphthalene where negative spin density occurs in the nodal plane of the unpaired electron.

2.1.1. Other Factors Governing the Magnitude of the Coupling Constant

The stronger the solvation of the cation, the smaller will be the coupling constant provided increased solvation does not alter the geometry of the ion pair. This arises for at least two reasons: on the one hand, strong coordination by the solvent will decrease the effective electron affinity of the cation; on the other hand, there may be ultimately an insertion of solvent between the cation and the anion. This may be quantized in some cases, but in others it may appear as a continuous "pulling-off" process (Section 1.4). These two mechanisms are graphically illustrated in Schemes III and IV,

III

IV

respectively. Factors favoring III include specific bonding between cation and anion and the use of a solvent which "bonds" to both anions and cations (such as water and the alcohols). Process IV would be favored by large anions with no specific bonding sites and by solvents which specifically solvate the cations (such as the glymes).

The extent to which the vibration depicted in Scheme II occurs depends upon such factors as the mass of the cation and the extent of its solvation, the bulk of orthosubstituents, and the nature of solvent molecules. Another possible controlling factor is the rate of migration between equivalent sites as discussed in Section 4.

2.2. Trends with Nature of Cation, Solvent, Temperature, and Anion

2.2.1. Cation

In an earlier discussion of this writer on the variation of the spin density on the cation with its nature, it was concluded that in general no well defined trends exist [25]. This is because the splittings are usually very small and often of ambiguous sign. However, a careful consideration of recent results suggests that at least some trends are reasonably well-defined, as can be seen from the results plotted in Fig. 1. In most cases there is a decrease in spin density as the electron affinity of the cation decreases. The nitroaromatic anions are a notable exception. Alternatively, we may say that there is a systematic decrease in spin density as the radius of the cation increases, and

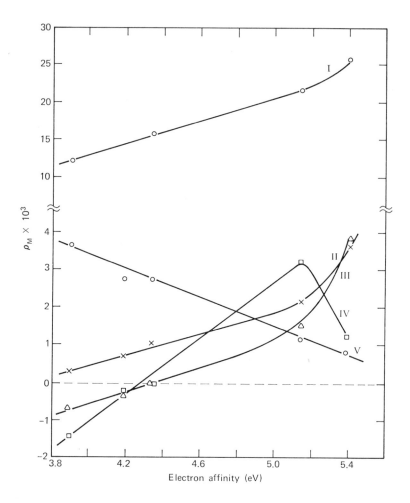

Figure 1. Spin-density on the cation in ion pairs as a function of electron affinity. (I) *o*-Dimesitoylbenzene in DME. (II) Acenaphthenequinone. (III) 1,2-Naphthoquinone. (IV) Naphthalene in THF. (V) Nitrobenzene in DME.

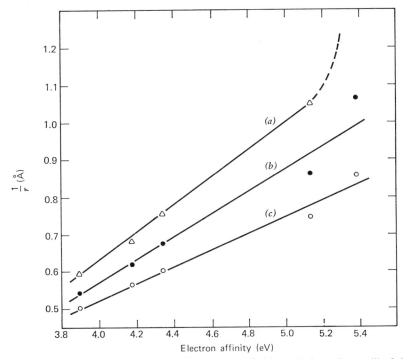

Figure 2. The correlation between the electron affinities and the cation radii of the alkali metals. The electron affinities of the cations are calculated from the gas-phase ionization potentials of the metal atoms (electron affinity = −ionization potential). (*a*) Pauling radii. (*b*) Experimental crystallographic radii [26]. (*c*) Calculated values of the cation radii *in vacuo* [27].

this is the parameter that has been usually selected by others. The relationship between the experimental gas-phase electron affinity, which is accurately known, and the reciprocal of the ionic radius is shown to be approximately linear (Fig. 2). Here we place reliance on the ionic radii obtained from electron diffraction maps [26], but we also include results for the radii given by Pauling and those calculated for the gaseous ions [27]. Hence, in a qualitative sense, when a decrease in spin density with increase in radius has been observed, the controlling factor is the electron affinity.

2.2.2. Solvent

With respect to solvent effects, it is unfortunate that almost all workers have used ethers as solvents; thus most of the results in Table 2 are for ion pairs in THF, MTHF, DME, and THP. The major conclusion that can be

Table 2 Solvent and Cation Dependence of Metal Hyperfine Couplings (in gauss) for Ion Pairs at Room Temperature

Anion	Cation	Metal Hyperfine Coupling					Ref.
		DME	THF	MTHF	THP	Solvent	
Benzophenone	Li$^+$	0.673(4.7)	0.32(2.2)				a, b, c
	Na$^+$	1.125(3.6)	1.18(3.7)				
	K$^+$	0.39(4.7)	0.24(2.9)				
4,4'-Dimethyl-benzophenone	Li$^+$		0.33				c
	K$^+$		0.25				
2,6,2'6'-Tetramethyl-benzophenone	Li$^+$		0.07				c
	Na$^+$		0.16				
	K$^+$		0.04				
Benzil	Na$^+$		0.61				d
2,2,5,5-Tetramethyl-hexane-3,4-dione	Na$^+$		0				d
Fluorenone	Na$^+$	0.35(10°C)					b
Xanthene-9-one	Na$^+$		0.9				e
o-Dimesitoylbenzene	Li$^+$	3.75(26.1)					f
	Na$^+$	6.95(22.0)					
	K$^+$	1.33(16.1)					
	Cs$^+$	10.2(12.4)					
Xanthene-9-thione	Na$^+$		4.0				g
	K$^+$		0.77				
Thioxanthene-9-thione	Na$^+$		3.78				g
3,5-Dimethylpyridine	Na$^+$	0.39	0.47				h, i
	K$^+$	0.14	0.15				
2,2'-Bipyridyl	Na$^+$	0.58(−8°C)					j
5,5,10,10-Tetramethyl-5,10-dihydro-silanthrene	Cs$^+$	2.17	2.16			2.44(DEE)	k

190

Compound	Ion						Ref
Flavin	$^{67}Zn^{2+}$					2.6(DMF/CHCl$_3$)	l
	$^{113}Cd^{2+}$					17 DMF/CHCl$_3$	m
Catechol	$^{89}Y^{3+}$					0.65(H$_2$O)	n
	$^{139}La^{3+}$					2.07(H$_2$O)	
di-Isopropyl-o-phthalate	Li^+		2.02				n
	Na^+	3.67					
	K^+	0.76					
di-Isopropyl-p-phthalate	K^+	0.095					o
Terephthalonitrile	Li^+	0.30		0.055	<0.03		
	Na^+		0.38	0.44	0.43		
	K^+			0.12	0.12		
Phthalonitrile	Na^+	0.30	0.26		0.19		p
Acenaphthene	Na^+	1.08					q
Cyclooctatetraene	Li^+		0.2				r
	Na^+		0.9				
Biphenyl	Li^+			0.136(0.95)			s, t
	$Na^+(K^+?)$	0		0.079(0.25)			
	K^+		0.043	0.083(1.0)			
	$^{85}Rb^+$			0.331(0.92)			
	$^{87}Rb^+$			1.113(0.91)			
	Cs^+			1.166(1.42)			
Pyrazine	Li^+		0.70(4.9)				u, i
	Na^+	0.54	0.58(1.8)				
	K^+	0.09	0.11(1.3)				
	Cs^+		1.27(1.6)				
2,5-Dimethylpyrazine	Na^+	0.15	0.20				i
	K^+	<0.025	<0.035				

Table 2 (Continued)

Metal Hyperfine Coupling

Anion	Cation	DME	THF	MTHF	THP	Solvent	Ref.
2,6-Dimethylpyrazine	Na+	0.36	0.45				i
	K+	0.07	0.09				
Tetramethylpyrazine	Na+	0.72	0.72				i
	K+	0.31	0.25	0.21			
Naphthalene	Li+	0	0.18(1.3)		0.39	0.39(dioxane)	v, w, x,
	Na+	0.405(1.28)	1.036(3.28)	0.46, 0.28	1.26	1.295(dioxane)	y, z,
	Na+			1.115		0.39(tetraglyme)	a',
	K+	0	0		0	0.109(TMED)	z
	^{85}Rb+	0.095(0.26)					
	^{87}Rb+	0.316(0.26)	0.290(0.24)		0.327		
	Cs+	1.071(1.31)	1.117(1.43)				
2,6-di-t-Butyl-naphthalene	Na+		1.5	2.0			b'
Anthracene	Na+		1.50			2.083(TMED)	c', d'
	K+		0.1			0.157(TMED)	
	K+					0.262(DEE)	
Azulene	Cs+	0.47	0.55				c'
	Li+		0.174				
	Na+	0.371	0.538				
	K+		0.2				

		DME	t-BuOH	t-AmOH	Solvent	Ref.
p-Benzoquinone	Li+	1.09		0.14		e', f', g'
	Na+		0.2	0.28		
2,6-Dimethyl-p-benzoquinone	Li+	0.10		0.21(1.5)		g', h'
	Na+			0.31(1.0)		

Compound	Cation	DME	THF	Diglyme	Solvent	Ref.
2,6-Dichloro-p-benzoquinone	Cs+				<0.1	h'
	Na+				0.17(0.21) 0.35, 0.25	h'
2,6-di-t-Butyl-p-benzoquinone	Na+				0.46	h'
Duroquinone	6Li+				0.22	f', g', h', i'
	7Li+				0.59	
	Na+	0.387			0.85	
	K+	0.05			0.36(THF)	
Anthraquinone	Na+	0.41	0.346			j'
Acenaphthenequinone	Li+	0.53(3.7)				k'
	Na+	0.69(2.2)				
	K+	0.09(1.1)				
	85Rb+	0.27(0.75)				
	87Rb+	0.87(0.71)				
	Cs+	0.29(0.35)				
1,2-Naphthoquinone (at −1°C)	Li+	0.54(3.8)				k'
	Na+	0.49(1.6)				
	K+	0(0)				
	85Rb+	0.10(0.27)				
	87Rb+	0.29(0.24)				
	Cs+	0.56(0.68)				
Nitrobenzene	Li+	0.125(0.87)			0.40(CH$_3$CN)0.55(acetone)	l', m'
	Na+	0.39(1.2)	0.36	0.50(1.6)		
	K+	0.23(2.8)	0.25	0.22(2.7)		
	87Rb+	3.45(2.8)		3.38(2.7)		
	Cs+	2.99(3.7)	3.23	2.82(3.4)		
m-Iodo-nitrobenzene	Li+	0.105(0.73)				n'
	Na+	0.37(1.2)				
	K+	0.18(2.2)				

Table 2 (Continued)

Metal Hyperfine Coupling

Anion	Cation	DME	THF	Diglyme	Ref.
	^{85}Rb$^+$	0.95(2.6)			
	^{87}Rb$^+$	3.20(2.6)			
o,o'-Dinitrobiphenyl	Cs$^+$	2.05(2.5)			o'
	Li$^+$	0.35			
	Na$^+$	0.18			
m-Dinitrobenzene	Na$^+$	0.32(1.0)			
	K$^+$	0.17(2.1)			
	^{85}Rb$^+$	0.9(2.5)			
	^{87}Rb$^+$	2.1(1.7)			
	Cs$^+$	2.45(3.0)			p'
o-Dinitrobenzene	Na$^+$	0.38			
	K$^+$	0.22			m'
	Cs$^+$	3.30			

		i-PrOH	t-BuOH	t-AmOH	t-HexOH	Ref.
Nitrobenzene	Na$^+$	0.58(1.8)	0.45(1.4)	0.36(1.2)	0.30(1.0)	
	K$^+$	0.17(2.1)	0.14(1.7)	0.13(1.6)	0.10(1.2)	
	^{87}Rb$^+$	2.46(2.0)	2.25(1.8)	2.27(1.8)	2.15(1.7)	l'
	Cs$^+$	1.73(2.1)	1.42(1.7)	1.32(1.6)	1.12(1.4)	

Values in brackets after metal hyperfine couplings are $\rho_M \times 10^3$.

[a] Ref. 55.
[b] Ref. 60.
[c] Ref. 45.
[d] Ref. 47.
[e] Ref. 94.
[f] Ref. 46.

i Ref. 44.

j J. dos Santos Veiga, W. L. Reynolds, and J. R. Bolton, *J. Chem. Phys.*, **44**, 2214 (1966).

k Ref. 77.

l A. Ehrenberg, L. E. Goran Eriksson, and F. Muller, *Nature*, **212**, 503 (1966).

m D. R. Eaton, *Inorg. Chem.*, **3**, 1268 (1964).

n M. Hirayama, *Bull. Chem. Soc. Japan*, **40**, 2234 (1967).

o Ref. 74.

p Ref. 36.

q Ref. 91.

r H. L. Strauss, T. J. Katz, and G. K. Fraenkel, *J. Am. Chem. Soc.*, **85**, 2360 (1963).

s H. Nishiguchi, Y. Nakai, K. Nakamura, K. Ishizu, Y. Deguchi, and H. Takaki, *J. Chem. Phys.*, **40**, 241 (1964).

t Ref. 34.

u Ref. 37.

v Ref. 7.

w M. Iwaizumi, M. Suzuki, T. Isobe, and H. Azumi, *Bull. Chem. Soc. Japan*, **41**, 732 (1968).

x Ref. 88.

y Ref. 87.

z Ref. 23.

a' Ref. 28.

b' Ref. 86.

c' Ref. 9.

d' Ref. 82.

e' E. A. C. Lucken, *J. Chem. Soc.*, 1964, 4234.

f' K. Khakhar, B. S. Prabhananda, and M. R. Das, *J. Am. Chem. Soc.*, **89**, 3100 (1967).

g' Ref. 42.

h' Ref. 100.

i' Ref. 92.

j' Ref. 98.

k' Ref. 121.

l' Ref. 29 and 30.

m' Ref. 33.

n' Ref. 32.

o' Y. Nakai, K. Kawamura, K. Ishizu, Y. Deguchi, and H. Takaki, *Bull. Chem. Soc. Japan*, **39**, 847 (1966).

p' Ref. 99.

drawn is that in these solvents the lowest metal coupling is almost always obtained with DME. This is presumably because of the ability of this solvent to chelate to the cations, thus increasing their overall degree of solvation. This effect is magnified still further in diglyme, triglyme, and tetraglyme [28]. In a few cases we can compare the results for various alcohols with those for the ethers, and it is found that the coupling constants are generally slightly larger in the alcohols [29]. In general, in aprotic solvents of relatively high dielectric constant, such as acetone, DMF, and MeCN, metal ion hyperfine coupling has not been detected. However, fairly large coupling constants have recently been measured for nitrobenzene anions in both acetone and MeCN [30]. These surprising results, which run contrary to earlier work, seem to be fairly clear; the important point is the method of preparation, which comprised dissolution of the solid alkali metal salts into highly purified solvent in vacuo. Similar experiments with semiquinones gave what appeared to be "free-ion" spectra [31]. Gradual changes in the coupling constants in mixed solvents resulting from increased proportion of the "better" solvent were reported for naphthalenides [40, 107, 136].

2.2.3. *Temperature*

An outstanding factor is that in nearly all cases the magnitude of the metal hyperfine coupling falls on cooling. This result is so general that when an increase has been observed, as in the example depicted in Fig. 3 (where the magnitude of the metal coupling approaches zero and then on further cooling increases), it has been concluded that the coupling constant must be negative. That negative coupling constants are possible has been confirmed by NMR studies (see Section 6), and it seems safe to conclude that whenever the magnitude of the metal hyperfine coupling increases on cooling, the metal hyperfine coupling is negative [21]. Examples include the lithium-nitrobenzenides [29, 32, 33], sodium (or potassium?), rubidium and cesium salts of the biphenyl anion in THP [34], the ion tetramer of the sodium ketyl of hexamethylacetone in THF or MTHF [35], the lithium salt of the phthalonitrile anion in DME [36], the lithium, sodium, and potassium salts of the pyrazine anion in DME or THF [37, 38], the cesium-*p*-xylene ion pair in DME [39], rubidium and cesium salts of naphthalene and anthracene in most of the common ether solvents [23, 40] (see Fig. 3), and the lithium and potassium salts of anthracene in DEE [40].

In examining Fig. 3, we see that the curves for Na/naphthalene/THF, DEE are *s*-shaped. Possible reasons for this are discussed in Sections 2.4 and 4.1.1.

2.2.4. *Anion*

We would expect to find a dependence of the alkali metal hyperfine splitting on the electron affinity of the anion and this is observed in some instances.

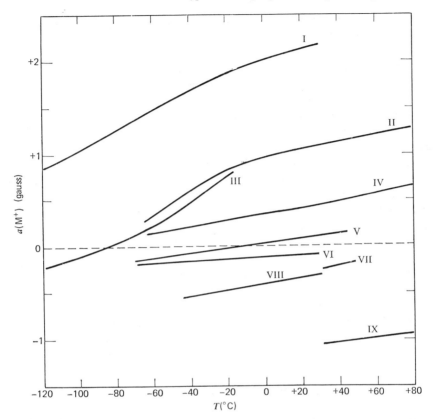

Figure 3. The temperature dependence of the metal hyperfine coupling in ion pairs. (I) Na/naphthalene/DEE. (II) Na/naphthalene/THF. (III) Cs/pyracene/THF. (IV) Na/fluorenone/DME. (V) Li/fluorenone/THF. (VI) ^{85}Rb/naphthalene/DME. (VII) ^{87}Rb/naphthalene/THF. (VIII) ^{87}Rb/naphthalene/DME. (IX) Cs/naphthalene/DME. (Data from references [21–23, 82].)

For example, on going from benzosemiquinone to durosemiquinone anions [41, 42], the magnitude of the metal splitting increases.* This is also found for tropenide dianion radicals, the metal splitting decreasing in the order tropenide > benzotropenide > dibenzotropenide [43]. This decrease in metal HFC (hyperfine coupling) is, however, also a function of the increased delocalization as the size of the aromatic system is increased. Unusual behavior has been observed with substituted pyrazine anions. As expected,

* However, this could be caused by steric hindrance, which favors out-of-plane location of the cation (Editor).

the alkali metal hyperfine coupling increases on going from pyrazine to tetramethylpyrazine, but it *decreases* on going to 2,5- and 2,6-dimethyl-pyrazine [44].

Steric effects also must be considered. For example, the alkali metal splitting increases slightly on going from benzophenone to 4,4′-dimethyl-benzophenone, as expected, but decreases markedly on going to 2,6,2′,6′-tetramethylbenzophenone [45]. This decrease is probably a result of steric crowding around the carbonyl group, preventing close approach of the cation.

Another significant observation is that when an anion can chelate to a cation via two bonding sites, the metal coupling is relatively large. This is particularly marked for diketyl structures such as *o*-dimesitoyl benzene [46] and benzil [47], but is not found for 2,2,5,5-tetramethylhexane-3,4-dione, where steric effects force the anion into a *trans*-configuration and prevent the formation of the *cis*-chelate structure presumably adopted by the other diketyls [47].

There is an unusually large increase in the metal hyperfine splitting for what appears to be a relatively subtle change in structure on going from xanthene-9-one to xanthene-9-thione [48]. This is caused partially by an increase in spin density on the sulfur atom relative to the oxygen atom, and this is supported by the decrease in the ring proton splittings. However, it is also possible that covalency effects are greater for sulfur.

2.3. Information from Solid-State Studies

Before discussing these results for paramagnetic anions in solution, we consider briefly the following solid-state data, because the results shed important light on the structure of ion pairs:

1. The formation of $M^+CO_2^-$ by reaction between alkali metals and carbon dioxide [49].

2. The formation of similar species by γ-irradiation of alkali metal formates [50].

3. The formation of the species NaH^+ and KH^+ by γ-irradiation of various solids [51, 138].

The alkali metal splittings derived from the ESR spectra of these systems are included in Table 3. One important result is that the anisotropic coupling to the alkali metal ion is small and in the particular case of NaH^+ can be accounted for entirely in terms of the dipolar interaction arising from spin on the adjacent hydrogen atom. This mechanism probably accounts for a large part of the anisotropy in all cases and hence we can conclude that the orbital used by the cation is almost entirely *s* in character, any *p*-contribution

Table 3   Some Relevant ESR Solid-State Data

Parent	Anion	Cation	Temperature ($^{\circ}K$)	Alkali Metal Hyperfine Coupling		$\rho = \dfrac{a_{iso}}{A_0} \times 10^2$	Ref.
				a_{iso}	a_x, a_y, a_z		
CO_2	CO_2^-	Li^+	77	11.4		7.95	a
		Na^+	77	22.7	22.8, 21.6, 23.6	7.18	
		K^+	77	5.5	5.5, 5.3, 5.8	6.7	
		$^{85}Rb^+$	77	23.6		6.53	
		$^{87}Rb^+$	77	79.4		6.52	
		Cs^+	77	53.0		6.47	
$HCOO^-M^+$	CO_2^-	Li^+	RT and 77	3.3		2.3	b
		Na^+	RT and 77	8.1	7.5, 7.5, 9.3	2.6	
		K^+	RT and 77	1.4	1.3, —, 1.4	1.7	
		Cs^+	77	16.0		1.9	
		Cs^+	RT	14.5		1.8	
CS_2	CS_2^-	Li^+	77	5.2		3.6	c
		Na^+	77	17.3	17.0, 17.4, 17.4	5.5	
		K^+	77	3.5	3.4, 3.4, 3.7	4.2	
		Cs^+	77	45.1		5.5	
$BaSO_4/$ $NaNO_3$	NaH^+	Na^+	77	17.2	21.4, 15.0, 15.1	5.44	d,e

[a] Ref. 49.
[b] Ref. 50 and D. W. Ovenall and D. H. Whiffen, *Mol. Phys.*, 4, 135 (1961).
[c] J. E. Bennett, private communication; J. E. Bennett, B. Mile, and A. Thomas, *Trans. Faraday Soc.*, 63, 262 (1969).
[d] Ref. 51.
[e] Ref. 138.

being very small. This information cannot be derived from liquid-phase studies, but if this conclusion is generalized to the liquid phase, we can conclude that the isotropic coupling is indeed a good measure of the actual spin density on the cation, and that spin polarization, which would be expected to affect the filled p-level of the cations, is relatively insignificant. Another important observation is that in the case of the γ-irradiated formates the location of the cation in relation to the anion is fairly well established. Unfortunately, this factor is not as useful as it might be since CO_2^- has an unpaired electron in what may be described as a σ-orbital [52] rather than a normal π-type orbital characteristic of radical anions. Nevertheless, it does suggest that spectra should be obtained from ion pairs in salts which contain radical anions having an unpaired electron in such a σ-orbital.

Another important aspect of these results is the large magnitude of the alkali metal coupling, especially for $M^+CO_2^-$ in CO_2 matrixes, compared with the values in alkali metal formates. This large difference is, we believe, a consequence of "solvation." In the alkali metal formates, the cation is strongly coordinated (solvated) to *several* formate ions. This, like ordinary solvation, reduces the electron affinity of the cation. No such coordination is available for $M^+CO_2^-$ in solid carbon dioxide, except that involving the paramagnetic anion. These results support our prognostication in Section 2.1 that the metal coupling should fall as the solvation increases.

These studies have been extended recently [139]. Labile species, such as $(Ag-NO_3)^-$ or $(Cd-CO_3)^-$, were detected by ESR in γ-irradiated salts at $70°K$.

2.4. Discussion of Trends

The purpose of this section is to take a further look at the trends listed in Section 2.2 in the light of the mechanisms discussed in Section 2.1. Of the many factors controlling the sign and the magnitude of the metal hyperfine coupling, the following are clearly important:

1. The electron affinity of the cation.
2. The effective spin density at the point of contact.
3. The nature of the contact site (direct or indirect spin transfer).
4. The extent of solvation.

In several cases in which the metal hyperfine coupling is large, the first effect appears to dominate (see Fig. 1 and the data for $M^+CO_2^-$ in solid CO_2 given in Table 3). In other cases where the extent of spin transfer is small, this trend with electron affinity is still approximately followed provided allowance is made for negative spin density in certain cases (Fig. 1).

For many anions, however, there is no clear trend in the metal hyperfine coupling, presumably because there is a competition between effects 1 to 4.

In the particular case of the aromatic nitroanions the normal trend is totally reversed. The easiest way to explain this is to recall that these anions are very readily perturbed, as is shown by the sensitivity of the ^{14}N hyperfine coupling constant to environmental charges. This could mean that effect 2 predominates. For example, lithium gives the greatest perturbation, thereby attracting negative charge to the oxygen with which it is associated, and hence causing loss of spin density on the ^{14}N. This effect might also be found for the ketyls but the information is scanty. One possible difference between the two systems is that there are two oxygen atoms on the nitro-group and if the cation is associated primarily with one of these at a time, the other oxygen may acquire relatively high spin density when the spin density on the oxygen associated with the cation is low (see Section 3.3). This is merely surmise, of course, since effect 3 may well play an important role.

We would expect steric factors to be particularly important in the third effect. This problem is made difficult by our frequent lack of knowledge of the sign of the coupling constant which is not given directly by ESR. The very large increase in coupling constant on going from benzosemiquinone to durosemiquinone or from pyrazine to tetramethylpyrazine anions may well reflect a steric resistance to the in-plane structure (see Scheme II).

Lowering the temperature can again have a variety of results. The most general is the increase in ionic solvation. This is illustrated by the fact that the spectra of supposedly "free ions" often appear on cooling, in addition to those of the ion pairs. This may result in a steady decrease in the metal hyperfine coupling, or there may be a dependence of the type shown in Fig. 3 (sodium naphthalenide in THF or DEE), which is more typical of an equilibrium between rapidly interchanging distinct species that is, $(I.P.)_1 \leftrightharpoons (I.P.)_2$. Of course, both effects can operate simultaneously. Reference to Fig. 3 shows that in general the evidence for a clear-cut equilibrium between distinct species is not strong. For further discussion of these equilibria see Section 4.1.

Another factor to consider is that an increase in cation solvation may result in a smaller perturbation of the anion (cf. effect 2 above). Effect 3 also must be invoked. Lowering the temperature will reduce the amplitude of vibration about the minimum energy position.* If this position is a nodal plane, then positive coupling constants will fall, whereas negative coupling constants will increase in magnitude and in certain cases the coupling constant may go through zero and become negative (Fig. 3).

It seems to us that of the various solvent effects discussed in Section 2.2 at least one important factor, the difference between the alcohols and the ethers, can be understood in the light of the discussion in Section 2.1. At first sight, it is somewhat surprising that alcohols, even such good solvents as iso-propanol, can give ion pairs with relatively large alkali metal hyperfine couplings. We would like therefore to propose the following model. Let us consider for simplicity a ketyl, $R_2\dot{C} = O^-$, although the same argument can be extended to semiquinones, nitroanions, etc. We envisage a situation in which, in the competition between cation and alcohol for the bonding site of the anion, the proton of the alcohol gets pride of place. Our model is depicted in Scheme V. The proton takes the in-plane position and the cation,

---

* The position occupied by the cation could be calculated, as shown by recent studies of Goldberg and Bolton [137]. For further discussion of this subject see Chapter 8, page 345. [Editor].

V

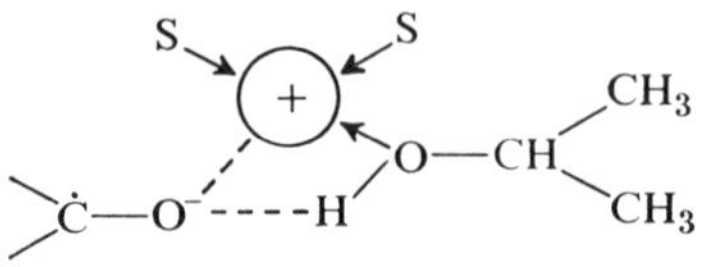

VI

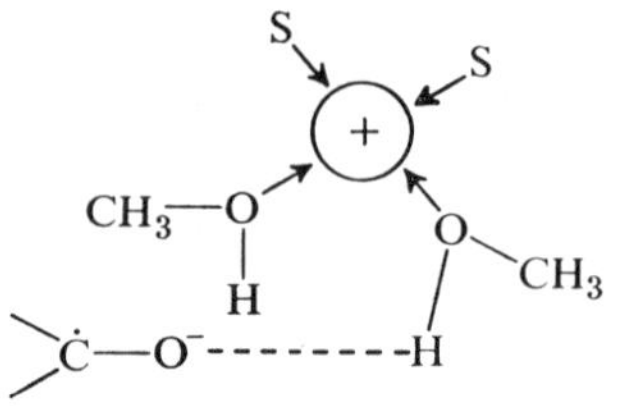

sharing the same solvent molecule, is forced into the out-of-plane position which, as we have concluded, should give rise to a relatively large positive spin density on the cation. In terms of the naive classification in Section 1.1, this could be described as a solvent-shared contact ion pair. As the R-group becomes more bulky, this unit becomes too crowded and the cation is forced away from the anion, thus causing a decrease in metal hyperfine coupling. As the R-group diminishes in size (Me- and Et-), the solvent molecules are small enough to surround the cation completely, thus insulating it from the anion and forming a true solvent-separated ion pair (depicted in Scheme VI).

Solvent-shared ion pairs of these types are expected to be favored only by protic solvents and by anions capable of forming good hydrogen bonds. They are expected to be energetically stabilized because there is a mutual reinforcement of bonding: the anion bonded to the proton makes the alcohol more basic with respect to the cation and the cation bonded to the oxygen of the alcohol makes the proton more acidic.

2.5. Hyperfine Coupling to Two Cations

Occasionally hyperfine coupling characteristic of two equivalent cations has been detected (Table 4), the most obvious system being that in which the anion has two negative charges. Usually, the dianion of a hydrocarbon is diamagnetic but in the specific cases cited previously, paramagnetic dianions occur either because the levels are degenerate or because the monoanion is diamagnetic. The former ions are ground-state triplets and are discussed in Section 5. In many instances the metal hyperfine coupling for the latter group of anions is comparatively large, which is to be expected in view of the high electron-donating power of the dianion. Another obvious reason for detecting two cations is in radiation damage work where the parent material has two cations symmetrically disposed about it. This has been

Table 4 Hyperfine Coupling to Two Equivalent Cations (in Solution)

Species	Cation	Solvent (RT)	$a(M^+)$	Ref.
Benzophenone	Li$^+$	DME	0.673	a
ketyl	2Li$^+$	DME	1.125	a
	2Na$^+$	MTHF	0.3	b
Hexamethylacetone	2Na$^+$	THF	1.58	c
ketyl	2Na$^+$	MTHF	0.65	d
Naphthalene	Na$^+$	TMED	1.593	a
anion radical	2Na$^+$	TMED	0.488	a
Tropenide	2Na$^+$	THF, MTHF	1.74	e
dianion radical		and DME		
Benzotropenide	2Na$^+$	THF	1.04	e
dianion radical				
Dibenzotropenide	2Na$^+$	THF	0.70	e
dianion radical				

^{a} Ref. 55.
^{b} Ref. 94.
^{c} N. Hirota and S. I. Weissman, *J. Am. Chem. Soc.*, 82, 4424 (1960).
^{d} Ref. 35.
^{e} Ref. 43.

observed in γ-irradiated cesium formate and seems to be purely a quirk of the crystal structure [50].

Cationic ion triplets, $A^-(M^+)_2$, which until recently have been thought of only as unstable intermediates in the cation displacement process (discussed in Section 4.2), have recently been unambiguously detected in the case of the semiquinones by Gough and Hindle [53, 140]. In all cases, addition of an equivalent of sodium tetraphenylboron in THF gave the triple ion, and when insufficient salt was added both the ion pair and the triple ion gave superimposed spectra. The cations were equivalent for the symmetrical semiquinones but quite inequivalent for the 2,6-dimethyl derivatives. The ion pairs of the symmetrical semiquinones showed the normal reduction in symmetry (Section 3.4), but symmetry was restored on addition of the second cation. This shows clearly that the two cations are bonded to the two oxygen atom sites, as depicted in Scheme VII. These results show that triple-ion

VII $\qquad$ $M^+, O = \!\!=\!\!\!\left\langle \begin{matrix} \cdots \\ (\,-\,) \\ \cdots \end{matrix} \right\rangle\!\!\!=\!\!= O, M^+$

formation can be of very great significance. Presumably the requirements for ESR detection would include (1) two well separated binding sites, each having high electron density, (2) an anion for the added salt having a very low cation affinity, and (3) low temperature and solvent of low dielectric constant. It is perhaps for these reasons that studies with added iodide or with the dinitrobenzene anions failed to give triple ions.

Extention of these studies [145, 146] revealed the existence of triple ions derived from pyrazine and tetramethyl-pyrazine radical anions.

Certain ketyls, particularly the penta- and hexamethylacetone ketyls, also give spectra characteristic of one anion bonded to two equivalent cations. The possibility that this is a triple ion had been discarded [22, 25] on the grounds that such a species is unlikely to have a long lifetime. However, Gough and Hindle's new results [53, 140] make this conclusion less compelling and it seems quite possible that the results in Table 4 for the anion radicals are best explained in terms of triple-ion formation. In the case of the aliphatic ketyls, the species showing coupling to two equivalent cations is in equilibrium with another species which Hirota and Weissman [54] have concluded is a paramagnetic ion tetramer, $(M^+)_2(A^-)_2$. This species, which exists in a triplet spin-state, is discussed in greater detail in Section 5.

2.6. Hyperfine Coupling to Diamagnetic Anions

Thus far, we have considered only paramagnetic anions associated with diamagnetic cations. It is at first sight surprising that no such interaction has been detected for paramagnetic cations associated with diamagnetic anions. Reasons for this failure probably include the following: (1) most solvents which favor the formation of radical cations have fairly high dielectric constants; (2) the most likely mechanism for the transfer of spin density to diamagnetic anions such as the halide ions would involve an unprobable electron transfer from the anion leaving a hole in the p-shell. A further step involving polarization of inner s-electrons is then required before any isotropic coupling could be detected.

3. INDUCED CHANGES IN MAGNETIC PARAMETERS OF PARAMAGNETIC ANIONS

3.1. g-Value Shifts on Ion-Pair Formation

Departure in g-values from 2.0023, characteristic for the free electron, generally are very small and have not been monitored by workers in the field. This is because the orbital magnetism of the unpaired electron is practically zero for organic radicals. In principle, induced changes in the three principal values of the g-tensor should be considered. In practice, the g-tensor is rarely known and it is $g_{av} = \frac{1}{3}(g_{xx} + g_{yy} + g_{zz})]$ that is measured. The g-value may be considered as being made up of various increments from different parts of the molecule, and if a particular atom or group has a structure which encourages orbital angular momentum, then this will be largely responsible for the deviations from the "free-spin" g-value. In particular, if heavy atoms are present under these circumstances, their large spin orbit coupling constants may magnify the effect considerably. A good example of this situation is the anion of benzophenone where the positive g-shift (i.e., g_{expt} is greater than the free spin value) stems from spin on oxygen. As the spin density on oxygen falls (e.g., on ion pairing), the g-shift decreases [55].

A second way in which the g-value can change on ion-pair formation is via transfer of spin onto the cation. If the cation has a large spin orbital coupling parameter, orbital motion may result. This is, however, a second-order effect since spin in the outer s-orbital cannot contribute and we would expect that it is the small population of the outer $p(\sigma)$-orbital which causes a g-shift, via coupling with the $p(\pi)$-orbitals. If this is the case, then negative shifts would be expected.

In contrast, ion pairing between a paramagnetic cation and a halide ion such as iodide might be expected to give a large positive g-shift as a result of charge transfer from the halide ion [25]. As far as we are aware, this effect has not been observed.

Some experimental values are given in Table 5. In the main they accord with the preceding principles and in particular cesium ion pairs of pyracene and naphthalene show the expected negative increment. The negative increment seen for the rubidium naphthalenide ion pair decreases as the temperature is lowered [23], as is expected for increased dissociation of the ion pair on cooling. Results for quinones should illustrate both of the foregoing principles. However, for various quinones [56, 57], cesium gives a small positive shift relative to the free ion. The sodium and potassium ion pairs give the normal negative increment, and we would certainly expect cesium to have a smaller effect on the spin density on oxygen. Gill and Gough [56] proposed that the small positive shift observed is a direct effect of the large cesium spin orbit coupling constant, but we do not at present understand why the g-shift is positive rather than negative.

Extensive studies of the effect of ion pairing on the g-value have been reported recently by Fraenkel et al. [141]. Using the technique described earlier [142] they succeeded in determining, with accuracy of about 5×10^{-6}, the variations of the g-values of naphthalenides resulting from their pairing with different alkali-metal cations in various solvents and over a wide temperature range. The results are given relative to the standard-potassium pyrenide in DME.

For all the naphthalenides that show metal coupling, g-factors had negative temperature coefficients, whereas the species exhibiting no metal splitting had higher g-values which were temperature independent.

Table 5 *g*-Shifts in Ion Pairs

Anion	Cation	Solvent	Temperature (°C)	g	Ref.
Pyracene	Free ion	DME	−70	2.00267	a
	Li+	Hexane/ MTHF	−113	2.00271	
	Na+	MTHF	−80	2.00265	
	Cs+	THF	−80	2.00247	
Naphthalene	Free ion	DME	26	2.002737	b
	Li+	DME	26	—	
	Na+	DME	26	2.002741	
	K+	DME	26	2.002745	
	Rb+	DME	26	2.002705	
	Cs+	DME	26	2.002409	
	Free ion	THF	26	2.002748	
	Li+	THF	26	2.002750	
	Na+	THF	26	2.002734	
	K+	THF	26	2.002737	
	Rb+	THF	26	2.002693	
	Cs+	THF	26	2.002533	
Terephthalonitrile	Free ion	DME, THF, THP, MTHF	20	2.00255	c
	Li+	DME, THF, THP, MTHF	20	2.00251	
	Na+	DME, THF, THP, MTHF	20	2.00253	
	K+	DME, THF, THP, MTHF	20	2.00253	

				Δg^d	
Phthalonitrile	Free ion	DME, THF, THP, MTHF	20	0	e
	Li+	DME, THF, THP, MTHF	20	-6.5×10^{-5}	
	Na+	DME, THF, THP, MTHF	20	-4.5×10^{-5}	
	K+	DME, THF, THP, MTHF	20	-3×10^{-5}	

				ΔH (gauss)^f	
Duroquinone	Na+	DME	−80	+0.16	g
	K+	DME	−80	+0.10	
2,3-Dimethyl- benzoquinone	K+	DME	−86	+0.14	
	Cs+	DME	−85	−0.05	

Table 5   (Continued)

2,6-Dimethyl-benzoquinone				
	Li^+	DME	−88	+0.11
	K^+	DME	−85	+0.12
	Cs^+	DME	−85	−0.08

Ref. 21.

Ref. 23.

[c] Ref. 74.

[d] Δg = g-value ion pair − g-value free ion.

[e] Ref. 36.

[f] ΔH = separation between center of ion-pair spectrum and center of "free" ion spectrum. A positive value of ΔH is equivalent to a negative g-shift.

[g] Ref. 56. Similar results were obtained by Warhurst et al. [57] for benzoquinone.

The difference between the g-value of an ion pair and that of the free ion is related to the spin-orbit coupling constant of the lowest unfilled atomic p-orbital of the cation. This is clearly seen from inspection of Table 5a.

Table 5a   Dependence of Naphthalenides g-Factor on Spin-Orbit Coupling Constant

Cation	$g - (g$ free-ion$) \times 10^5$ at 20°C		Spin-Orbit Coupling Constant (cm^{-1})
	DME	THF	
$^7Li^+$	−0.07	+0.07	0.29
$^{23}Na^+$	−1.13	−0.56	11.46
$^{39}K^+$	−1.19	−0.32	38.48
$^{85}Rb^+$	−4.84	−4.73	158.40
$^{133}Cs^+$	−18.66	−20.69	369.41

Spin-orbit coupling constant in the lowest unoccupied p-atomic orbital of the cation. Calculated from the data reported by C. E. Moore, Nat. B. of Standards Circ., 467 (1949).

3.2. Proton Hyperfine Coupling in Ion Pairs

Ring protons in aromatic anions generally are remarkably insensitive to major perturbations such as those involved in ion-pair formation. In most cases, only careful measurement will reveal these changes and it is probably not very safe to draw major conclusions from the trends revealed. The results for some systems are given in Table 6.

Some general comments would seem to be in order. With anthracene, for example, in solvents where contact ion pairs are formed [9, 40], as the perturbation by the cation increases, spin density is acquired by the central

Table 6 ^{1}H, ^{13}C, and ^{14}N Hyperfine Coupling in Ion Pairs

Anion	Cation	Solvent	Temperature	Proton Hyperfine Coupling				Ref.
				a_o	a_m	a_p	$a(^{13}C)^a$	
Benzophenone								
	Mg^{2+}	DME	25°C	2.87	1.06	3.46	15.8	b
	Ca^{2+}	DME	25°C	2.76	0.98	3.46	13.2	
	Ba^{2+}	DME	25°C	2.64	0.93	3.44	12.1	
	Na^{2+}	DME	25°C	2.60	0.87	3.45	—	
	K$^+$	DME	25°C	2.58	0.86	3.44	9.3	
	Free ion	DME	25°C	2.55	0.85	3.38	—	
				$a_{1,8}$	$a_{3,6}$	$a_{4,5}$	$a(^{13}C)$	
Fluorenone								
	Li$^+$	DME	RT	—	—	—	6.20	c
	Na$^+$	DME	RT	2.08	3.15	0.66	4.85	
	K$^+$	DME	RT	2.03	3.12	0.66	4.20	
	Free ion	DMF	RT	1.88	3.07	0.64	2.75	
				a_2	a_3	$a(R_\alpha)$		
Di-isopropyl o-phthalate								
	Li$^+$	THF	RT	0.46	3.03	0.29		d
	Na$^+$	DME	RT	0.43	2.90	0.34		
	K$^+$	DME	RT	0.55	3.19	0.37		
	Free ion	CH$_3$CN	RT	0.82	3.47	0.32		

Anthracene [g]

			a_γ	a_α	a_β	$a_\alpha + a_\beta$
Li$^+$/Na$^+$	DME	RT	5.337	2.740	1.509	4.249
K$^+$	DME	RT	5.346	2.726	1.513	4.239
Bu$_4$N^{+e}	DME	RT	5.384	2.662	1.532	4.194
Li$^+$	THF	26°C	5.314	2.732	1.508	4.240
Na$^+$	THF	26°C	5.310	2.745	1.503	4.248
K$^+$	THF	26°C	5.294	2.757	1.498	4.255

Naphthalene [h]

			a_α	a_β
Li$^+$	Dioxane	24°C	4.792	1.788
Na$^+$	Dioxane	24°C	4.871	1.857
K$^+$	Dioxane	24°C	4.861	1.837
Free ion	DMF	24°C	4.906	1.812

Azulene [g]

			a_6	$a_{4,8}$	a_2	$a_{5,7}$
Li$^+$	THF	26°C	9.385	6.796	3.536	1.670
Na$^+$	THF	26°C	9.368	6.792	3.593	1.634
K$^+$	THF	26°C	9.148	6.542	3.709	1.512
Free ion	DMF	26°C	8.829	6.219	3.948	1.338

2,6-Dimethyl-p-benzoquinone [i]

			a(Me)	a(H)
Li$^+$	t-AmOH	RT	2.91	1.13
Na$^+$	t-AmOH	RT	2.77	1.26
K$^+$	t-AmOH	RT	2.56	1.51
Rb$^+$	t-AmOH	RT	2.56	1.51
Cs$^+$	t-AmOH	RT	2.56	1.51
Bu$_4$N$^+$	t-AmOH	RT	2.02	2.02

Table 6 (Continued)

Anion	Cation	Solvent	Temperature	Proton and ^{14}N Hyperfine Coupling					Ref.
1,2-Naphthoquinone				a_4	a_1	a_2	a_3	a_5	
	Li$^+$	DME	RT	4.29	1.57	1.31	0.27	0.21	
	Na$^+$	DME	RT	4.11	1.57	1.20	0.29	0	
	K$^+$	DME	RT	4.01	1.48	1.14	0.34	0	
	Rb$^+$	DME	RT	4.03	1.53	1.17	0.36	0.07	
	Cs$^+$	DME	RT	4.01	1.48	1.13	0.35	0.09	
Duroquinone				$a_{2,6}$(Me)	$a_{3,5}$(Me)	$a_{2,6}+a_{3,5}$			k
	Li$^+$	t-AmOH	RT	2.91	0.93	3.84			
	Na$^+$	t-AmOH	RT			3.86			
	K$^+$	t-AmOH	RT			3.86			
	Hex$_4$N$^+$	t-AmOH	RT			3.86			
	Free ion	i-PrOH	RT	1.95	1.95	3.90			
2,5-di-t-Butyl-1,4-benzoquinone				a_3	a_6	a_3+a_6			
	Na$^+$	DME	17°C	1.53	2.95	4.48			
	K$^+$	DME	17°C	2.28	2.28	4.56			
	Rb$^+$	DME	17°C	2.28	2.28	4.56			
Nitrobenzene				a_0	a_m	a_p	$a(^{14}N)$		m
	Li$^+$	DME	0°C	3.49	1.135	3.94	11.55		
	Na$^+$	DME	23°C	3.56	1.12	4.08	11.60		
	K$^+$	DME	23°C	3.55	1.12	4.12	11.05		
	Cs$^+$	DME	0°C	3.54	1.12	4.16	10.80		
	Free ion	DME	23°C	3.54	1.12	4.29	10.15		
	Free ion	DMSO	23°C	3.52	1.11	4.18	10.40		

210

m-Iodonitrobenzene

			a_0	a_m	a_p	$a(^{14}\mathrm{N})$	n
Sr^{2+}	DME, THF	RT	3.50	1.15	3.84	11.84	
Ba^{2+}	DME, THF	RT	3.46	1.13	3.89	11.24	
Li^+	DME, THF	RT	3.40	1.10	4.01	10.05	
Na^+	DME, THF	RT	3.36	1.10	4.05	10.07	
K^+	DME, THF	RT	3.36	1.05	4.13	9.72	
Rb^+	DME, THF	RT	3.33	1.03	4.15	9.5	
Cs^+	DME, THF	RT	3.33	1.03	4.15	9.3	

m-Dinitrobenzene

			a_5	a_2	a_4	a_6	$a(^{14}\mathrm{N})$	o
Na^+	DME	RT	1.10	3.30	3.85	4.45	9.85, 0.29	
K^+	DME	RT	1.10	3.10	3.84	4.55	9.00, 0.22	
Cs^+	DME	RT	1.08	3.16	4.24	4.24	4.66, 4.66	
Free ion	DME	RT	1.12	2.95	4.66	4.66	4.20, 4.20	

o-Dinitrobenzene

			$a_{3,6}$	$a_{4,5}$	$a(^{14}\mathrm{N})^p$	q
Na^+	DME	23°C	0.81	1.62	4.05	
K^+	DME	23°C	0.62	1.65	3.60	
Cs^+	DME	23°C	0.43	1.64	3.30	
Free ion	DME	23°C	0.21	1.71	2.79	

Table 6 (Continued)

Anion	Cation	Solvent	Temperature	Proton and ^{14}N Hyperfine Coupling	
				$a(H)$	$a(^{14}N)^r$
s-Trinitrobenzene[s]					
	Li^+	DMSO	RT	4.22	2.23
	Na^+	DMSO	RT	4.06	2.56
	K^+	DMSO	RT	4.16	2.30
	Cs^+	DMSO	RT	4.25	2.12
	Pr_4N^+	DMSO	RT	4.20	2.08
	Bu_4N^+	DMSO	RT	4.20	2.10

[a] $a(^{13}C)$ is ^{13}C hyperfine coupling for the carbonyl carbon.
[b] Ref. 59.
[c] Ref. 22 and N. Hirota in *Radical Ions* E. T. Kaiser and L. Kevan, Eds., Interscience, 1968, p. 35.
[d] H. Hirayama, *Bull. Chem. Soc. Japan*, 40, 2234 (1963).
[e] Concentration of Bu_4N^+ is approximately 50 times that of the alkali metal cations.
[f] Ref. 58.
[g] Ref. 9.
[h] M. Iwaizumi, M. Suzuki, T. Isobe, and H. Azumi, *Bull. Soc. Chem. Japan*, 41, 732 (1968).
[i] Ref. 42.
[j] Ref. 121. Proton couplings were not assigned and so numbering is arbitrary.
[k] Ref. 41.
[l] C. Chippendale and E. Warhurst, *Trans. Faraday Soc.*, 64, 2332 (1968). See also ref. 57 for a similar study with 1,4-benzoquinone.
[m] Ref. 33. Similar results were obtained by Gross et al. [30].
[n] Ref. 32.
[o] C. Ling and J. Gendell, *J. Chem. Phys.*, 46, 400 (1967). Similar results were obtained by Symons et al. [99].
[p] Two equivalent nitrogens.
[q] Ref. 33.
[r] Three equivalent nitrogens.
[s] Ref. 65.

ring at the expense of the outer rings. This result is expected if the cation is located above the central ring, as indicated by Bolton's calculations [137] Similarly with azulene anions [8, 9] spin density is acquired by the larger ring at the expense of the smaller. The picture presented by naphthalene anions is less clear-cut [23, 40]. The results of Bolton and Fraenkel [58] for anthracene anions in DME may appear strange since they show a marked difference from the normal ion-pair behavior in that the biggest perturbation is given by Bu_4N^+ ions, whereas lithium ions are ineffective. This is attributed to the fact that cation-solvent interactions dominate with the smaller ions and the anthracene anion with its very dispersed charge and complete absence of bonding sites is unable to compete with DME molecules, that is, Li^+ forms large solvated ions. However, the weaker cation solvators MTHF and DEE give the normal behavior in which lithium gives the greatest perturbation [9, 40].

For molecules such as the ketyls, which contain one strong binding site, the ring protons reflect the fact that as negative charge is pulled onto the oxygen by the cation, so spin is gained by the ring, but the effect is very small (see also Section 3.3). This effect is illustrated by the results for fluorenone [22] and benzophenone [59, 60]. A similar effect would be expected for substituted nitrobenzenes. The results in Table 6 show that as the perturbation by the cation increases, so also does the ^{14}N hyperfine coupling and hyperfine coupling to the ortho- and meta-protons in the benzene ring. However, the hyperfine coupling from the para-protons decreases with increase in perturbation and it would be difficult to assess the effect of the perturbation on the overall spin density in the ring. The perturbation of the nitro-group by the cation is further discussed in Section 3.3.

The total spin densities in the benzene rings in the p-benzoquinones appear to be independent of the perturbation by the cation (Table 6). The p-benzoquinones and the dinitrobenzenes are molecules containing two binding sites and the differences in the behavior of these two systems on perturbation by the cation are further discussed in Section 3.4.

In some systems, the proton hyperfine coupling is temperature dependent. In the case of anthracene ion pairs, the decrease in the sodium or potassium hyperfine coupling on cooling is roughly proportional to the increase in the α-proton hyperfine coupling, which in turn is proportional to the changes in the β and γ proton hyperfine coupling [40]. The proton hyperfine coupling in lithium-azulene ion pairs is similar to the free-ion values at low temperatures, but as the temperature is raised the proton hyperfine coupling for the larger ring increases and that for the smaller ring decreases as a result of conversion of loose ion pairs into tight ones [8]. The temperature dependence of the proton hyperfine coupling in acenaphthylene ion pairs [10] has been interpreted in a similar manner.

3.3. ^{13}C Hyperfine Coupling in Ion Pairs

The isotropic coupling to nuclei other than protons in π-radicals also arises from a spin polarization mechanism, but the situation is more complicated than that for protons, since the isotropic coupling from a given carbon nucleus is influenced not only by spin on that nucleus but by spin on all adjacent nuclei as well. Generally, however, this effect serves simply to magnify the perturbation causing the change. The situation can be seen by reference to the spin densities on the oxygen and carbon of a ketyl. As the field of a cation is imposed, electrons are drawn on to the oxygen, thus reducing the spin density on the oxygen and increasing the spin density on the carbon. Since positive spin density on carbon gives a positive contribution to $a_{iso}(^{13}C)$, whereas positive spin density on ^{17}O oxygen gives a negative contribution, both the gain of spin by carbon and the loss of spin by oxygen cause an increased positive coupling to carbon. This is well illustrated by the studies of the carbonyl ^{13}C and oxygen ^{17}O of p-benzosemiquinone as a function of solvent [61]. Both these nuclei are remarkably sensitive to the cation perturbation, in contrast to the protons.

We would therefore expect to find that the ^{13}C isotropic coupling for the carbonyl carbon in ketyls is strongly dependent upon the nature of the cation in ion pairs. This is indeed the case [22, 59]; the perturbation from the value of the free ion in DMF increases steadily as the radius of the cation is reduced. The perturbation is reduced in all cases on cooling, which is consistent with the reduction in the metal hyperfine coupling [22]. This is also consistent with the changes in the proton coupling for fluorenone and benzophenone discussed in Section 3.2. The only other detailed study of ^{13}C splittings in ion pairs is that of Bolton and Fraenkel [58] for the anthracene anion. Their results show the same trends as the proton splittings discussed in Section 3.2.

3.4. Cation and Solvent Dependence of ^{14}N Hyperfine Coupling

The simple nitrogen heterocyclic anions such as pyrazine have been studied extensively with respect to two-site binding (Section 3.4), but not at all with respect to perturbation of the ^{14}N isotropic coupling. In contrast, aromatic nitroanions have been studied from this point of view in great depth. This is probably because of the great sensitivity of $a_{iso}(^{14}N)$ to small changes in solvent polarity and ion pairing. This sensitivity stems partly from the reinforcing effect discussed for ^{13}C, but it also may be associated with two geometrical factors, (1) the out-of-plane twisting of the nitro-group which is sensitive, for example, to ortho-substitution, and (2) deviations from planarity in which the nitrogen atom becomes the apex of a shallow pyramid.

Effect (1) is important because, in the limit of a 90° twist, the π-system of the benzene ring no longer overlaps with that of the nitro-group and hence the unpaired electron is forced to make a choice between the two sites. Apparently, for mono-nitro-compounds this choice is the nitro-group, and hence on twisting the spin density on the nitro-group steadily increases [62].

Effect (2) also has a strong influence on $a_{iso}(^{14}N)$ because in the pyramidal molecule there is a direct mixing of the $2s(N)$ orbital into the wavefunction and hence there is a dramatic increase in $a_{iso}(^{14}N)$ on bending [63]. That such a deformation should be significant and proportional to the spin density

Table 7 Spin Populations of Nitrogen Atomic Orbitals

Radical	Shape	$\rho_{2p}{}^a$	$\rho_{2s}{}^a$	ρ_p/ρ_s	$\rho_p + \rho_s$	Ref.
Me_3N^+	Flat	~1	0.035	~30	~1.0	b
ϕNO_2^-	—		0.411	0.025	16.6	0.44	c
$(CH_3)_2CHNO_2^-$	—	0.575	0.050	11.4	0.62	d
$NO_3{}^{2-}$	Pyramidal	0.64	0.082	7.8	0.72	e

^{a} ρ_{2s}: Spin density in $2s(N)$ calculated from the isotropic ^{14}N hyperfine coupling. ρ_{2p}: Spin density in $2p(N)$ calculated from the anisotropic ^{14}N hyperfine coupling.
^{b} T. Cole, *J. Chem. Phys.*, 35, 1169 (1961).
^{c} Ref. 63.
^{d} Ref. 64.
^{e} R. S. Eachus and M. C. R. Symons, *J. Chem. Soc. (London)*, A, 790 (1968).

on the nitro-group is evidenced by the pyramidal character of the structurally similar ion $NO_3{}^{2-}$ [13]. The limit of this pyramidal deformation would be reached when the spin density on the nitro-group is unity and this must be the case for the aliphatic nitroanions. Recent solid-state data for such ions [64] can be analyzed to give both s and p character of ion nitrogen [63], and the results of such calculations are compared with those for aromatic nitro-compounds in Table 7. These results confirm that, as the pyramidal character becomes more pronounced, the spin density on the nitro-group increases.

The effect of solvent polarity on $a(^{14}N)$ in the aromatic nitroanions is extremely large and tends at times to mask the effect of the cation in ion pairs. The chart (Fig. 4) compiled by John Slater of these laboratories illustrates this sensitivity and brings out the fact that solvent perturbation may well be much larger than that caused by cations. In all cases the normal rule that small ions perturb more strongly than large ions is obeyed. This behavior is that expected for anions with strong bonding sites (as stressed in Section 3.2) which can actively compete with the solvent for cations.

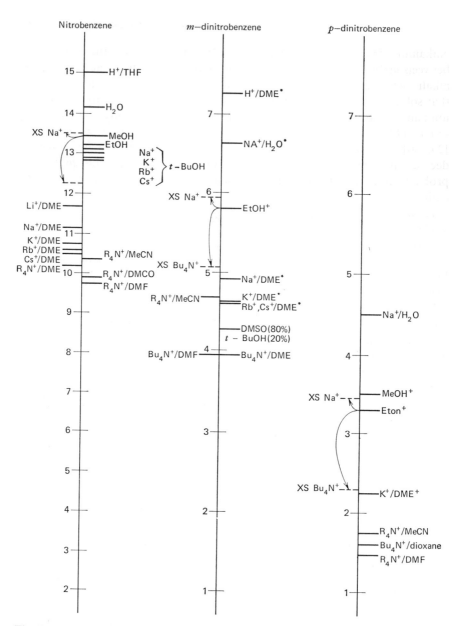

Figure 4. The effect of solvent and cation on the ^{14}N hyperfine coupling in aromatic nitroanions. For *m*-dinitrobenzene, when the anion was asymmetric (marked by *) the hyperfine coupling recorded is the average of the two inequivalent ^{14}N hyperfine couplings. For meta- and para-dinitrobenzene, in cases (marked by †) where line-width alternation was sufficiently rapid that alternate lines disappeared from the spectrum, the hyperfine coupling recorded is half the apparent coupling to one ^{14}N. Where the cation is not speci-fied, the ^{14}N hyperfine coupling is independent of the nature of the gegenion.

216

Nakamura [32] and Gross et al. [29, 30] have found a linear correlation between $a(^{14}N)$ and $Z/(R + 0.6)$, where R is the Pauling ionic radii of the alkali metal cations and Z is the charge on the cation. Others have found that solvents such as DMSO and DMF [65, 66] which normally do not give ion pairs, nevertheless give the same trend when an excess of the appropriate salt is added. When the effect of temperature on $a(^{14}N)$ has been studied [29, 30], $a(^{14}N)$ has been found to decrease approximately linearly with decrease in temperature. One of the factors contributing to this decrease is probably a general increase in the solvation of the ions, particularly the cation.

It has been commonly assumed that the use of alkylammonium salts in studies of anions of this type serves to overcome the difficulties involved in ion-pair formation [66, 67]. This is a good conclusion when aprotic solvents of high dielectric constant such as DMF or MeCN are used, as is evidenced by the results of Adams et al. [66], who found that $a(^{14}N)$ for p-chloronitrobenzene anions was independent of the concentration of added tetraethylammonium salt. However, this is not a safe generalization. For example, if alcoholic solvents are used, added tetraalkylammonium salts can induce very large changes in $a(^{14}N)$ for various nitroanions [11, 68]. A typical example is given in Fig. 5 together with the effect of added sodium and potassium salts. Qualitatively, these results can be understood in terms of the following model.

Ion-pair formation with the alkali metal ion is virtually complete in the 0.001 mole fraction region. In contrast, ion-pair formation with Bu_4N^+ is incomplete even in the 0.04 mole fraction region. The greater affinity for the

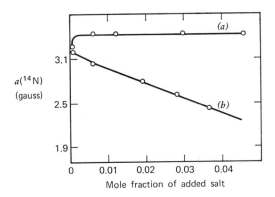

Figure 5. Dependence of a (^{14}N) for p-dinitrobenzene anions in ethanol on the concentration of added salts. (*a*) Sodium or potassium ethoxide. (*b*) Tetra-*n*-butylammonium bromide or perchlorate.

alkali metal ions has been brought out by competition studies which show that relatively low concentrations of alkali metal salts compete favorably with relatively high concentrations of tetraalkylammonium salts [68]. The perturbation caused by the alkali metal ions is very small in alcohols compared with that found in aprotic solvents such as the ethers and is practically independent of the nature of the alkali metal ion. We believe that the ions involved are best described as solvent shared ion pairs, in which the major perturbation of the nitro-group is caused by hydrogen bonding, the extra effect of the cation being to increase slightly the hydrogen-bond strength by forming units such as Scheme V. This mechanism is not available to the very weakly solvated tetraalkylammonium ions, which seem to behave rather like an added co-solvent. Indeed, the trend shown in Fig. 5 is similar in form to that obtained on the addition of dioxane [68]. Clearly, perturbation by the tetraalkylammonium ion is less than that caused by alcohol molecules. The lack of effect observed by Adams and others [66, 67] can then be understood by postulating that the perturbing effect of the tetraalkylammonium ion is approximately equal to that of solvents like DMF of MeCN. If this is correct, then it seems very likely that the spectra of tetraalkylammonium ion pairs in such solvents as the ethers would be virtually indistinguishable from those of the free ions, and hence the spectra often labeled "free-ion" spectra may in reality be those of tetraalkylammonium ion-pairs.

3.5. *Cation and Solvent Dependence of* ^{17}O *Hyperfine Coupling*

When a perturbation applied to a nitro- or carbonyl-group causes the spin density on nitrogen or carbon to increase, it is expected simultaneously to cause a corresponding decrease in the spin density on oxygen. Measurements of ^{13}C and ^{14}N hyperfine coupling have confirmed that the spin density on nitrogen or carbon does increase on perturbation by a cation or a highly polar solvent; however, the ^{17}O results are less clear-cut.

Apart from the results mentioned previously for semiquinones [61], which fulfill all our expectations, the most relevant results for ^{17}O appear to be those of Gulick and Geske [69, 70], who labeled the oxygen atoms in substituted nitrobenzene anions with ^{17}O. Some of their results are given in Table 8. These authors have shown that the sign of the ^{17}O hyperfine coupling is negative, but because ^{17}O has a negative nuclear magnetic moment the spin density on oxygen is positive (the spin density on nitrogen is also positive). The results given in Table 8 are surprising in that there is a slight *increase* in the magnitude of $a_{iso}(^{17}O)$ as $a_{iso}(^{14}N)$ increases on perturbation except for pentamethylnitrobenzene, which has a sterically crowded nitrogroup. As was pointed out by Gulick and Geske, we would have expected a decrease in the magnitude of $a(^{17}O)$ as negative charge density was pulled

Table 8 Effect of Cation and Solvent on $a(^{17}O)$ Hyperfine Coupling

Anion	Cation	Solvent	$a(^{14}N)$	$(-)a(^{17}O)$	Ref.
Nitrobenzene	Et_4N^+	DMF	9.67	8.84	a
	Et_4N^+	DMF + 0.2% H_2O	9.84	8.84	
	Et_4N^+	DMF + 0.6% H_2O	10.15	8.85	
	Et_4N^+	DMF + 1.0% H_2O	10.39	8.93	
	Et_4N^+	DMF + 5.0% H_2O	11.17	8.99	
	Et_4N^+	DMF + 10.0% H_2O	11.78	8.99	
	$Na^+(0.1M)$	DMF	10.98	8.94	
	$K^+(0.1M)$	DMF	10.43	9.83	

Anion	Mole Fraction H_2O in DMF (Et_4N^+ cation)	$a(^{14}N)$	$(-)a(^{17}O)$	Ref.
p-Dinitrobenzene	0.000	1.41	3.82	b
	0.079	1.56	4.29	
	0.149	1.66	4.36	
Pentamethyl-nitrobenzene	0.000	20.31	11.54	b
	0.125	20.85	11.17	
	0.223	21.12	11.12	
	0.364	21.30	11.11	
	0.463	21.36	11.04	
	0.588	21.43	10.90	
	0.696	21.50	10.86	
	0.788	21.66	10.80	

[a] Ref. 69.
[b] Ref. 70.

onto it by, for example, cations or a more highly polar solvent. Ling and Gendell [33] have suggested that this difficulty can be overcome by simulating the perturbation by an increase in the Coulomb parameters of both oxygen and nitrogen.

It is difficult to understand the physical significance of this model and the results are certainly in marked constrast with those of the carbonyl compounds discussed earlier. We suggest that the difference lies in the tendency for the nitro-groups to become pyramidal. If an increase in spin density on nitrogen causes a slight increase in bending, this will cause an increase in s-character on oxygen and hence an increase in the magnitude of $a_{iso}(^{17}O)$. This arises because of the delocalization effect through the nitrogen-oxygen bond, which has been reviewed in some detail elsewhere [71]. The $p(\pi)$-orbital of the oxygen contributes s-character to the oxygen nucleus by spin polarization of the nitrogen-oxygen σ-bond and if the effect of a decrease in spin density in the $p(\pi)$-orbital is outweighed by the increase in s-character on the oxygen caused by bending, then a small net increase in the magnitude of $a_{iso}(^{17}O)$, such as that observed, would be expected. In the case of the pentamethylnitrobenzene anion, the electron is forced out of the ring and onto the nitro-group by steric twisting, and $a(^{14}N)$ is close to the value expected for nitro aliphatic anion radicals (23–25 gauss [62]). The large spin density on the nitro-group gives rise to a pyramidal distortion which must be close to the upper limit. Hence, in this case, increased perturbation of the nitro-group will not lead to much further distortion, and so the decrease in spin density on oxygen controls the value of the ^{17}O hyperfine coupling, resulting in a decrease in the magnitude of $a(^{17}O)$ as the perturbation is increased. This is a possible explanation which seems to accommodate the facts, but the overall situation appears to be highly complex.

3.6. Ions Containing Two or More Binding Sites

When benzosemiquinone is monoprotonated, its symmetry is lost and the spectrum changes from a quintet to a triplet of triplets. The difference between the hyperfine coupling of the two sets of protons has been taken as a measure of the perturbation induced by the proton. The same applies to the change induced on protonating (or alkylating) an aromatic dinitroanion; however, for these anions the differences in the coupling constants to ^{14}N are usually considered, rather than those observed for the protons.

If, in an ion pair, the cation has a sufficient affinity for one binding site and a long enough residence time there, a similar perturbation is induced. For semiquinones, this is proportional to the size and charge of the cation in the manner expected for contact ion pairs, as seen in Fig. 6. In marked contrast, the perturbation induced in the anion of metadinitrobenzene remains as large as that for protonation and is independent of the size of the cation.

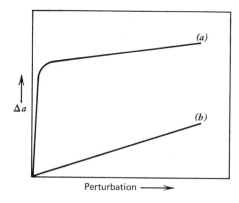

Figure 6. Qualitative representation of the effect of an asymmetric perturbation on the ESR spectra of (a) m-dinitro-benzene anions and (b) semiquinones. Δa is the difference in proton hyperfine coupling for the semiquinones as well as the difference in the ^{14}N hyperfine coupling for m-dinitrobenzene anions.

The first type of behavior is useful for comparing the effect of cations, solvent, temperature, and substituents on the structure of ion pairs [41, 42]; the second is useful because its "all-or-nothing" character enables us to study different types of perturbations such as those induced by solvent molecules [72].

The reason for these differences probably lies in the electronic structures of the anions, although it is not clear to what extent the ability of the nitro-group to be twisted out of the aromatic plane may contribute. The electronic problem may be illustrated by considering an anion, such as that of *sym*-trinitro-benzene, having the unpaired electron in a twofold degenerate orbital. Such an ion has a natural tendency to distort its symmetry (the Jahn-Teller effect), and this will be aided, and "fixed" by a cation (or strongly bonded solvent molecule) near to a site of high negative charge density. For meta-dinitrobenzene anions no such orbital degeneracy exists, but nevertheless there must be two relatively low-lying unsymmetrical equivalent orbitals which concentrate negative charge on one or the other nitro-group (Fig. 7). Although this orbital pair is less stable than the symmetrical orbital, the *total* energy is lowered if one of these levels is utilized, because of the strong ion pairing or solvation, which serves to fix the distortion. Since a completely

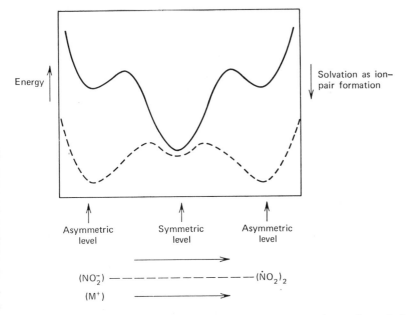

Figure 7. A diagramatic representation of the effect of solvation or ion-pair formation on the structure of *m*-dinitrobenzene anions. The solid line represents the weakly solvated or "free" ion and the dashed line represents the strongly solvated anion or ion pair.

different molecular orbital has been used, this effect is not a simple pertur- bation such as that found for the semiquinones. No doubt the new level is also more or less perturbed, but the major effect is the movement of the electron out of the symmetrical into one of the unsymmetrical orbitals. [The minor perturbation of the new level shows up in some of the trends in $a(^{14}N)$ shown in Fig. 4 and Table 6.]

Many other systems show these effects notably the pyrazines [37, 38, 73], various dicyano aromatic anions [36, 74] the pyracene anion [21, 75], the acenaphthene anion [76], and the anion of 5,5,10,10-tetramethyl-5,10- dihydrosilanthrene [77].

Paradinitrobenzene anions resemble the meta-compounds, but it is not yet clear to what extent the perturbation is of the "all-or-nothing" kind. This is partly because the anion is less stable and partly because spectra characterized by line-width alternation are generally obtained, rather than those characteristic of tied-down species [78]. The significance of this state- ment will be outlined in detail in the next section.

4. CHANGES IN LINE-WIDTH

One of the most useful—and to many of us unexpected—rewards of ESR studies is the information available from the widths of the component lines. This width may be a result of unresolved hyperfine interactions, but in the liquid phase it more commonly stems from various magnetic field modulations caused by the motions of the molecules in the investigated solution. Thus the width is essentially a kinetic parameter.

In any complete discussion it is convenient to divide all the possible effects into two major classes: those that govern the lifetime of a given spin state, and those that cause fluctuations in the energies of these states. The first is controlled by a characteristic time, T_1, which is often termed the spin-lattice relaxation time, or the longitudinal relaxation time. The second class is controlled by T_2, which is the spin-spin or transverse relaxation time. However, since our concern here is with the chemical implications of the studied effects, we prefer to discuss them simply in terms of the line-width, which is controlled by a characteristic time, T_2', containing possible contri- butions from both T_1 and T_2.

Of the various results of molecular tumbling, perhaps the most important for the radicals of the type presently under consideration is an asymmetric broadening which stems from the anisotropy in their g- and A-tensors. These are not of great importance for hydrocarbon radical anions, or for most radicals showing only proton hyperfine coupling. The effect does appear, however, when hyperfine coupling to such nuclei as ^{13}C, ^{14}N, or

^{19}F is important. The way in which the line-widths behave can be most readily understood by reference to Fig. 8, in which a coupling to a single ^{14}N in a typical π-radical is depicted. In the solid state, the envelope spectrum of the frozen solution is characterized (if the tensors have axial symmetry, as is often the case) by a shoulder (the parallel feature) and a peak (the perpendicular feature) for each hyperfine component. The greater the field separation between these parallel and perpendicular components, the greater will

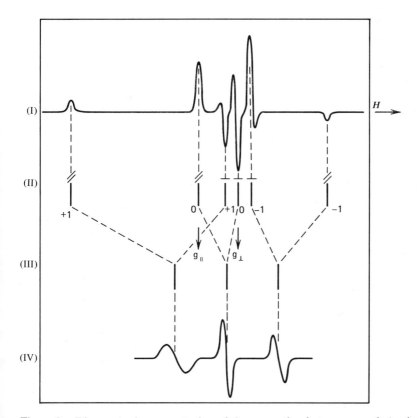

Figure 8. Diagrammatic representation of the connection between g- and A-anisotropy and the line-width of the solution ESR spectrum for a radical showing coupling to one ^{14}N nucleus. (i) Solid state (powder spectrum). (ii) Anisotropic hyperfine coupling, parallel —perpendicular—$\perp$. (iii) Isotropic hyperfine coupling. (iv) Liquid-phase spectrum. This figure illustrates the way in which line-widths of solution spectra can be used to derive the sign of the isotropic coupling. In the case illustrated, since $g_{\parallel} > g_{\perp}$ and the ^{14}N anisotropic coupling is assumed to be $+2B$ ($\parallel$) and $-B$ ($\perp$) the form of the broadening ($+1 > -1$) shows that $(A_{\parallel}) > (A_{\perp})$ and hence A_{iso} is positive. [In this and subsequent figures showing ESR spectra, we display first derivatives of the absorption as a function of field H. The maxima and minima are points of maximum and minimum slope.]

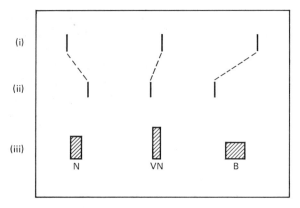

Figure 9.   Effect of rapid interconversion on the spectrum of a paramagnetic anion containing one ^{14}N and existing in two states (A and B) having different g-values which could, for example, be two types of ion pair. The figure represents the situation when the concentration of A and B are equal. (i) Species A. (ii) Species B. (iii) Averaged spectrum. (N = narrow, VN = very narrow, B = broad.)

be the width of the time-averaged line in the liquid-phase spectrum. This is, in fact, quite general. The greater the field swept during the tumbling motion, the greater the opportunity for spin relaxation and hence the broader the line. With increase in temperature, the tumbling rate increases and the line becomes narrower.*

Any other factor that modulates the position of a resonance line at a rate which is fast compared with the inverse of the difference between the extreme positions of the line will give such a broadening. For example, solvent fluctuations could have this effect, or rapid equilibria between types of ion pair or between loose ion pairs and free ions. Differential broadening would result if the individual species had different g-values as well as different A-values, as can be seen in Fig. 9. [This could be distinguished from the broadening due to rapid tumbling (Fig. 8) by reference to the solid-state spectrum.]

If we consider a simple case in which there is a rapid equilibrium between the two species, A and B, then, as the rate of equilibration increases (e.g., on heating), the individual line (or lines) associated with A and B will

---

* If the solid-state spectrum is known, the hyperfine A and g-tensors can be derived [13]. Again, if we can predict the sign of the purely anisotropic part of the A-tensor, then that of a_{iso} generally can be deduced. Similarly, if the form of the g-tensor can be safely predicted, the sign of a_{iso} can be deduced from the form of the broadening in the liquid-phase spectrum as shown in Fig. 8. This illustrates, in a simple manner, the rather complicated formalism given by others [79, 80, 81].

broaden, coalesce, and become a weighted-mean single line (or lines), which will ultimately become very narrow. This situation is depicted in Fig. 10. In the slow-exchange region, the broadening is approximately given by $\delta_{slow} \propto h\omega^2$, and is independent of Δ. However, in the fast-exchange region, the width depends upon Δ, and is given by $\delta_{fast} \propto \Delta^2/\omega$. In the intermediate region the dependence is more complicated, but conditions usually can be arranged so that the slow or fast ranges are being monitored.

Clearly, if a range of different nuclei is involved, different Δ values will occur. This will not affect the broadening in the slow-exchange region, but it will markedly affect the relative widths in the fast-exchange region. Some idea of the complexity that can arise can be gathered from Fig. 11.

Various equilibria that give rise to such changes are

$$(M^+A^-)_t \underset{\longleftarrow}{\overset{a}{\longrightarrow}} (M^+A^-)_t \underset{\longleftarrow}{\overset{b}{\longrightarrow}} M^+ + A^- \tag{1}$$

$$M^+ + A^-M^+ \rightleftharpoons M^+A^- + M^+ \tag{2}$$

$$M^+A^- \rightleftharpoons A^-M^+ \tag{3}$$

$$A + M^+A^- \rightleftharpoons A^-M^+ + A \tag{4}$$

$$A^- + M^+A^- \rightleftharpoons A^-M^+ + A^- \tag{5}$$

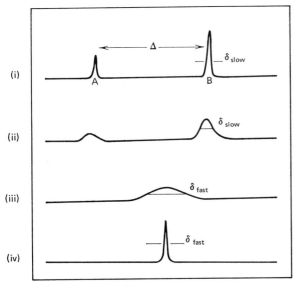

Figure 10. The effect of increasing the rate of exchange between two species, A and B, each of which, in the slow exchange region, gives a separate single ESR line. $\delta_{slow} \propto \omega^2$ when $\omega \ll \Delta$ and $\delta_{fast} \propto \Delta^2/\omega$ when $\omega \gg \Delta$. (i) Very slow exchange. (ii) Slow exchange. (iii) Fast exchange. (iv) Very fast exchange.

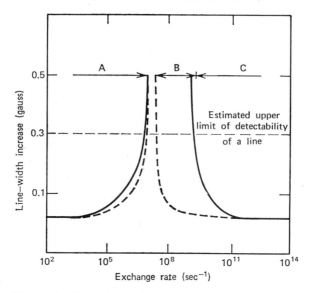

Figure 11. Dependence of the line-width on the rate of exchange between two species. The solid line represents the line-width increase for two lines separated by 9 gauss and the dashed line represents the line-width increase for two lines separated by 0.8 gauss. A is the slow-exchange region (static spectrum), C is the fast-exchange region (averaged spectrum), and B is an intermediate region where lines separated by 9 gauss or more will not be detected but lines separated by less than 9 gauss might be detected.

These equilibria are discussed in the following sections.

4.1. $(M^+A^-)_t \underset{\longleftarrow}{\overset{a}{\longrightarrow}} (M^+A^-)_l \underset{\longleftarrow}{\overset{b}{\longrightarrow}} M^+ + A^-$

The first part of this equilibrium represents equilibration between structurally different ion pairs. In the slow-exchange region, both pairs will be detected, and of the differences expected, that of the hyperfine coupling to M^+ is likely to be most marked. Hirota [40] found that anthracenide in DEE gave two superimposed spectra, one with a large lithium hyperfine coupling and the other without a lithium hyperfine coupling but with proton splittings that were different from the normal free-ion values. He suggested that this system comprised a slow equilibrium between a tight ion pair and a solvent-separated ion pair.

Hirota had reported an apparently more definitive result in which two sets of metal couplings were observed for sodium naphthalenide in DEE [82]; but later expressed considerable reservations regarding these results [40]. It is particularly significant that Tuttle and others [83, 84] have shown that, unless precautions are taken, potassium salts present as impurities on glass surfaces or in the metalic sodium, may exchange with ion pairs containing sodium to give appreciable concentrations of ion pairs containing potassium. It seems probable that in Hirota's earlier work [82] on sodium-naphthalenide in DEE, the small coupling assigned to loose sodium ion pairs came from potassium ion pairs.* Clearly care must be taken, and at present any results that show two sets of quartet couplings in which the ratios of the coupling constants are close to those expected for ^{23}Na and ^{41}K must be treated with caution.

The simultaneous presence of two types of ion pairs of sodium naphthalenide was demonstrated by Höfelmann, Jagur-Grodzinski, and Szwarc [28]. Sodium naphthalenide forms tight ion pairs in THP ($a_{Na^+} = 1.23$ G at 20°C), but loose, glyme separated ion pairs ($a_{Na^+} = 0.38$ G) are formed on addition of tetraglyme. Both pairs are seen in the spectrum at low-glyme concentrations; the intensity of the lines due to the latter species increases with increasing glyme concentration and eventually, for glyme concentration greater than 0.2M, the spectrum reveals the presence of loose pairs only.

The results made possible the determination of the equilibrium constant ($\sim 200M^{-1}$)

$$(Na^+, N^-)_t + glyme \rightleftarrows (Na^+, glyme, N^-)_{loose}$$

and of the rate constants of tight-pair solvation by glyme ($\sim 10^7$ M^{-1} sec^{-1}) and of collapse of loose pair into tight ($\sim 10^5$ sec^{-1}). The rate of solvation by external agent (glyme) is slower than for the solvation caused by the bulk of the solvent, provided the latter does occur.

The loose pair could not be attributed to traces of potassium. This is evident from the way of its preparation. Moreover, the experiments were repeated, with the same results, using sodium prepared by decomposing recrystallized sodium azide [136]. Finally, it was shown that ESR spectrum of potassium naphthalenide in THP does not reveal any splitting due to K^+ [136, 141] (see also Table 2).

Another piece of evidence for the presence of two types of ion pair is the line-width effect observed by Hirota [82, 85, 86]. In the fast-exchange region, the ESR spectra of two types of ion pair with different metal splittings should exhibit selective broadening of the metal hyperfine coupling as depicted in Fig. 12. The situation is more complicated if the g-values or anion a-values differ for the two species since then the form of this modulation (broad-narrow-narrow-broad; B-N-N-B) will cease to be symmetrical [87, 88]. Even if the metal hyperfine coupling is zero in $(M^+A^-)_t$, the same broadening pattern will result, provided the same cation is involved in the equilibrium between the two types of ion pair. Such broadening has been observed by

---

* When sodium was prepared by the decomposition of NaN$_3$ only one set of lines was observed in DEE even at −90°C [136]. [Edit.].

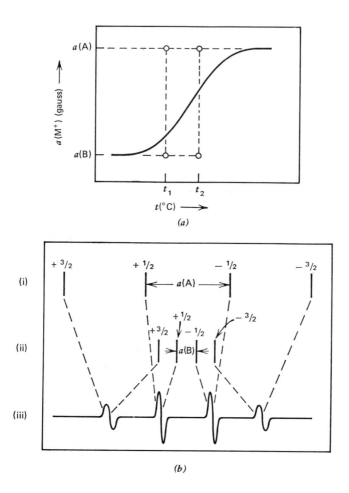

Figure 12. (*a*) Expected temperature dependence of the metal hyperfine coupling for two ion pairs A and B in rapid equilibrium. At low temperature *t*, the concentration of A [with metal hyperfine coupling $a(A)$] is small and the concentration of B [with metal hyperfine coupling $a(B)$] is large. The reverse occurs at higher temperature t_2. (*b*) Line-width modulation for two ion pairs in equilibrium and showing hyperfine coupling to a cation having $I = \frac{3}{2}$ and metal hyperfine coupling constants of $a(A)$ and $a(B)$. In this example, the concentrations are taken as equal. (i) Species A in slow-exchange region. (ii) Species B in slow-exchange region. (iii) Averaged spectra for A $\rightleftharpoons$ B in fast-exchange region.

Hirota for sodium anthacenide in MTHF or DEE [82, 86], sodium 2,6-di-*t*-butylnaphthalenide in THF [86], and sodium napththalenide in DEE [82] and by Szwarc and co-workers [28] for sodium naphthalenide in THP with added tetraglyme (no specific broadening was observed in the absence of tetraglyme). A rather different example is that for ion pairs of 2,6-dimethyl-*p*-benzosemiquinone, where the predicted equilibrium involves migration of the cation between the two oxygen atoms [42, 89]. These sites are inequivalent; thus the metal coupling is expected to vary and hence in the fast-exchange region selective broadening results (Fig. 13).

In Section 2.4 we mentioned that an *s*-shaped curve (see Fig. 3) for the temperature dependence of the metal hyperfine coupling suggests the occurrence of a rapid equilibrium between distinct species (compare Fig. 12). Hirota [85] has considered that the two plateaus that are linked by the *s*-shaped curve correspond to the metal hyperfine coupling constants of the two distinct ion pairs, A and B. The metal hyperfine coupling at any temperature, $a(M^+)$, is then related to the equilibrium constant K by

$$a(M^+) = a(A) + \frac{a(B)K}{1 + K}$$

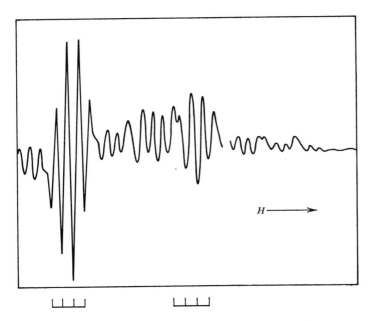

Figure 13. Half-field ESR spectrum of sodium 2,6-dimethylbenzosemiquinone in *t*-AmoH at 40°C showing selective broadening of the metal hyperfine coupling due to rapid equilibrium between the two ion-pair species [42].

where $a(A)$ and $a(B)$ are *temperature-independent* metal hyperfine couplings of the two ion-pair species A and B.

Hirota [40, 82, 85, 86] showed that plots of $\log K$ versus $1/T$ are linear, an observation which supports the proposed model. This model was supported by recent studies of the temperature dependence of g-values [141]. Assuming that the variation in the observed g-value results from the rapid equilibrium, tight pair $\rightleftarrows$ loose pair, in which the proportion of the two types of ion pair vary with temperature, one may show that $\Delta g(T)$ should be a linear function of $a_{\text{cation}}(T)$, provided that g of the tight pair and of the loose one are temperature independent. Such a linear relation was reported for sodium naphthalenide in THF but not for cesium naphthalenide [141]*. It was concluded, therefore, that the existence of two types of ion pair is responsible for the variation of a_{Na^+} and Δg in sodium naphthalenide system, but other factors are responsible for the variation of a_{Cs^+} in cesium naphthalenide.

The temperature independence of a_{cation} was questioned [28]. This assumption may affect the reported ΔH and ΔS values. For example, Hirota et al. [107] reported $\Delta H = -5.6$ kcal/mole and $\Delta S = -24$ eu for the conversion of tight sodium naphthalenide pairs in THF into loose ones, whereas Szwarc et al. [143] obtained $\Delta H = -6.9$ kcal/mole and $\Delta S = -32$ eu for the same process by utilizing data derived from the equilibrium studies of electron-transfer. Both groups were led to the same value of the equilibrium constant at $-70°C$. Consideration of the numerical value of ΔS [143] suggests that the thermodynamic results reported by Szwarc are more reliable than those of Hirota.

Another example of equilibrium involving different types of ion pair is provided by the sodium salt of triphenylene, which gives three types of solution ESR spectra [90] depending on the solvent and the temperature. These spectra have been interpreted in terms of rapid equilibria between three distinct species, the free ion, a solvent-separated ion pair (without sodium hyperfine coupling), and a contact ion pair (with sodium hyperfine coupling). The equilibrium between the contact ion pair and the solvent-separated ion pair gave the characteristic *B-N-N-B* line-width effect of the sodium hyperfine coupling, and analysis of the s-shaped temperature dependence of the sodium hyperfine coupling gave a linear plot of $\log K$ versus $1/T$. The analysis of the temperature dependence of the sodium hyperfine coupling in sodium acenaphthaleneion pairs in THF also gave a linear plot of $\log K$ versus $1/T$ [91] interpreted in terms of an equilibrium between tight and loose ion pairs.

---

* The linear relation is still expected if a (tight) and Δg (tight) are slightly temperature dependent. [Editor].

Further support for the preceding model comes from the marked temperature dependence of the ^{13}C hyperfine coupling of the carbonyl group in fluorenone ketyl ion pairs. Hirota [22] proposed that the decrease in ^{13}C hyperfine coupling on cooling arises from a change in the ion-pair structure, one structure being a tight ion pair and the other a looser ion pair with more solvation, the latter being favored at low temperatures. A decrease in the ^{13}C hyperfine coupling certainly suggests an increased separation between cation and anion.

If, on cooling or altering the medium, it is not possible to resolve the spectrum into separate lines of $(MA)_t$ and $(MA)_l$ (the slow-exchange region), then it is difficult to distinguish between equilibrium (reaction 1a) and a range of equilibria involving, in effect, an infinite number of possible species. This general perturbation system could also, in principle, give rise to the broadening observed (B-N-N-B), but on cooling lines would simply continue to broaden rather than resolve out into those due to two (or more) distinct ion-pair species in slow equilibrium. In the event that there are two limiting structures with a wide range of intermediate structures, the familiar S-shaped curve would still be obtained.

Differentiation between the dynamic model, postulating equilibrium between two types of ion pair, and a static model has been achieved in the system cesium biphenylide in glyme [144]. The results indicated that static system, continuous change of the structure of ion-pair, accounts better for the experimental observations (see Chapter 7).

The second half of reaction 1, dissociation of the ion pairs into separate solvated ions, will manifest itself in the slow-exchange region as separate ESR spectra of the ion pairs and A^-. Often the individual A^- lines appear between the two central lines of the M^+ multiplets, although if there are differences between the anion coupling constants or g-values of A^- and M^+A^-, these will not be centrally placed.

If the rates of dissociation and association are increased, the individual lines will all broaden and merge to give eventually lines characteristic of A^- but somewhat shifted to the weighted-mean positions (Fig. 14). In this case, although ion pairs which normally would be expected to show metal hyperfine coupling are present, no such coupling is detected and, indeed, it may be difficult to deduce their presence by inspection of the spectra. A good example is the lithium salt of azulene in DME [8], which shows no lithium hyperfine coupling, whereas the sodium salt shows coupling to a sodium nucleus [9]. By studying the temperature dependence of the proton hyperfine coupling Reddoch has inferred that the lithium-azulene ion pair is in fast equilibrium with the free ion. Similar studies have been made with the alkali metal salts of acenaphthylene [10] and of anthracene [9].

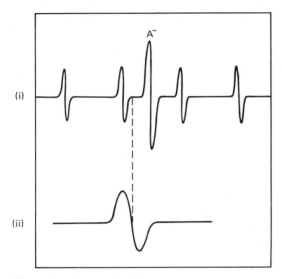

Figure 14. Hypothetical ESR spectra for the equilibrium $M^+A^- \rightleftharpoons M^+ + A^-$, the cation having $I = \frac{3}{2}$, In this example the concentrations of M^+A^- and A^- are equal but the species have different g-values. (i) Slow exchange. (ii) Fast exchange.

In practice, many simple salts in pure solvents appear to be in the slow-exchange region, so that separate sets of lines are detected. Some examples are sodium naphthalenide in THF [7], sodium durosemiquinone in DME [92], sodium nitrobenzenenide in DME/DMSO [33], and the cesium-pyracene ion pair in DME [21]. If an excess of inert electrolyte with the same cation is added, the fast exchange case can be achieved, but the process is then reaction 2, which is the bimolecular equivalent of reaction 1b (p. 225).

In certain systems both parts of reaction 1 may be occurring simultaneously, in which case these factors must all be combined. The real situation is often simplified, since reaction 1a is usually fast and 1b slow.* In that case, it is not possible to decide from the spectra which of the two ion pairs is actually responsible for the dissociation process, and the ion pairs can be treated as if they were one entity. [This is a good example of a reaction in which it would be unwise to calculate $\Delta S^{\ddagger}$ and $\Delta H^{\ddagger}$ for reaction 1b if proper allowance for 1a cannot be made. This is further discussed in Section 4.3.]

It may not be easy to distinguish between a spectrum due to loose ion pairs having no detectable metal coupling and one due to "free" ions. If it is not possible to infer the structure from the ESR data, then perhaps the best test

---

* Frequently the dissociation of loose, solvated ion pairs (1b) is faster than their collapse into tight pairs, which requires extensive desolvation. [Editor].

is simply the law of mass action. Equilibria between ion pairs will be independent of total concentration [93], whereas those between ion pairs and free ions will favor the free ions on dilution. This dilution effect has been observed in many instances [7–9, 92].

Alternatively, the presence of free ions may be established by conductance studies. Such investigations [135] cast doubts on earlier claims of free ions being present in some systems (e.g., [7]) and imply that, after all, loose ion pairs were confused with free ions.

4.2. $M^+ + A^-M^+ \rightleftharpoons M^+A^- + M^+$

Normally, provided the triple-ion $M^+A^-M^+$ is not stable, the major effect of this equilibrium is to cause a broadening and ultimately a loss of the hyperfine features associated with coupling to the metal nuclei as the concentration of M^+ is increased. For example, Hirota and Weissman [94] have shown that when sodium iodide is added to sodium xanthone ketyl, the spectrum of the ion pair broadens and ultimately changes to one characteristic of the "free" ketyl. This effect can be used as an aid to spectral interpretation in cases where many overlapping lines are involved. Further, if free A^- is present, then the concentration of A^- will of course fall as salt is added. This could be used as a method of distinguishing between A^- ions and M^+A^- pairs, which give no metal hyperfine coupling. Any kinetic study must include the fact that the actual concentration of "free" M^+ will be low in solvents that encourage ion pairing with A^-, and ion-pair equilibria involving diamagnetic anions from added salt must be taken into account. This is a rather difficult correction to apply [94, 95].

If two different cations, M_1^+, and M_2^+, are used, then, in slow exchange, the individual spectra for both ion pairs will be detected. As stressed previously, care must be taken to distinguish between this situation and that involving two different types of ion pair having a common cation. As the exchange rate is increased, all lines will broaden to give, ultimately single lines so that the spectrum again resembles that of A^-, with slightly modified a- and g-values. An example of slow exchange was observed by Adams and Atherton [96], who added sodiumtetraphenyl boron to cesium-m-dinitrobenzenide in DME and obtained superimposed spectra from both sodium and cesium-m-dinitrobenzenide ion pairs. It is significant that they did not obtain the mixed triple ion (Section 2.4). An example of fast exchange was observed by Ward and Weissman [97] when potassium iodide was added to sodium naphthalenide in DME.

If an anion has two or more binding sites, and triple ions are not detected, displacement via the symmetrical intermediate (Scheme VIIIa) is likely to be preferable to that involving only one of the anion sites (VIIIb), and this has indeed been established in certain cases [95, 98]. The test derives from the fact that displacement via Scheme VIIIa on the addition of salt would have

$$M^+ + O\!=\!\!\langle\ominus\rangle\!=\!OM^+ \;\rightleftharpoons\; M^+O\!=\!\!\langle\ominus\rangle\!=\!O + M^+$$

VIIIa

VIIIb

the same effect as an increase in the cation-migration rate (Section 4.3). This acceleration should and does [95, 98] parallel the rate of cation displacement as measured from the widths of the hyperfine components associated with the cation.

There is an alternative reaction, namely, $A^{\bar{}}$, $M_1^+ + M_2^+X^- \rightleftharpoons X^-M_1^+ + A^{\bar{}}$, M_2^+, that produces ESR changes similar to those anticipated for the reaction $M^+ + A^-M^+ \rightleftharpoons M^+A^- + M^+$. Differentiation between those alternatives was achieved by Rutter and Warhurst [147], who investigated the spectral changes resulting from the addition of NaI or NaBPh$_4$ to sodium-2,5-di-t-butyl-p-benzosemiquinone solution in THF. The original spectrum of doublet of doublets, characterizing the asymmetrical ion pair, broadens on the addition of NaI and eventually, at higher concentration of the salt, it collapses into a triplet spectrum expected for the symmetrical cationic field on the anion. The dissociation of NaI is minute in THF at concentrations of about 10^{-3} M or higher; hence, for the reaction involving the undissociated ion pair, the lifetime of the semiquinone pair should be proportional to [NaI], but is should be proportional to [NaI]$^{1/2}$ had the reaction involved free Na$^+$ ions. The results (for 40-fold variation in the concentration of the iodide) clearly demonstrated that the exchange involves the undissociated NaI. At 20°C the rate constant of exchange is 7.6×10^8 M^{-1} sec^{-1} and the Arrhenius parameters are E $\sim$ 2.5 kcal/mole and A $\sim 6 \times 10^{10}$ M^{-1} sec^{-1}.

When NaBPh$_4$ is added instead of NaI the spectrum of the semiquinone again broadens but, in addition, a triplet spectrum with another g-value appears. The latter is attributed to triple ions Na$^+$Q$^-$Na$^+$. It is proposed that the triple ion is an intermediate in the exchange process, that is,

$$Na^+Q^- + Na^+ \rightleftharpoons Na^+Q^-Na^+ \rightleftharpoons Na^+ + Q^-Na^+$$

4.3. $M^+A^- \rightleftharpoons A^-M^+$

This equation is meant to symbolize migration of the cation between different binding sites in the anion. If these sites are equivalent, as in *p*-benzosemiquinone and durosemiquinone, then the situation is particularly simple and, in the fast-exchange region, it is manifested by a broadening of alternate lines throughout the spectrum. The situation for two equivalent nitrogen atoms (as in *m*-dinitrobenzene) is illustrated qualitatively in Fig. 15. This is, of course, simplified by the omission of any ring-proton components. In fact, contributions from protons that are equivalent in the symmetrical anion will also show this alternation of widths in the ion pair when cation migration occurs at the appropriate rate [99] (see Fig. 11).

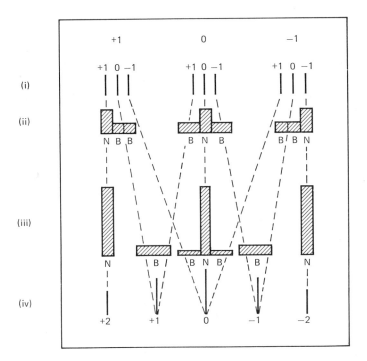

Figure 15. Hypothetical spectra for a radical anion with two equivalent ^{14}N nuclei ($I = 1$). However, on association with a cation which induces a fluctuation in the ^{14}N hyperfine coupling the coupling constants of the two ^{14}N become different. (i) Slow exchange. (ii), (iii) Intermediate exchange. (iv) Fast exchange. (N = narrow, B = broad.)

The situation is much more subtle when inequivalent sites are involved. The only example that has been thoroughly studied is that of 2,6-dimethyl-*p*-benzosemiquinone [42, 56, 89]. Two extra factors must be considered. One is the fact that the unhindered oxygen is strongly preferred by the cation and the other is that simple alternation can no longer be expected. In the general case, all but the central line (if any) is expected to broaden to a greater or lesser extent in the fast-exchange region. In practice, there may fortuitously be other lines whose position is hardly modified by the migration, which also appear narrow in the experimental spectra [42, 56, 89].

How, then, can we know that cation migration is actually occurring? The answer is that it is much more difficult to gauge this than the symmetrical case. The situation expected for the 2,6-dimethylsemiquinone is illustrated, relative to those for the two symmetrical anions, in Fig. 16. On cooling, spectra from the two different ion pairs might be resolved, and in one case, with the potassium salt, extra weak features, possibly stemming from the species having the cation at the hindered oxygen [100], have been detected. An alternative explanation for the various line-broadening effects observed has been proposed [56]; this envisages modulations caused by movement of the cation around the unhindered oxygen, with no tendency to migrate. For several reasons we are inclined to reject this suggestion, the most compelling being that the same movements or vibrations are free to occur for the symmetrical quinones, but no line-width effects stemming from such movements

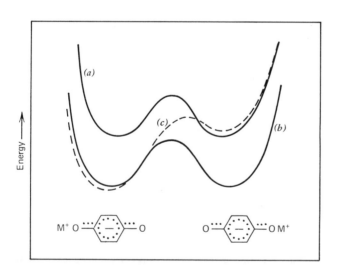

Figure 16. Potential energy contours for cation migration in semiquinone ion pairs. (*a*) Duroquinone. (*b*) Benzoquinone. (*c*) 2,6-Dimethylbenzoquinone.

have been detected. In fact, it seems most probable that these one-site fluctuations are too fast to be detected by ESR methods.

The way in which the cation moves is of interest, even if it is a somewhat esoteric question since it can only be the relative motion of the ions that is monitored by ESR. However, it has been pointed out [101] that there is no case for specifically picturing the effect as a movement of the cation alone. We might just as well fix the cation and imagine the anion revolving about one of its axes and, indeed, in cases where the cation is strongly bound into a viscous solvent network and the anion is not so held, this would probably be the most realistic extreme. Moreover, the relative movement could be either across the π-electron surface of the aromatic ring or around its periphery.

If the route is across the center of the ring, then there could well be another binding site of real stability with the cation in this position. This would result in an extra minimum in the configurational coordinate diagram in Fig. 16 and any contribution from this structure would need to be taken into consideration in the overall scheme. In the case of slow migration, all three species ought to be detected, that is, spectra from two inequivalent ion pairs ought to be found, one having a symmetrical anion and the other an unsymmetrical anion. The only instance that we know of in which this has been claimed is for a benzene derivative having two ortho-substituents [36], which seems to us to be the least likely structure to give such an effect. In fact, the authors claimed that one species was the symmetrical ion pair and the other was the unsymmetrical one under rapid-migration conditions. If, as seems likely for the ortho-dicyano anion being studied, the pathway for migration is via the envisaged symmetrical structure, how can one structure be long-lived while the other two are very short-lived? This could occur if the symmetrical species were much more stable than the others, but the relative intensities of the spectra do not support this concept. The spectra under consideration were extremely complicated and it may well be that there are equally acceptable alternatives which would not suffer from this drawback.

Rates of migration seem to be governed by a large number of factors, and it is not clear to what extent we can trust the breakdown of purely kinetic parameters obtained over a range of temperatures into enthalpies ($\Delta H^{\ddagger}$) and entropies ($\Delta S^{\ddagger}$) of activation. This is because a change in temperature influences more than just the rate under consideration. Such a problem is very frequently encountered in liquid-phase reactions, but it is underlined in ESR work by the fact that hyperfine constants are often changing markedly in the temperature range under consideration. This means that equilibria are changing and if these are in any way connected to the process whose rate is being studied, there will be a change in rate quite apart from the $\Delta H^{\ddagger}$ and $\Delta S^{\ddagger}$ terms being sought. For this reason, despite the extra insight gained in understanding rate processes by measuring $\Delta H^{\ddagger}$ and $\Delta S^{\ddagger}$, we prefer to discuss changes in rates at a fixed temperature.

The overriding factors governing the rate of migration of the cation between two binding sites is likely to be the strength of the binding. The solvent will play an important role in that it competes with the anion for binding to the cation and insofar as it weakens the cation-anion binding it will cause an acceleration in the migration rate. This is well illustrated by the results for semiquinones [100], which show that as the difference in proton hyperfine coupling (which is a measure of the perturbation) falls on changing the solvent so the rate of migration increases.

This is also illustrated by the frequently observed steady trend of an increase in migration rate on going from lithium to cesium. A range of typical results are summarized in Table 9. (Some examples of differences in proton hyperfine coupling induced by ion-pair formation were given in Table 6.)

Line-width effects from cation migration have been observed for the ion pairs of other anions with two binding sites such as pyracene [21, 75], 2,2'-dinitrobiphenyl [102], o-dinitrobenzene [99, 103], p-dinitrobenzene [99], terephthalonitrile [74], phthalonitrile [36], 5,5,10,10-tetramethyl-5,10-dihydrosilanthrene [77], and acenaphthene [76].

If line-width alternation is detected for an anion having two equivalent binding sites, but there is no detectable coupling to cation nuclei, then alternation may still be the result of cation migration, but the cation may be so well solvated that it is effectively insulated from the electron.

Extensive studies of Arrhenius parameters characterizing the jump of cations between two equivalent bonding sites have been recently reported by Warhurst [148]. Di-t-butyl-p-benzosemiquinone was chosen for these investigations because its ESR spectrum is so simple. The study of alkali salts (Li^+, Na^+, K^+, Rb^+, and Cs^+) in several ethereal solvents led to rather complex results. The ESR spectra did not show coupling to cations (Cs^+ being the exception); nevertheless, Warhurst believes they are virtually of contact type. The trend in the rate constant with increasing size of the cation was confirmed. This result arises mainly from increase in the A-factor which varies more extensively than the activation energy. The latter increases with size of the cation in THF but decreases in MeTHF.

The discussion of the transition state stresses the changes in the solvation of O^- sites, whereas it seems that the variation in the degree of solvation of cations is more important. It appears that it is too early to judge finally this situation and more data are needed. Only limited information on this subject was reported by other workers [164].

An interesting example of linewidth alternation caused by the movement of cations was reported for barium salt of acenaphthene semiquinone in DME [149]. Two semiquinone molecules are associated with one Ba^{2+}

cation which presumably is well solvated. Hence the interaction between both odd electrons is weak. The spectra show slow exchange at 0°C but fast at 80°C. It is suggested that the four oxygens form apexes of a tetrahedron, and Ba^{2+} ion moves from one of its faces to another.

Alternatively, the line-width effect could be caused by asymmetric solvation. This is rather a different phenomenon, since the movement is now caused by solvent fluctuation in which one set of solvent molecules becomes organized about one of the substituents while another set relaxes its orientation about the other substituent. Only in the case of symmetrical *m*-dinitrobenzene anions and some of its derivatives has this been observed [72, 150]. In alcohols the lifetime of the solvate is such as to give alternate broadening [72, 150], but in water this time is so great that a slow-exchange situation is achieved [104]. Added electrolytes have no further effect upon the line-widths, and there can be little doubt that this is purely a solvent effect, unaided by the cations.

This solvent barrier to interchange between the two unsymmetrical "excited states" of *m*-dinitrobenzene anions is very similar indeed to the barrier to intermolecular electron-transfer discussed in the following section, and, in some sense, it is identical with the barrier to intramolecular electron-transfer in such systems as bridged dinitrobiphenyl anions studied by Harriman and Maki [105] and Gupta and Narasimhan [106].

It is interesting to speculate on the way in which triple-ion formation ($M^+A^-M^+$) would affect migration phenomena. On any theory, migration would be greatly inhibited, and exchange via, for example, rotation of the anion would be difficult to detect since there is no asymmetry to monitor.

Gough and Hindle [53] did observe a new, interesting line-width effect, but this seems to be related more to the tumbling phenomena mentioned in Section 4 (cf. Fig. 8) rather than to cation migration. They suggest that the broadening found across the sodium septets stems from anisotropy in the *g*-tensor and in the ^{23}Na hyperfine tensor. This is the first time that such broadening of cation features has been observed. It is most unexpected since the outer *p*-orbital contribution on sodium is usually negligible. Indeed, as we have stressed in Section 2, anisotropy probably stems indirectly from spin on the anion. However, by assuming that $g_\parallel > g_\perp$ and that $a_{iso}(^{23}Na)$ is positive for durosemiquinone and negative for *p*-benzosemiquinone, the line-width effect suggests that the cations lie along the molecular axis.

4.4. $A + M^+A^- \rightleftharpoons A^-M^+ + A$

This well established electron-transfer process, first studied by Weissman and co-workers [6, 97], is discussed fully by Jagur-Grodzinski and Szwarc in the second volume of this book. However, since it serves to illustrate many interesting facets of ion-pair formation and solvation it deserves a brief mention here. It has been described as an atom ($M \cdot$) transfer, but since

Table 9 Cation Migration

Anion	Cation	Temperature (°C)	Solvent	Rate of Modulation[a]	ΔH^b	ΔN^c	Ref.
Benzoquinone	H+	25	t-AmOH, DME, THF	Static	5.62(t-AmOH)		d
	Li+	25	t-AmOH, t-BuOH	Slow	1.38(t-AmOH)		
	Na+	25	t-AmOH, t-BuOH, DME	Intermediate	—		
	K+	25	t-AmOH, t-BuOH, DME	Fast	0		
	Rb+	24	DME	Fast	0		
	Rb+	−17	DME	Slow	0.80		
	Rb+	−27	DME	Slow	0.80		
	Rb+	−60	DME	Static	0.80		
Duroquinone	H+	25	t-AmOH, THF	Static	5.70(t-AmOH)		d
	Li+	25	DME, THF	Static	2.22(THF)		
	Li+	25	t-AmOH	Slow	1.98		
	Na+	25	THF	Slow	1.66		
	Na+	25	t-AmOH, DME	Intermediate	—		
	K+	25	t-AmOH, DME, THF	Fast	0		
	K+	−20	THF	Slow	1.22		
	K+	−40	THF	Slow	1.18		
	K+	−60	THF	Static	1.11		
2,6-Dimethyl-p-benzoquinone	H+	25	t-AmOH, DME	Static			d
	Li+	25	t-AmOH	Static			
	Na+	25	DME	Static			
	Na+	25	t-AmOH	Slow			
	K+	25	DME	Intermediate			
	K+	25	t-AmOH	Fast			
	Rb+	25	DME, t-AmOH	Fast			
	Cs+	25	DME, t-AmOH	Fast			
	Cs+	−90	DME	Static			

Compound	Bu$_4$N$^+$	Temp	Solvent (DME, t-AmOH)	Modulation[a]	ΔH[b]	ΔN[c]
2,6-di-t-Butyl-p-benzoquinone	H$^+$	25	t-AmOH	Static		
	Li$^+$	25	t-AmOH	Static		
	Na$^+$	25	t-AmOH	Static		
	K$^+$	25	t-AmOH	Static		
	Rb$^+$	25	t-AmOH	Static		d
	Cs$^+$	25	t-AmOH	Static		
Pyrazine	Li$^+$	RT	THF	Static	1.35	3.32
	Na$^+$	RT	THF, DME	Fast	0	0
	Na$^+$	−32	THF, DME	Intermediate	—	—
	K$^+$	RT	THF, DME	Fast	0	0
	Rb$^+$	RT	THF, DME	Fast	0	0
	Cs$^+$	RT	THF, DME	Fast	0	0 e
m-Dinitrobenzene	Na$^+$	20	DME	Static	0.7	9.36
	K$^+$	20	DME	Slow	0.8	8.74
	Rb$^+$	20	DME	Completely averaged	0	0 f
	Cs$^+$	20	DME	Completely averaged	0	0

[a] Static: lifetime is long compared to induced changes in the hyperfine coupling. Slow: modulation at a rate just sufficient to broaden some of the static hyperfine lines. Intermediate: modulation at a rate such that lines are merging. Fast: modulation at a rate such that a coalesced spectra are observed with some lines broadened. Completely averaging: modulation at a rate sufficient for the observation of coalesced lines having equal widths.

[b] ΔH = difference in proton hyperfine couplings (for protons nearest to each bonding site).

[c] ΔN = difference between nitrogen hyperfine couplings.

[d] Ref. 100 and references therein.

[e] Ref. 37, 38, and 73.

[f] Ref. 33, 99, and C. Ling and J. Gendell, *J. Chem. Phys.*, **46**, 400 (1967).

M^+ is present continuously as a slightly modified cation, this seems to be a very restricting description. Furthermore, it draws too big a contrast between this reaction and the simpler electron-transfer

$$A + A^- \rightleftharpoons A^- + A \tag{4'}$$

which occurs side-by-side with reaction 4 in the appropriate media. [It might have been expected that the symmetrical species, A_2^-, would have some stability, especially since the corresponding cations such as $(C_6H_6)_2^+$ are readily formed. However, we know of no instance in which such dimer anions have been detected.]

In the slow-exchange region, all lines broaden, and the extra width increment gives the rate of electron transfer. In the limit of fast-exchange, a single line remains for reaction 4′ and this is split into a multiplet by M^+ in reaction 4. This result has an analytical use: ESR spectra are often very complicated and hard to analyze. By adding excess A, if metal coupling contributes, this will remain in the fast-exchange limit and can be estimated readily. This method of simplifying the spectrum is an alternative to that of adding an excess of salt which, under favorable circumstances, will remove the cation coupling, leaving the A^- features (Section 4.2).

These electron-transfer reactions must follow a pathway of the type depicted in Fig. 17. Once molecule A has reached the ion pair and there is a small overlap between the orbitals of A and A^-, the electron still cannot transfer until some solvent reorganization occurs to give the symmetrical

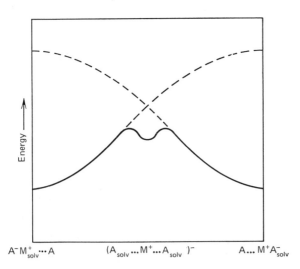

Figure 17. A diagrammatic representation of the movement of an electron from A^- to A in the collision complex A^-M^+A.

"intermediate," $(A_{solv} \cdots M^+ \cdots A_{solv})^-$, in which the electron can be on either of the A molecules. Any other mode of electron transfer, for example, one in which the state of solvation of the reagents remains unaltered, followed by relaxation of the solvent after the transfer is completed would contravene the law of microscopic reversibility. (This pathway might, however, be followed on optical excitation.)

We would therefore expect that, for a given A, the rate of electron transfer would increase as the anion-solvating power of the solvent decreased. Moreover, for the ion pairs, M^+ must necessarily be transferred, and this requirement may result in a lower rate, other things being equal. The results illustrate this point. When tight ion pairs are involved, the rate of electron transfer is usually at least a hundredfold less than that for the corresponding loose ion pairs or free ions [107].*

Electron transfer was observed also in a system involving triple ions, namely, Na^+, duro-semiquinone$^-$, Na^+ [151]. In the fast exchange limit the spectrum collapsed into $2Na^+$ septuplet indicating that both sodium ions are transferred. However, the sodium coupling constant increased by a factor of 3 on the addition of 2M duroquinone solution (from 0.66 MHz in the absence of duroquinone to 1.96 MHz). This was interpreted as an evidence for the formation of a complex $DQ - (Na^+, DQ^-, Na^+)$, where DQ denotes molecule of duroquinone. The rapid dissociation of this complex, coupled with even faster association, leads to a rapid exchange of quinone units.

4.5. $A^- + M^+A^- \rightleftarrows A^-M^+ + A^-$

It is of interest to compare reaction 4 or 4′ with that of spin-exchange (5). This occurs when two paramagnetic anions, A^-, approach to give slight overlap, under which conditions the electrons can exchange to give, at high enough concentrations, the same broadening of the hyperfine components of A^- and ultimately a single, central line. The major difference is that for A^- or M^+A^- units there is no solvent barrier, nor is there any need for M^+ to transfer, and hence the rate constants are far higher, being effectively diffusion controlled [108].

---

* This point deserves further comments. In a tight ion pair the cation is solvated on its "outside." In ethereal solvents this solvation is much more powerful than the solvation of the anion, especially for small cations such as Li^+ or Na^-. In the transition state of an electron transfer process either the cation becomes fully solvated, as in a loose pair, or the "outside" solvation shell is stripped off to permit a close contact of the cation with the acceptor (transition state of an exchange should be symmetric in respect to the donor and acceptor). In either case the additional reorganization of solvent molecules hinders the process. The reorganization is not required for a loose pair where the cation is fully solvated and, hence, this exchange may be as fast (or perhaps even faster) as the exchange involving the free ions (see refs. 28 and 18). [Editor].

In the particular case of A^- and M^+A^- interacting M^+ could well transfer, even though this is *not* a condition for spin-exchange. The triple ion $A^-M^+A^-$, which would exist as a spin-triplet or singlet, normally has too low a lifetime for detection. However, the aggregates $(M^+)_2(A^-)_2$ and $M^{2+}(A^-)_2$ frequently have long lives, and these are discussed in more detail in the following section.

5. ION CLUSTERS AND TRIPLET STATE SPECIES

Various aspects of ion-clustering have already been discussed, especially in Section 2.5. Here we are specifically concerned with spin-triplet species. Structures that have been studied include $(M^+)_2(A^-)_2$, $M^{2+}(A^-)_2$, and $(M^+)_2A^{2-}$, where A^{2-} contains two unpaired electrons. As with so much of the work discussed in this chapter, Weissman and co-workers pioneered this field. In contrast with most of the work thus far discussed, it is the solid state which generally provides the most useful information, and if this is to be extrapolated to give information about the solutions, care must be taken to ensure that no phase separation occurs on freezing. This can generally be avoided by the use of clear glasses.

The type of solid-state spectrum obtained, and its analysis, are indicated in Fig. 18. The interaction responsible for this is an asymmetric dipole coupling between the spins of the electrons, which lifts the degeneracy of the ± 1 and 0 levels within the triplet manifold. The general ESR properties of such triplet states have recently been reviewed [109].

Usually, hyperfine features are not resolved in these studies, although, by application of ENDOR techniques it ought to be possible to derive useful information. However, the overall splitting, labeled $2D$ in Fig. 18, is related to the parameter D in the equation for the zero-field splitting between the levels, and this, in turn, is related to the mean separation between the spins, $\langle r^{-3}\rangle^{-1/3}$, by $D = \frac{3}{2}g^2\beta^2\langle 1/r^3\rangle$. If the perpendicular features are split, as in curve b of Fig. 18, then this splitting is related to the E term of the spin Hamiltonian, which is a measure of the deviation from axial symmetry. The field values at which resonance occurs are given by the following equations:

$$H_{\parallel} = H_z = \frac{g_e}{g_z}(H_0 - D)$$

$$H_{\parallel'} = H_{z'} = \frac{g_e}{g_z}(H_0 + D)$$

$$H_{\perp}^2 = H_{xy}^2 = \left(\frac{g_e}{g_{xy}}\right)^2 H_0(H_0 - D)$$

$$H_{\perp}^2 = H_{x'y'}^2 = \left(\frac{g_e}{g_{xy}}\right)^2 H_0(H_0 + D)$$

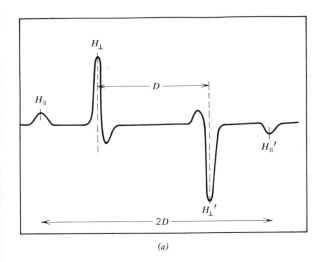

(a)

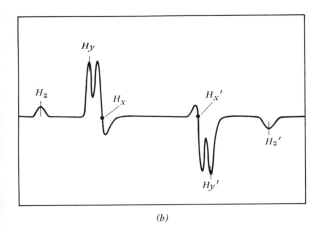

(b)

Figure 18. Typical solid-state spectra in the $g = 2$ region for triplet-state species. (a) $E = 0$. (b) $E > 0$.

For nonaxial triplets the equations are far more complicated, but if $E/D \ll$ 1 and $g_x = g_y = g_z$, then they simplify:

$$H_{y'}^2 - H_{x'}^2 \simeq 2E(3H_0 + 2D)$$
$$H_x^2 - H_y^2 \simeq 2E(3H_0 - 2D)$$

In addition to the allowed $\Delta m_z = 1$ transition in the $g = 2$ region, there is another formally forbidden line which can sometimes be detected in the $g = 4$, i.e. half-filed region. This absorption can loosely be described as the $\Delta m_z = 2$ transition. The intensity of this feature falls rapidly as the mean separation increases, and it is often undetectable in the work described here. Its main utility in this work is to prove that a triplet state is being studied especially when, as is often found, the high-field lines are seriously broadened.

Normally, no hyperfine coupling is resolved in the rather broad solid-state spectra. However, in some cases structure is detected for liquid-phase species. Two limiting cases arise with respect to hyperfine coupling. If spin exchange (Section 4.5) is rapid ($J \gg a$), each electron "sees" both anions and spends 50% of its time on each. Hence a hyperfine pattern is obtained which is characteristic of the A_2 unit with hyperfine coupling constants half the normal values.

If overlap is poor ($J \ll a$), exchange will be slow and then the hyperfine pattern is normal for A^-. Under exceptional circumstances, in which the exchange energy and the hyperfine energies are comparable, ($J \approx a$) extra features associated with transitions between the triplet and singlet states, which are normally forbidden, may be detected. These appear as satellite lines in the wings of the main $g = 2$ features [110–112], but they have not yet been found for the investigated radical clusters.

Results from studies on radical clusters fall into two main groups. One group is characterized by relatively strong coupling corresponding to the "effective" spin separation of about 5–7 Å, and well-defined solid-state lines are obtained of the type shown in Fig. 18. These give very broad single lines in the liquid phase because there is an efficient spin-relaxation mechanism which can be related to the range of field swept by the components of the fine-structure tensor as the unit tumbles. (This splitting averages to zero, but as with g- and A-anisotropy, the modulation induced by tumbling is a function of the magnitude of D.)

The other group gives no resolved features but simply a broad singlet in the solid state whereas the liquid-phase spectra are well resolved. This means (if the species really contains two A^- ions) that D must be very small (<15 G) and the spins must be separated by >10 Å.

Systems that exhibit a small dipole interaction are the alkaline earth metal salts of the aromatic ketyls [54] and the alkali metal salts of naphthalene, anthracene, and terphenyl [113]. The alkali metal salts of the aliphatic ketyls [22] give two dimer species in equilibrium, one species with a large dipole interaction (see Table 10) and the other with a small dipole interaction. On warming, the spectrum of the species with a large dipole interaction decreases in intensity and the peaks come closer together. The central broad line resolves out into a spectrum showing coupling to two equivalent alkali metal cations.

Hirota [22] suggested that species with a large dipole interaction are contact ion clusters and species with a small dipole interaction are solvent-separated ion clusters. The former species is well-defined. The large D term shows that they must contain two A^- ions relatively close together. The symmetry requirement for $E \approx 0$ is satisfied by a structure in which the planes of the negative ions are perpendicular to each other. The r_{av} values given in Table 10 are somewhat misleading since the electrons are delocalized and, since the dipolar coupling goes as r^{-3}, it is the spin on the closest atom that really controls the magnitude of D. For example, if two ketyls approach along a common C—O axis, as depicted in Scheme IX (cf. Hirota [22]), D

will stem partly from spin on oxygen and partly from spin on carbon, but spin delocalized into the R groups will be largely ineffective. Since the spin density on oxygen is far lower than that on carbon, both will give appreciable contributions. The situation is well illustrated by the results for hexamethylacetone ketyl (the spin in this case being localized on the carbonyl groups). In a simple model involving one or two cations (X and IX), the 5.6 Å separation between the electrons (calculated on a point-dipole model from the observed D term) would fall roughly on the midpoints of the carbon-oxygen bonds, in accord with expectation. Thus it seems that a unit of this sort, with no intervening solvent molecules, will satisfactorily accommodate the results for the aliphatic ketyl clusters. The larger separation between the unpaired electrons calculated for the aromatic ketyl clusters should then be understandable in terms of delocalization into the aromatic rings. If the same model (IX) is used and the same separations between the carbonyl groups assumed (5.6 Å), then from the measured D term for sodium fluorenone ketyl we can calculate an approximate spin density of 0.7 on each carbonyl group.

This is a surprisingly large spin density since the calculations of Fraenkel and Rieger [114] for benzophenone free-ion ketyl predict a spin density on

Table 10 Triplet-State ESR Parameters for Ion Clusters

Anion	Cation	Solvent	D	E	$\langle r^{-3}\rangle^{-\frac{1}{3}}$	Ref.
Hexamethylacetone	Li^+	MTHF	225	0	5.0	a
	Na^+	MTHF	167	0	5.6	
	K^+	MTHF	179	0	5.4	
Pentamethylacetone	Na^+	MTHF	161	0	5.6	a
	K^+	MTHF	150	0	5.7	
Benzophenone	Li^+	MTHF	137	0	5.9	a
	Na^+	MTHF	103	0	6.5	
Fluorenone	Li^+	MTHF	120	0	6.2	a
	Na^+	MTHF	99	0	6.6	
	K^+	MTHF	78	0	7.1	
Xanthone	Li^+	MTHF	90	—	6.8	b
	Na^+	MTHF	79	—	7.1	
4,7-Diphenyl-1,10-phenanthroline	Be^{2+}	MTHF	95	$\sim$0	6.8	c
	Mg^{2+}	MTHF	98	$\sim$0	6.7	
	Zn^{2+}	MTHF	113	$\sim$0	6.3	
	Ca^{2+}	MTHF	162	$\sim$0	5.5	
2,2'-Biquinoline	Be^{2+}	MTHF	84	$\sim$0	7.0	c
	Mg^{2+}	MTHF	86	$\sim$0	7.0	
	Zn^{2+}	MTHF	99	$\sim$0	6.7	
	Ca^{2+}	MTHF	73	$\sim$0	7.3	
	Sr^{2+}	MTHF	77	$\sim$0	7.2	
	Ba^{2+}	MTHF	76	$\sim$0	7.1	
2,2'-Bipyridine	Be^{2+}	MTHF	119	$\sim$0	6.2	c
	Mg^{2+}	MTHF	120	$\sim$0	6.2	
	K^+	MTHF	82	$\sim$0	7.0	
	Na^+	MTHF	94.0	$\sim$0	6.7	d
	Cs^+	MTHF	57	$\sim$0	7.9	
	K^+	DME	81	$\sim$0	7.0	
	K^+	TMED	86	$\sim$0	6.9	
	K^+	THF	87	$\sim$0	6.8	
Dibenzoylmethane	Li^+	MTHF	106	4	6.4	e
	Na^+	MTHF	85	5.8	6.9	
	K^+	MTHF	77	4	7.2	
	Li^+	MTHF	97	0	6.6	
	Na^+	MTHF	74	0	7.2	
	K^+	MTHF	67.5	0	7.4	
Dibenzamide	Na^+	MTHF	94.8	6	6.6	f
	K^+	MTHF	86	6.2	6.9	
Benzoylacetone	Na^+	MTHF	90	6	6.8	f
	K^+	MTHF	82	1	7.0	

[a] Ref. 22.
[b] Ref. 54.
[c] Ref. 116.
[d] J. D. W. Van Voorst, W. G. Zijlstra, and R. Sitters, *Chem. Phys. Letters*, 1, 321 (1967).
[e] Ref. 115.
[f] F. W. Pijpers and H. van Willigen, *Recueil*, 86, 511 (1967).

the carbonyl group of only 0.4. (Unfortunately, although the ^{13}C carbonyl isotropic hyperfine couplings are known for hexamethylacetone and fluorenone ketyls, it is very difficult to make any estimates of the spin-density changes because of the many factors involved which include possible nonplanarity.) If, indeed, a spin density of 0.7 on each carbonyl group is too high, then the value of the D term for aromatic ketyl clusters requires that the carbonyl groups are closer than the carbonyl groups in the aliphatic ketyl clusters. If the structure of the cluster is as depicted in Scheme IXb, then a closer approach could be understood, especially since the anion repulsion effect is reduced on delocalization into the benzene rings. Our approximate calculations suggest that a decrease of 1.8 Å between the carbonyl groups would be required to accommodate the expected spin density of about 0.4 on each carbonyl group.

In general, for the alkali metal salts of the ketyls, the larger the cation, the larger the effective separation between the electrons, but this is not a very marked effect, and the results for the alkaline earth metal chelates (of bipyridine, biquinoline, and diphenylphenanthroline) do not fit in with this statement (Table 10). This may be because two cations are probably involved in the alkali metal-ketyl clusters (although this is not known for certain for the strongly interacting species) and hence the approach of the two negative ions is somewhat closer than that required by a single intervening divalent cation (as depicted in X). Moreover, the greater polarization effect of the divalent cation will lower the spin density on oxygen, thus increasing the effective separation of the electrons.

The paramagnetic dianion (its monoanion is diamagnetic) of dibenzoylmethane also gives strongly coupled triplet-state clusters and spectra from two dimeric species have been observed [115]. The results are given in Table 10. One species, with $E > 0$, is assigned a planar structure with the two dianions coordinated to two alkali metal cations; the other species, $E = 0$, is assigned to a structure with the four oxygens of the two dianion molecules tetrahedrally disposed about a single alkali metal cation. A similar structure has been proposed [116] for the alkaline earth metal chelates of bipyridine, biquinoline, and diphenylphenanthroline and for the dimers of the dianions of dibenzamide and benzoylacetone [117]. As with the alkali metal-ketyl clusters, the larger the alkali metal cation, the larger the effective separation between the unpaired electrons.

Extremely strong interactions lead to the formation of ionic diamagnetic clusters [152, 153] which should be distinguished from covalently bonded associates such as pinacoles [94, 154]. These ionic clusters are linked by π—π bonds which are reinforced by cation-anion attraction. The latter substantially adds to the stability of such clusters because free anions do not associate [153]. Of interest is the fact that some details of structure of such associates were obtained from kinetic studies of electron transfer processes [153].

5.1. Weak Coupling

The weakly interacting species are far less well defined since the solid-state spectra give only broad single lines and the liquid-state spectra are characteristic of one anion only, showing that exchange is slow. This slow exchange can be understood if the two π-systems are held in mutually perpendicular positions which would give zero overlap, for example [149] (cf. the contact ion cluster IX). Unfortunately, this means that there is practically nothing to distinguish the spectra from those of the simple anions. Perhaps the only indicative feature is that hyperfine coupling to two equivalent cations is detected. This could well be the triple ion, $M^+A^-M^+$, discussed in Sections 2.5, 4.2, and 4.3. However, for this to be correct, there would have to be extra cations fortuitously present, possibly because of decomposition, since no "free" anions were detected. As mentioned earlier, Hirota [22], who assumes the structure $(M^+A^-)_2$, has suggested that solvent molecules participate in such a way as to force the anions apart (as depicted in Scheme XI).

Rather different triplet-state systems have been studied recently by de Boer and co-workers. The first antibonding π-orbitals of triphenylene [90] and 1,3,5-triphenylbenzene [118] are doubly degenerate and hence the dianions are triplet molecules, as proven by the solid-state ESR spectra. A similar situation is encountered for dianions derived from other molecules having a three- or sixfold symmetry axis, for example, coronene, 2,4,6-triphenyl-s-triazene and decacyclene. Trinaphthylene is an exception; the ground state of its dianion is singlet [155]. The interesting feature of these results (Table 11) is the dependence of the D term upon the nature of the cation and solvent. The results can be understood in term of a perturbation of the spin distribution by the cations. We consider two extremes. One is an axially symmetric dianion, with two cations at either end (Scheme XII). The other is a planar dianion with the cations centrally placed (Scheme XIII). As either the proximity or the effective charge density of the cations is increased, the cations will attract negative charge, thus reducing or increasing the dipolar coupling between the unpaired electrons for XII and XIII, respectively. De Boer and co-workers [90] obtained three distinct types (I, II, and III) of solid-state triplet spectra for triphenylene, one giving way to the other, on the addition of successive amounts of polyethers to the triplet dianions in MTHF glasses. The polyethers (glymes) preferentially solvate the cations, thus reducing their perturbing effect (a change from a to b in XII or XIII). The decrease in cation perturbation caused a striking decrease in the zero-field splitting parameter D. The fact that a

Table 11 Triplet-State ESR Parameters for Dianion-Alkali Metal Ion Pairs

Dianion	Cation	Solvent	D	E	Species	Ref.
Triphenylene	K⁺	MTH-DME (2:1)	466	94	I	a
	K⁺	MTHF-THF (2:1)	480	92	I	
	K⁺	MTHF	492	94	I	
	K⁺	MTHF-diglyme (20:1)	402	82	II	
	K⁺	MTHF-diglyme (5:1)	304	0	III	
	K⁺	Diglyme	304	0	III	
	Rb⁺	MTHF-DME (2:1)	442	89	I	
	Rb⁺	Diglyme	304	0	III	
	K⁺	MTHF + tetraglyme	492–471	96–92	I	b
	K⁺	MTHF + tetraglyme	395	73	II	
	K⁺	MTHF + tetraglyme	310	0	III	
1,3,5-Triphenyl-			498	0 ⎫		
benzene	K⁺	MTHF	488	0 ⎬	I	b
			463	0 ⎭		
	K⁺	MTHF + tetraglyme	408	0	II	b
	K⁺	MTHF + tetraglyme	356	0	III	
	Na⁺	MTHF	523	0	I	
	Na⁺	MTHF + tetraglyme	480	0	II?	
	Na⁺	MTHF + tetraglyme	359	0	III	

[a] Ref. 90.
[b] Ref. 118.

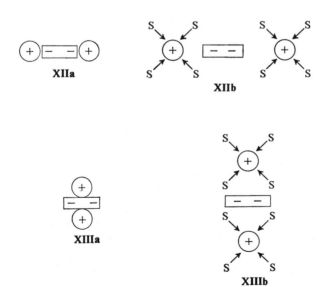

251

decrease was observed means that model XIII is preferable to model XII. The authors suggest that the three species (species I, II, and III in Table 11) are a dianion with two counterions (in the absence of glyme), an intermediate species with the dianion coordinated with one counterion, and the unperturbed dianion (in an excess of polyether). The two ion-pair species show marked distortion of the trigonally symmetric conformation ($E > 0$) but, in the presence of excess polyether, the spectrum is that of a triplet species with trigonal symmetry ($E = 0$) and is assigned to the unperturbed dianion.

1,3,5-Triphenylbenzene gives very similar behavior except that three dimer species are seen in MTHF and also there is no loss of symmetry on ion-pair formation (i.e., $E = 0$ for the ion pair and the free dianion). Again there is a marked decrease in D when the cation perturbation is reduced by the addition of tetraglyme.

Studies of these paramagnetic dianions have been extended to their solutions. In powerfully solvated media, like glymes, their ground state was found to be a triplet [156]. However, in poorer solvents, like MTHF, the perturbation caused by cations makes the singlet the state of lowest energy [157, 158]. This reversal of energy levels arises from a lowering of the symmetry of the aggregate due to noncentric location of cations. The problem of reversal of singlet and triplet states in these systems has been comprehensively discussed in a recent paper by de Boer et al. [159] and again in Section 6 of Chapter 8 of this book.

6. SOME NEGLECTED TOPICS

To round off this chapter, we make brief references to topics which have been omitted from the main body of the review. In Section 6.1 we recall that workers in this field must bear in mind the tendency of certain anion radicals to disproportionate. This is discussed in detail by Jagur-Grodzinski and Szwarc in the second volume of this book. We then consider briefly the complementary role played by nuclear-magnetic resonance, which is again described in detail elsewhere (Chapters 7 and 8). In Section 6.3, the possibility of pairing between like-charged ions, normally ignored, is stressed, and finally in Section 6.4 some properties of metal solutions are described. This is done partly because these solutions are often used in the preparation of radical anions and partly because some sort of ion-pair formation between the cation and the solvated electron appears to be important in these solutions also.

6.1. Disproportionation and Related Topics

The difference in the standard reduction potentials of the neutral molecule and the monoanion is a measure of the stability of the monoanion to the disproportionation reaction. This difference is reduced by the solvation energies of the ions (the dianion generally having a higher solvation energy than that of the two monoanions [119]), and by the free energy of the ion-recombination reaction if the dianion forms ion pairs with cations [120].

Three disproportionation equilibria need to be considered:

$$2A^- \rightleftarrows A + A^{2-} \tag{6}$$

$$A^-M^+ + A^- \rightleftarrows A + A^{2-}M^+ \tag{7}$$

$$2A^-M^+ \rightleftarrows A + M^+A^{2-}M^+ \tag{8}$$

Equilibrium 6 is well defined. The solvation energy of the dianion is greater than the sum of the solvation energies of the two monoanions and so disproportionation is favored by increasing solvent polarity. Equilibria 7 and 8, however, depend in a complicated manner on the solvent, temperature, and cation (see, e.g., ref. 18 or 19, page 354).

ESR studies of the temperature, gegenion, and solvent dependence of systems thought to involve ion pairs may be complicated by disproportionation reactions. Two instances have recently appeared in the literature. De Boer [21] found that the intensity of the ESR signal from "loose" lithium-pyracene ion pairs in hexane/MTHF decreased as the temperature was increased and eventually disappeared. The signal reappeared on cooling. The ESR signal also decreased on the addition of the corresponding alkali metal halide to the sodium or potassium-pyracene ion pair. De Boer proposed that the driving force of the disproportionation reaction is the reduced cation-solvating power of the solvent at high temperatures, which favors the formation of dinegative ions strongly associated with gegenions. (This system is probably an example of equilibrium 8.) An opposite reversible temperature dependence has been observed by Warhurst and Wilde [121] for the lithium-acenaphthenequinone ion pair in DME. As the temperature was lowered, the signal decreased in intensity. The disproportionation equilibrium constant was dependent on the gegenion and decreased with increasing cationic radius.*

In general, quantitative aspects of the disproportionation equilibria of ion pairs have been studied by UV spectrophotometry [122–124]. The ESR spectra are easier to interpret but less informative since only the spectrum of

---

* In this system the loss of the ESR signal is probably due to the formation of diamagnetic dimer and not to disproportionation (see ref. 153). (editor).

the monoanion is obtained, whereas in UV spectrophotometry spectra of the neutral molecule, the monoanion, and the dianion are observed.

Other chemical reactions may also be encountered in studies of pair formation. In ether solvents, the paramagnetic ketyl dimers (discussed in Section 5) exist in equilibrium with the monomer species. However, in nonpolar solvents, the major equilibrium is between the paramagnetic dimer and a diamagnetic dimer (pinacolate) [54]. The dissociation of the paramagnetic dimer into monomer species is solvent and cation dependent with the dissociation constant increasing with increasing solvent polarity (DME < THF < MTHF) and increasing cation radius [54].

Another example of a dimerization reaction that is promoted by ion-pair formation is the dimerization of the radical anions of quinoline and other nitrogen heteroaromatics [125]. In HMPA the radical anions are stable, but in THF they dimerize to form the dimeric dianion, the driving force of the reaction being the stabilization of the dianion by strong association with gegenions. In the case of pyridine, dimerization was extensive even in HMPA and the dimer eventually is aromatized and further reduced by excess sodium metal to the bipyridyl radical anion.

We feel that it is important to stress that reactions such as these can occur in studies of ion-pair formation by radical anions and may drastically affect the conclusions. In particular, disproportionation reactions may interfere with ESR studies of the cation, solvent, and temperature dependence of ion pairs and dimerization reactions may lead to misinterpretation of ESR spectra.

6.2. Relation between ESR and NMR Spectra of Paramagnetic Ion Pairs

Although our prime concern is with ESR, this review would not be complete without brief reference to the complementary technique of NMR. The technique is complementary because it can be used only to study nuclei which give no resolved hyperfine contribution to the corresponding ESR spectra. That is, as far as the nucleus under consideration is concerned, the electron spin must appear to be rapidly inverting. If this is not the case, then, provided there is an isotropic hyperfine coupling between the nucleus and the unpaired electron, the nuclear resonance is split to such a great extent that no resonance is observed. When inversion is rapid, the averaging will not return the resonance to the normal value because of the small difference in the population of the spin states in a magnetic field. This difference in population is responsible for the "Knight shift" or "contact shift" that is observed. This is a function of the s-character of the unpaired electron's orbital on the atom in question, and the shift is positive or negative according to the sign of the spin density in that orbital. This ability to measure the sign of the spin density is one of the most important advantages of this technique over ESR.

Rapid electron-spin relaxation can be achieved either directly, by increasing the concentration until spin-spin relaxation is effective, or indirectly, by contriving rapid electron-nuclear exchange. Examples of the indirect method are the rapid movement of solvent molecules in and out of the solvation shells of the anions and the rapid exchange of cations induced by adding an excess of an appropriate salt to the solution. Thus solvent nuclei or cation nuclei may experience Knight shifts. A disadvantage of the former approach is that concentrated solutions are often very difficult to obtain. Further, if electron nuclear exchange is occurring, the shift is a weighted average of resonances from the few nuclei in contact, and for the many not in contact. If, as is often the case, appropriate concentrations are unknown, then the magnitude of the coupling constant cannot be obtained.

Another difficulty that may arise is that equilibria can occur at such rates that some of the anions make no contribution to the shift. Consider, for example, the equilibria 1 and 2 (Section 4): if, in the presence of added diamagnetic salt, $(IP)_l$ were to exchange rapidly with M^+, but $(IP)_t$ exchanged slowly, then only $(IP)_l$ would make a contribution to the shift, the cations in $(IP)_t$ giving no detectable resonance.

6.2.1. NMR of Cation Nuclei

The results reported by de Boer and his co-workers are given in Fig. 19. In the particular case of the lithium salt of fluorenone, even in the presence of about 1 M lithium bromide in THF over a wide temperature range, no shift was observed [126]. This is in marked contrast with the ESR results and has been interpreted to mean that Li^+ exchange was too slow to give the required averaging (i.e., slow compared with the lithium hyperfine coupling). There was, however, a marked broadening, which was assigned to a quadrupole effect resulting from a large field gradient at the lithium nucleus from the anion.

The other results [127, 128] gave shifts, both positive and negative, but these were not always in line with the results of ESR studies (Fig. 19). One obvious reason for this is that the solutions used for NMR studies were far more concentrated than those required for ESR work. Under these conditions ion clusters are expected, and the results discussed in Section 2.5 show that this can give rise to an increase in the metal hyperfine coupling. This problem is discussed in greater detail in Chapter 7.

6.2.2. Pseudo-Contact Shifts

Before concluding this brief section on NMR, mention should be made of the effect that an anisotropic g-tensor can have on the NMR shifts. If

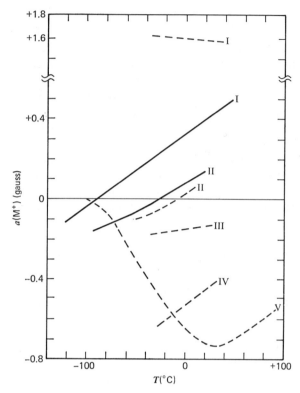

Figure 19. A comparison between the metal hyperfine coupling derived from comparable NMR and ESR experiments. – – – –, NMR results; ————, ESR results. (I) Na^+/fluorenone/THF. (II) Na^+/biphenyl/MTHF. (III) $^{85}Rb^+$/fluorenone/THF. (IV) $^{87}Rb^+$/fluorenone/THF. (V) Cs^+/biphenyl/diglyme.

Δg $(= g_{\parallel} - g_{\perp})$ is large, then there will be a contribution to the isotropic coupling, and hence to the Knight shift, other than that arising from direct electron-nuclear contact. This is because a g-shift is derived from orbital motion and this, in turn, is associated with a corresponding magnetization. If the g-tensor is anisotropic, this extra source of magnetization will not average to zero, and due allowance must be made.

If the nucleus under consideration is not the nucleus around which the orbital motion is occurring, then the effect falls off as r^{-3} when r is the mean separation between the center "responsible" for the g-shift and the nucleus being studied. This has been put to good use in some recent studies of ion pairing betwen tetraalkyl ammonium ions, R_4N^+, and various transition-metal complex anions [129, 130].

6.3. Pairing between Like-Charged Ions

Although ion clusters are considered in this review, we have no section on pairing between like-charged ions. This is a phenomenon which is found especially for large, planar cations or anions whose structures minimize charge repulsion and maximize dispersion forces. The effect is markedly solvent dependent, being especially significant in water [131].

6.4. Metal Solutions

Although we cannot strictly classify the species obtained when alkali metals are dissolved in solvents such ethylamine as "ion pairs"—since their structures are unknown—nevertheless, they must be closely related. These species have well-resolved hyperfine features from one cation with markedly temperature-sensitive separations. These range from about 3% at low temperatures to nearly 45% of the atomic values at the highest temperatures so far studied. In addition to the amines [132, 133], very similar spectra have recently been detected for potassium in various ethers [134].

These species are in equilibrium with another paramagnetic entity having a single narrow line and dilution tests [133] have shown that the stoichiometry of this equilibrium is well represented by

$$M^+ + e^- \rightleftarrows M^+e^-$$

where e^- is the solvated electron. Recently, equilibrium constant of such a dissociation was determined for $M = Na^+$ in THF by flash-photolysis technique [160].

Two models have been proposed for the nature of M^+e^- and the temperature dependence of the alkali metal hyperfine coupling. One is very similar to the two-state model proposed by Hirota (discussed in Section 4.1) with M^+e^- being the time average of a loose ion pair. (Low a value) and a centro-symmetric unit (high a value) sometimes called a "monomer" or "solvated atom" [133]. Increasing temperature favors the solvated atom. The second model [132] pictures the "monomer" as a solvated cation with the additional unpaired electron distributed between the cation and the solvent. The occupancy of the cation s-orbital decreases as the solvation shell tightens on cooling.

For heavy metals such as cesium, there is an appreciable negative g-shift, which increases as the species becomes more "atomlike" (i.e., on raising the temperature). This represents a deviation from the gas-phase atomic g-value and can best be understood in terms of a mixing with the outer p-orbitals induced by the surrounding solvent molecules.

Other systems of similar nature have been observed in γ-irradiated glasses [161] or in flash-photolysis studies [162]. Interesting comments on this subject have been published by Tuttle [163].

REFERENCES

1. J. Burgess and M. C. R. Symons, *Quart. Rev.*, **22**, 276 (1968).
2. A. Hudson and G. R. Luckhurst, *Chem. Revs.*, **69**, 191 (1969).
3. S. I. Weissman, E. de Boer, and J. J. Couradi, *J. Chem. Phys.*, **26**, 963 (1957).
4. T. R. Tuttle, R. L. Ward, and S. I. Weissman, *J. Chem. Phys.*, **25**, 189 (1956).
5. A. Carrington, F. Dravnieks, and M. C. R. Symons, *J. Chem. Soc. (London)*, **1959**, 947.
6. P. J. Zandstra and S. I. Weissman, *J. Am. Chem. Soc.*, **84**, 4408 (1962).
7. N. M. Atherton and S. I. Weissman, *J. Am. Chem. Soc.*, **83**, 1330 (1961).
8. A. H. Reddoch, *J. Chem. Phys.*, **41**, 444 (1964).
9. A. H. Reddoch, *J. Chem. Phys.*, **43**, 225 (1965).
10. M. Iwaizumi, M. Suzuki, T. Isobe, and H. Azumi, *Bull. Chem. Soc. Japan*, **40**, 1325 (1967).
11. J. Oakes and M. C. R. Symons, *Chem. Comm.*, **1968**, 294.
12. A. Carrington and A. D. McLachlan, *Introduction to Magnetic Resonance*, Harper International Edition, 1967.
13. P. W. Atkins and M. C. R. Symons, *The Structure of Inorganic Radicals*, Elsevier, Amsterdam, 1967.
14. P. B. Ayscough, *Electron Spin Resonance in Chemistry*, Methuen, London, 1967.
15. H. M. McConnell, *J. Chem. Phys.*, **24**, 762 (1956); *Ann. Rev. Phys. Chem.*, **8**, 105 (1957).
16. T. F. Hunter and M. C. R. Symons, *J. Chem. Soc. (London)*, A, 1770 (1967).
17. T. R. Griffiths and M. C. R. Symons, *Trans. Faraday Soc.*, **56**, 1125 (1960).
18. M. Szwarc, *Accts. Chem. Res.*, **2**, 87 (1969).
19. M. Szwarc, *Carbanions, Living Polymers and Electron-Transfer Processes*, Interscience, New York, 1968.
20. S. Aono and K. Oohashi, *Prog. of Theor. Phys.*, **32**, 1 (1964).
21. E. de Boer, *Rec. Trav. Chim.*, **84**, 609 (1965).
22. N. Hirota, *J. Am. Chem. Soc.*, **89**, 32 (1967).
23. C. L. Dodson and A. H. Reddoch, *J. Chem. Phys.*, **48**, 3226 (1968).
24. M. C. R. Symons, *Nature*, **224**, 685 (1969).
25. M. C. R. Symons, *J. Phys. Chem.*, **71**, 172 (1967).
26. B. S. Gourary and F. J. Adrian, *Solid State Physics*, **10**, 127 (1960); M. J. Blandamer and M. C. R. Symons, *J. Phys. Chem.*, **67**, 1304 (1963).
27. R. H. Stokes, *J. Am. Chem. Soc.*, **86**, 979 (1964).
28. K. Hofelmann, J. Jagur-Grodzinski, and M. Szwarc, *J. Am. Chem. Soc.*, **91**, 4645 (1969).
29. J. M. Gross and J. D. Barnes, *J. Chem. Soc. A*, **1969**, 2437.

30. J. M. Gross, J. D. Barnes, and G. N. Pillams, *J. Chem. Soc. (London)*, **A**, 109 (1969).
31. J. M. Gross and J. D. Barnes; *J. Phys. Chem.*, **74**. 2936 (1970).
32. K. Nakamura, *Bull. Chem. Soc. Japan*, **40**, 1 (1967).
33. C. Ling and J. Gendell, *J. Chem. Phys.*, **47**, 3475 (1967).
34. H. Nishiguchi, Y. Nakai, K. Nakamura, K. Ishizu, Y. Deguchi, and H. Takakai, *Mol. Phys.*, **9**, 153 (1965).
35. G. R. Luckhurst, *Mol. Phys.*, **9**, 179 (1965).
36. K. Nakamura and Y. Deguchi, *Bull. Chem. Soc. Japan*, **40**, 705 (1967).
37. N. M. Atherton and A. E. Goggins, *Trans. Faraday Soc.*, **62**, 1702 (1966).
38. J. dos Santos Veiga and A. F. Neiva Correia, *Mol. Phys.*, **9**, 395 (1965).
39. E. de Boer and C. MacLean, *J. Chem. Phys.*, **44**, 1334 (1966).
40. N. Hirota, *J. Am. Chem. Soc.*, **90**, 3603 (1968).
41. T. E. Gough and M. C. R. Symons, *Trans. Faraday Soc.*, **62**, 269 (1966).
42. T. A. Claxton, J. Oakes, and M. C. R. Symons, *Trans. Faraday Soc.*, **64**, 596 (1968).
43. N. L. Bauld and M. S. Brown, *J. Am. Chem. Soc.*, **87**, 4390 (1965); **89**, 5417 (1967).
44. C. A. McDowell and K. F. G. Paulus, *Can. J. Chem.*, **43**, 224 (1965).
45. R. Tanikawa, K. Maruyama, and R. Goto, *Bull. Chem. Soc., Japan*, **37**, 1893 (1963) **38**, 2538 (1964).
46. B. J. Herold, A. F. Neiva Correia, and J. dos Santos Veiga, *J. Am. Chem. Soc.*, **87**, 2661 (1965).
47. G. R. Luckhurst and L. E. Orgel, *Mol. Phys.*, **7**, 297 (1964).
48. E. G. Janzen and C. M. DuBose, *J. Phys. Chem.*, **70**, 3372 (1966).
49. J. E. Bennett, private communication; J. E. Bennett, B. Mile, and A. Thomas, *Trans. Faraday Soc.*, **61**, 2357 (1965).
50. J. H. Sharp and M. C. R. Symons, *J. Chem. Soc. (London)*, *A*, 3075 (1970).
51. M. B. D. Bloom, R. S. Eachus, and M. C. R. Symons, *Chem. Comm.*, **1968**, 1495.
52. M. C. R. Symons, *Ann. Rev. Phys. Chem.*, **20**, 219 (1969).
53. T. E. Gough and P. R. Hindle, *Can. J. Chem.*, **47**, 1698, 3393 (1969).
54. N. Hirota and S. I. Weissman, *J. Am. Chem. Soc.*, **86**, 2537 (1964).
55. A. H. Reddoch, *J. Chem. Phys.*, **43**, 3411 (1965).
56. P. S. Gill and T. E. Gough, *Trans. Faraday Soc.*, **64**, 1997 (1968).
57. D. H. Chen, E. Warhurst, and A. M. Wilde, *Trans. Faraday Soc.*, **63**, 2561 (1967).
58. J. R. Bolton and G. K. Fraenkel, *J. Chem. Phys.*, **40**, 3307 (1964).
59. N. Hirota, *J. Chem. Phys.*, **37**, 1884 (1962).
60. P. B. Ayscough and R. Wilson, *Proc. Chem. Soc.*, **1962**, 229.
61. Ref. 1, Figure 6.
62. Y. Nakai, *Bull. Chem. Soc. Japan*, **39**, 1372 (1966).
63. W. M. Fox, J. M. Gross, and M. C. R. Symons, *J. Chem. Soc. (London)*, A, 448 (1966).
64. C. Chachaty, *Compt. Rend., Ser. C.*, **262**, 680 (1966).
65. S. H. Glarum and J. H. Marshall, *J. Chem. Phys.*, **41**, 2182 (1964).
66. T. Kitagawa, T. Layloff, and R. N. Adams, *Analyt. Chem.*, **36**, 925 (1964).
67. D. H. Geske and A. H. Maki, *J. Am. Chem. Soc.*, **82**, 2671 (1960).

68. J. Oakes, J. Slater, and M. C. R. Symons; *Trans. Faraday Soc.*, **66**, 546 (1970).

69. W. M. Gulick and D. H. Geske, *J. Am. Chem. Soc.*, **87**, 4049 (1965).

70. W. M. Gulick, W. E. Gieger, and D. H. Geske, *J. Am. Chem. Soc.*, **90**, 4218 (1968).

71. H. J. Bower, M. C. R. Symons, and D. J. A. Tinling, *Radical Ions*, E. T. Kaiser and L. Kevan (Eds.), Interscience, New York, 1968, p. 417.

72. C. J. W. Gutch and W. A. Waters, *Chem. Comm.*, **1966**, 39.

73. N. M. Atherton and A. E. Goggins, *Trans. Faraday Soc.*, **61**, 1399 (1965).

74. K. Nakamura, *Bull. Chem Soc. Japan*, **40**, 1019 (1967).

75. E. de Boer and L. Mackor, *Proc. Chem. Soc.*, **1963**, 23; *J. Am. Chem. Soc.*, **86**, 1513 (1964).

76. M. Iwaizumi, M. Suzuki, T. Isobe, and H. Azumi, *Bull. Chem. Soc. Japan*, **40**, 2754 (1967).

77. E. G. Janzen and J. B. Pickett, *J. Am. Chem. Soc.*, **89**, 3649 (1967).

78. J. M. Gross and M. C. R. Symons, *Trans. Faraday Soc.*, **63**, 2117 (1967).

79. J. W. H. Schreurs and D. Kivelson, *J. Chem. Phys.*, **36**, 117 (1962).

80. A. Carrington and H. C. Longuet-Higgins, *Mol. Phys.*, **5**, 447 (1962).

81. E. de Boer and E. L. Mackor, *Mol. Phys.*, **5**, 537 (1962).

82. N. Hirota, *J. Phys. Chem.*, **71**, 127 (1967).

83. P. Graceffa and T. R. Tuttle, *J. Chem. Phys.*, **50**, 1908 (1969).

84. G. E. Werner and W. H. Bruning, *J. Chem. Phys.*, **51**, 4170 (1969).

85. N. Hirota and R. Kreilick, *J. Am. Chem. Soc.*, **88**, 614 (1966).

86. A. Crowley, N. Hirota, and R. Kreilick, *J. Chem. Phys.*, **46**, 4815 (1967).

87. P. B. Ayscough and F. P. Sargent, *J. Chem. Soc. (London)*, **B**, 900 (1966).

88. N. M. Atherton, *Chem. Comm.*, **1966**, 254.

89. T. A. Claxton and J. Oakes, *Trans. Faraday Soc.*, **64**, 607 (1968).

90. H. van Willigen, J. A. M. van Broekhoven, and E. de Boer, *Mol. Phys.*, **12**, 533 (1967).

91. A. M. Hermann, A. Rembaum, and W. R. Carper, *J. Phys. Chem.*, **71**, 2661 (1967).

92. P. S. Gill and T. E. Gough, *Can. J. Chem.*, **45**, 2112 (1967).

93. R. F. Adams, N. M. Atherton, A. E. Goggins, and C. M. Goold, *Chem. Phys. Letters*, **1**, 48 (1967).

94. N. Hirota and S. I. Weissman, *J. Am. Chem. Soc.*, **86**, 2537 (1964).

95. R. F. Adams and N. M. Atherton, *Trans. Faraday Soc.*, **64**, 7 (1968).

96. R. F. Adams and N. M. Atherton, *Trans. Faraday Soc.*, **65**, 649 (1969).

97. R. L. Ward and S. I. Weissman, *J. Am. Chem. Soc.*, **79**, 2086 (1957).

98. A. W. Rutter and E. Warhurst, *Trans. Faraday Soc.*, **64**, 2338 (1968).

99. T. A. Claxton, W. M. Fox, and M. C. R. Symons, *Trans. Faraday Soc.*, **63**, 2570 (1967).

100. J. Oakes and M. C. R. Symons, *Trans. Faraday Soc.*, **66**, 10 (1970).

101. T. A. Claxton, *Trans. Faraday Soc.*, **65**, 2289 (1969).

102. J. Subramanian and P. T. Narasimhan, *J. Chem. Phys.*, **48**, 3757 (1968).

103. J. Gendell, *J. Chem. Phys.*, **46**, 4152 (1967).

104. W. E. Griffiths, C. J. W. Gutch, G. F. Longster, J. Myatt, and P. F. Todd, *J. Chem. Soc. (London)*, **B**, 785 (1968).

105. J. E. Harriman and A. H. Maki, *J. Chem. Phys.*, **48**, 2453 (1967).

106. R. K. Gupta and P. T. Narasimhan, *J. Chem. Phys.*, **48**, 3757 (1968).

107. N. Hirota, R. Carraway, and W. Schook, *J. Am. Chem. Soc.*, **90**, 3611 (1968).

108. T. A. Miller and R. N. Adams, *J. Am. Chem. Soc.*, **88**, 5713 (1966).

109. C. Thomson, *Quart. Rev.*, **22**, 45 (1968).

110. R. Briere, R.-M. Dupeyre, H. Lemaire, C. Morat, A. Rassat, and P. Rey, *Bull. Soc. Chim. Fr.*, **1965**, 3290.

111. S. H. Glarum and J. H. Marshall, *J. Chem. Phys.*, **47**, 1374 (1967).

112. R. H. Dunhill and M. C. R. Symons, *Mol. Phys.*, **15**, 105 (1968).

113. P. Biloen, R. Prins, J. D. W. van Voorst, and G. J. Hoijtink, *J. Chem. Phys.*, **46**, 4149 (1967).

114. G. K. Fraenkel and P. H. Rieger, *J. Chem. Phys.*, **37**, 2811 (1962).

115. H. van Willigen and S. I. Weissman, *J. Chem. Phys.*, **37**, 2811 (1962).

116. I. M. Brown, S. I. Weissman, and L. C. Snyder, *J. Chem. Phys.*, **42**, 1105 (1965).

117. F. W. Pijpers, H. van Willigen, and J. J. Th. Gerding, *Recueil*, **86**, 511 (1967).

118. J. A. M. van Broekhoven, H. van Willigen, and E. de Boer, *Mol. Phys.*, **15**, 101 (1968).

119. N. S. Hush and J. Blackledge, *J. Chem. Phys.*, **23**, 514 (1955).

120. G. J. Hoijtink, E. de Boer, P. H. van der Meij, and W. P. Weijland, *Recueil*, **75**, 487 (1956).

121. E. Warhurst and A. M. Wilde, *Trans. Faraday Soc.*, **65**, 1413 (1969).

122. J. E. Bennett, A. G. Evans, J. C. Evans, E. D. Owen, and B. J. Tabner, *J. Chem. Soc.*, **1963**, 3954; A. G. Evans and B. J. Tabner, *J. Chem. Soc.*, **1963**, 4613.

123. J. F. Garst and E. R. Zabolotny, *J. Am. Chem. Soc.*, **87**, 495 (1965); J. F. Garst, E. R. Zabolotny, and R. S. Cole, *J. Am. Chem. Soc.*, **86**, 2257 (1964); E. R. Zabolotny and J. F. Garst, *J. Am. Chem. Soc.*, **86**, 1645 (1964).

124. R. C. Roberts and M. Szwarc, *J. Am. Chem. Soc.*, **87**, 5542 (1965).

125. J. Chaudhuri, S. Kume, J. Jagur-Grodzinski, and M. Szwarc, *J. Am. Chem. Soc.*, **90**, 6421 (1968).

126. G. W. Canters, H. van Willigen, and E. de Boer, *Chem. Comm.*, **1967**, 566.

127. G. W. Canters, E. de Boer, B. M. P. Hendriks, and H. van Willigen, *Chem. Phys. Letters*, **1**, 627 (1968).

128. G. W. Canters, E. de Boer, B. M. P. Hendriks, and A. A. K. Klaassen, *Colloque Ampere XV*, North-Holland, Amsterdam, 1969, p. 242.

129. H. M. McConnell and R. E. Robertson, *J. Chem. Phys.*, **29**, 1361 (1958).

130. D. W. Larsen, *J. Am. Chem. Soc.*, **91**, 2920 (1969).

131. M. J. Blandamer, J. A. Brivati, M. F. Fox, M. C. R. Symons, and G. S. P. Verma, *Trans. Faraday Soc.*, **63**, 1850 (1967).

132. K. Bar-Eli and T. R. Tuttle, *J. Chem. Phys.*, **40**, 2508 (1964).

133. R. Catterall, M. C. R. Symons, and J. Tipping, *J. Chem. Soc.* (*London*), A, 4342 (1966), 1234 (1967); L. R. Dalton, J. D. Rynbrandt, E. M. Hansen, and J. L. Dye, *J. Chem. Phys.*, **44**, 3969 (1966); J. L. Dye and L. R. Dalton, *J. Phys. Chem.*, **71**, 184 (1967).

134. R. Catterall, J. Slater, and M. C. R. Symons, Colloque Weyl II, Cornell University, Ithaca, N.Y. June 1969. *J. Chem. Phys.*, **52**, 1003 (1970).

135. R. V. Slates and M. Szwarc, *J. Phys. Chem.*, **69**, 4124 (1965); P. Chang, R. V. Slates, and M. Szwarc, *J. Phys. Chem.*, **70**, 3180 (1966).
136. L. Lee, R. Adams, J. Jagur-Grodzinski, and M. Szwarc, *J. Am. Chem. Soc.*, **93**, 4149 (1971).
137. I. B. Goldberg and I. R. Bolton, *J. Phys. Chem.*, **74**, 1965 (1970).
138. M. D. B. Bloom, R. S. Eachus, and M. C. R. Symons, *J. Chem. Soc.*, **A833** (1971).
139. M. C. R. Symons and R. S. Eachus, *Chem. Comm.*, 70 (1970).
140. T. E. Gough and D. R. Hindle, *Trans. Faraday Soc.*, **66**, 2420 (1970).
141. W. G. Williams, R. J. Pritchett, and G. K. Fraenkel, *J. Chem. Phys.*, **52**, 5584 (1970).
142. B. G. Segal, M. Kaplan, and G. K. Fraenkel, *J. Chem. Phys.*, **43**, 4191 (1965).
143. Y. Karasawa, G. Levin, and M. Szwarc, *J. Am. Chem. Soc.*, **93**, 4614 (1971); *Proc. Roy. Soc.*, in press.
144. G. W. Canters, Thesis, Nijmegen, The Netherlands (1969).
145. S. A. Al-Baldawi and T. E. Gough, *Can. J. Chem.*, **48**, 2799 (1970).
146. S. A. Al-Baldawi and T. E. Gough, *Can. J. Chem.*, **49**, 2059 (1971).
147. A. W. Rutter and E. Warhurst, *Trans. Faraday Soc.*, **66**, 1866 (1970).
148. E. Warhurst and A. M. Wilde, *Trans. Faraday Soc.*, **67**, 605 (1971).
149. E. Warhurst and A. M. Wilde, *Trans. Faraday Soc.*, **66**, 2124 (1970).
150. (a) C. J. W. Gutch, W. A. Waters, and M. C. R. Symons, *J. Chem. Soc.*, **B1261** (1970). (b) D. Jones and M. C. R. Symons, *Trans. Faraday Soc.*, **67**, 961 (1971).
151. R. F. Adams, T. L. Staples, and M. Szwarc, *Chem. Phys. Lett.*, **5**, 474 (1970).
152. K. Maruyama, *Bull. Chem. Soc. Japan*, **37**, 553 (1964).
153. T. L. Staples and M. Szwarc, *J. Am. Chem. Soc.*, **92**, 5022 (1970).
154. N. Hirota, in *Radical-Ions*, E. T. Kaiser and L. Kevan, Eds., Interscience, New York (1968), Chapter 2.
155. J. L. Sommerdijk, E. de Boer, F. W. Pijpers, and H. van Willigen, *Z. Phys. Chem. (Frankfurt)*, **63**, 183 (1969).
156. J. A. M. van Broekhoven, Thesis, Nijmengen, The Netherlands (1970).
157. M. Glasbeek, J. D. W. van Voorst, and G. J. Hoijtink, *J. Chem. Phys.*, **45**, 1852 (1966).
158. J. A. M. van Broekhoven, B. M. P. Hendriks, and E. de Boer, *Molec. Phys.*, **20**, 993 (1971).
159. B. M. P. Hendricks, A. A. K. Klaassen, and E. de Boer, *Int. J. Pure and Appl. Chem.*, in press.
160. M. Fisher, G. Rämme, S. Claesson, and M. Szwarc, *Chem. Phys. Lett.*, **9**, 309 (1971).
161. F. W. Froben and J. E. Willard, *J. Phys. Chem.*, **75**, 35 (1971).
162. L. J. Giling, J. G. Kloosterboer, R. P. H. Rettschnick, and J. D. W. van Voorst, *Chem. Phys. Lett.*, **8**, 457 (1971).
163. T. R. Tuttle and P. Graceffa, *J. Phys. Chem.*, **75**, 843 (1971).
164. K. S. Chen and N. Hirota, to be published.

6

Nuclear Magnetic Resonance Studies of Carbon-Lithium Bonding in Organolithium Compounds

L. DENNIS McKEEVER

Physical Research Laboratory, The Dow Chemical Company, Midland, Michigan

INTRODUCTION

Elucidation of the relationship between molecular structure and chemical reactivity is of substantial importance in chemistry. Systematic studies of the highly reactive organolithium compounds reveal substantial variations in their reactivity with a common reagent in a common solvent [1, 2]. An

263

understanding of this reactivity behavior requires a detailed knowledge of the structure and the bonding character of these reactants. Studies of structure and the nature of carbon-lithium bonding in organolithium compounds has received serious attention only recently [3–5] and among the physical methods available for their elucidation, nuclear magnetic resonance techniques are among the most fruitful. Since all the nuclei in the more common organolithium reagents possess isotopes with magnetic moments, proton, carbon, and lithium resonance can be examined. Owing to facile observation, proton magnetic resonance has received the widest attention.

The most important parameter obtained in the NMR experiment is the chemical shift. The density of electron charge surrounding the nucleus being studied as well as the charge distribution in neighbor atoms determines the magnitude of the shift. Both of these factors may cause a diamagnetic or paramagnetic shift of the nuclear resonance.

In the NMR experiment, a diamagnetic moment arises from the induced orbital motion of electrons caused by the presence of an external magnetic field. The local magnetic field at the nucleus is thus proportional to the external field, the propitionality coefficient being the "screening constant." In the absence of hybridizational changes, variations of chemical shifts can be understood in terms of charge polarization; for example, shifts to high field (relative to a given reference) can be attributed to increased electron charge density at a given nucleus. Consideration of the diamagnetic contribution to the chemical shift is useful in assessing relative charge densities at all nuclei endowed with a magnetic moment.

Paramagnetic contributions to the chemical shift are difficult to assess quantitatively. They arise from the induced mixing by the magnetic field of the ground-state electronic configuration with low-lying excited states. For example, an excited state contributes to the local current on a given atom, if it corresponds partly to the transfer of an electron from one p-orbital to another on the same atom [6].

p-Orbitals are not important for hydrogen atoms; thus, paramagnetic currents generally may be neglected when dealing with the NMR chemical shift of this nucleus. A sizable paramagnetic contribution can, however, arise from the neighbor anisotropy effect in protons as well as carbon. The paramagnetic contribution is inversely proportional to the average electronic excitation energy [6]. As a result, the chemical shift changes in a direction such that the total effective field determining the resonance frequency is reduced.

Quantitative relationships between chemical shifts and charge density are available for several nuclei. For example, a conversion constant of 10 ppm/ electron is found for hydrogen [7] and 160–200 ppm/electron for carbon [8, 9]. In using these constants to predict the charge polarization, it is usually assumed that the state of hybridization remains unchanged.

Carbon chemical shifts and ^{13}C-^{1}H coupling constants are both useful for assessing the state of carbon hybridization. $J_{^{13}C^{-1}H}$-values range from $\sim$125Hz for sp^3 to $\sim$165 Hz for sp^2 hybridized carbon [10]. Similarly, extensive data indicate that carbon chemical shifts of sp^3 hybrids are in the range 160–170 ppm upfield from carbon disulfide, while the chemical shifts of sp^2 hybridized species generally are in the 60–70 ppm range. Rehybridization from sp^2 to sp^3 would thus be expected to shift the ^{13}C resonance to higher field.

2. NATURE OF THE CARBON-LITHIUM BOND

Bonding in organolithium compounds varies from polar covalent in alkyllithiums,

$$R^{\delta-} - Li^{\delta+}$$

to essentially ionic in "delocalized" or resonance stabilized organolithiums such as triphenylmethyllithium,

$$R^-, Li^+$$

Depending on solvent and concentration, the ionic species can exist in the form of free ions, tight ion pairs, solvent-separated ion pairs, or higher aggregates. The nature of the solvent and the presence of strong chelating donor reagents is known to drastically affect the reactivity of organolithiums [5]; because their structure and the character of the carbon-lithium bond are profoundly affected by these factors.

2.1. Alkyllithiums

In contrast to other organoalkali metal derivatives, alkyllithium reagents apparently have polar covalent carbon-lithium bonds. In accord with Fajans' concepts of covalent bonding [11], the small size and high polarizing power of the lithium cation favors this bonding character to a greater degree than in other organoalkali metal compounds. Alkyllithiums are uniquely soluble in both aromatic and aliphatic hydrocarbons, methyllithium being an exception. Furthermore, organolithiums have a higher vapor pressure and lower reactivity than other organoalkali metal compounds. These properties reflect the high degree of covalent character of the carbon-lithium bonds. Colligative studies reveal that the simple alkyllithium compounds are polymeric in solution, generally forming tetramers or hexamers [3, 5]. This association implies p-orbital participation in the overall bonding favored by the low $s \rightarrow p$ promotion energy. X-ray studies suggest that the tetramer assumes a tetrahedral geometry (Fig. 1) with the lithium atom situated at the

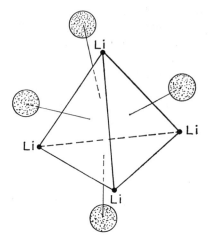

Figure 1.   Tetrahedral structure of the methyl-lithium tetramer. See also Fig. 5 for an alternative representation of the same structure.

apexes of a tetrahedron and the alkyl groups located above each facial plane [12].

The exchange reaction between methyllithium and ethyllithium in ether solution was studied by utilizing the lithium-7 NMR technique [13]. The results imply that in solution, as in the solid state [12], the tetramer is tetrahedral. The ^{7}Li NMR spectra of mixtures of ethyl- and methyllithium in ether at $-80°$ are shown in Fig. 2. Solutions of the individual species exhibit single, sharp ^{7}Li resonances at room temperature, their respective chemical shifts differing by ~0.6 ppm. The ^{7}Li NMR spectrum of the mixed solutions at room temperature consists of a single line, the chemical shift being the weighted average of the methyllithium and ethyllithium values. However, as temperature is lowered, the line broadens and eventually a

$C_2H_5Li/CH_3Li = 0.20$ 0.33 1.0

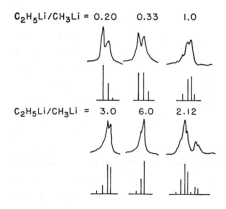

$C_2H_5Li/CH_3Li = $ 3.0 6.0 2.12

Figure 2. ^{7}Li resonance spectra of mixtures of ethyllithium and methyllithium in ether at $-80°$ [13]. The spectrum at lower right represents a solution containing lithium ethoxide in addition to methyl- and ethyllithium. The calculated spectra, based on the "local environment" hypothesis, are shown beneath the observed spectra.

reasonably well resolved spectrum is obtained again at about $-80°$. The relative intensities of the four moderately defined ^7Li resonance lines depend on the ethyl/methyl ratio. If it is assumed that the lithium atoms occupy the apexes of a regular tetrahedron and further that each of them interacts only with the three nearest alkyl neighbors (see Fig. 1), then there are four possible environments for lithium atoms in the mixed tetrameric species, as shown in Table 1. A lithium atom may interact with three methyls, three ethyls, two

Table 1   Number of Lithium Atoms in Each Type of Local Environment for All Possible Tetrameric Species in Methyllithium-Ethyllithium Mixtures [13]

	3Me 0Et	2Me 1Et	1Me 2Et	0Me 3Et
Li_4Me_4	4	0	0	0
Li_4Me_3Et	1	3	0	0
$Li_4Me_2Et_2$	0	2	2	0
Li_4MeEt_3	0	0	3	1
Li_4Et_4	0	0	0	4

methyls and one ethyl, or two ethyls and one methyl. For random distribution of methyl and ethyl groups in the mixed species, the relative abundances of these four types of environments can be readily calculated, and the results agree with the observed spectra if the four consecutive ^7Li resonances shown in Fig. 2 are attributed to the environments having increasing numbers of ethyl groups. Indeed, the chemical shifts of the extreme lines observed in the quartets are those of the pure methyllithium and ethyllithium, respectively. Thus the "local environment" hypothesis, advocated by Brown [14], seems to be fully justified. The preceding results show also that the alkyl groups in a mixed species do not undergo a rapid, intramolecular exchange. Had that been the case, five distinct NMR lines should have been observed in the spectrum of a mixture—each corresponding to a different type of tetramer.

Further evidence for the tetrahedral structure of the tetramer and support for the "local environment" hypothesis was provided by McKeever, Waack, Doran, and Baker [15, 16], who used ^{13}C-^7Li spin-spin coupling as a probe. On the basis of the "local environment" hypothesis a seven-line NMR spectrum is anticipated for the ^{13}C-enriched $(CH_3Li)_4$, the lithium resonance being split by one, two, or three neighboring ^{13}C atoms. However, had a rapid, intramolecular exchange of methyl groups taken place, each lithium atom would have interacted with four carbon atoms and the spectrum should have revealed nine lines.

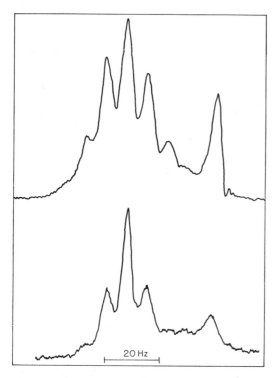

Figure 3. ^{7}Li NMR spectrum of ^{13}C-enriched methyllithium and lithium iodide in THF solution at $-80°$: upper, 51 % ^{13}C enrichment; lower, 25 % ^{13}C enrichment. The high-field resonance line is that of lithium iodide added for calibration purpose [16].

Table 2 Line Intensity Ratios from ^{7}Li NMR of CH$_3$Li [16]

Mole % ^{13}C	Solvent	Intensity Ratios[a]		
		Observed	Calculated 7-Line	Calculated 9-Line
25	THF	0.48, 0.10	0.44, 0.07 (0.53, 0.15)[b]	0.54, 0.13 (0.61, 0.21)[b]
51	THF	0.74, 0.34, 0.10	0.76, 0.31 0.05 (0.78, 0.38, 0.12)[b]	0.81, 0.41, 0.12, 0.03 (0.83, 0.47, 0.18, 0.07)[b]

[a] Calculated for a stick model. Intensity ratios are relative to center line of the multiplet, which has a relative defined intensity 1.00.
[b] Calculated spectrum using Lorentzian line shapes and a half-width of 5.0 Hz. The approximation, analogous to the stick model, favors the seven-line spectrum.

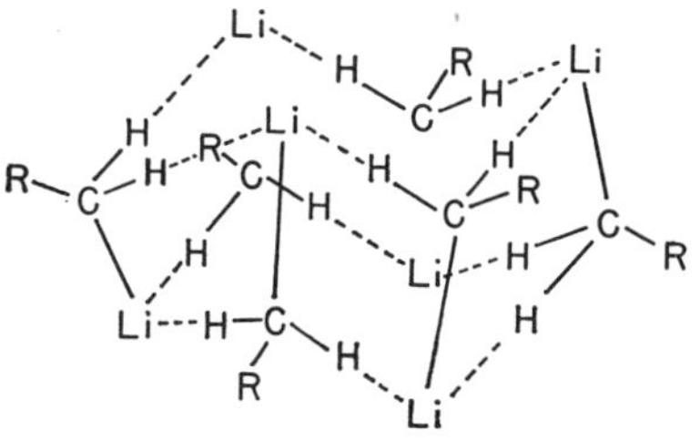

Figure 4. Hydrogen bridge-bonded structure for the *n*-butyllithium hexamer [21].

The ^{7}Li NMR spectrum of ^{13}C-enriched methyllithium in tetrahydrofuran solution at $-80°$C is shown in Fig. 3, illustrating its distinct dependence on the ^{13}C concentration. As seen from Table 2, the line intensity ratios, calculated on the basis of the "local environment" hypothesis (cf. Table 1), are in good accord with those observed. The alternative hypothesis leads to a nine-line coupling pattern with intensities which are in poor agreement with the observed ratios.

Similar ^{13}C-^{7}Li spin-spin coupling patterns have been observed for the *t*-butyllithium tetramer in hydrocarbon solution [17]. Interpretation of the spectral data again points to a tetrahedral structure for the tetramer. ^{13}C-^{7}Li scalar coupling was not observed for hydrocarbon solutions of α-^{13}C *n*-butyllithium [17]. In such solvents, *n*-butyllithium is reported to be hexameric [18] within a wide concentration range [19] and if its structure is octahedral [20] then $J_{^{13}C-^7Li}$ is expected to be in the 10–15 Hz range. The absence of scalar coupling may, however, be due to rapid interaggregate exchange. Alternatively, if the provocative hydrogen bridge-bonded hexameric structure suggested by Carubner [21] and illustrated in Fig. 4 is correct, then a distinctly different (if any) ^{13}C-^{7}Li coupling pattern would be expected.

Brown, Seitz, and Kimura [22] have reported that no ^{6}Li-^{7}Li scalar coupling is observed in alkyllithium compounds, and this implies a virtually zero lithium-lithium bond order in the tetramer. Such findings indicate that most of the bonding electrons are distributed in the regions adjacent to the bridging alkyl carbons and that the aggregation results from the multiple bridge bonds between lithium-carbon-lithium, and not from the hypothetical lithium-lithium bonding [12]. In accord with an earlier suggestion by Weiner, Vogel, and West [23], the cubic structure illustrated in Fig. 5 is the best representation of the lithium-carbon-lithium bonding in the alkyllithium tetramer.

Proton nuclear magnetic resonance studies by Brown, Dickerhoof, and Bafus [14] and Fraenkel, Adams, and Williams [24] revealed that the protons alpha to the lithium in alkyllithiums resonate at high field. These data reflect the partial polar character of the carbon-lithium bond leading to the substantial concentration of negative charge in the organic moiety causing the shielding of the alpha protons. However, recent studies of the ^{13}C resonance of alkyllithiums suggest that the carbon-lithium bond is less polar than previously thought [16]. The pertinent data for various alkyllithiums are summarized in Table 3. In each case, the carbon resonance of the organo-lithium is shifted only slightly upfield from that of the parent hydro-carbon.

In previous studies of arylmethyllithium reagents [25], in which there is appreciable electron delocalization, substantial differences were observed between the ^{13}C chemical shifts of the lithium reagents and that of the corresponding hydrocarbon. The downfield shift of the α-carbon-13 of the organolithiums relative to the corresponding hydrocarbons was explained by substantial sp^2 hybridization of the carbon bound to lithium. In contradis-tinction, the slight upfield ^{13}C chemical shift of the alkyllithiums suggests a predominant sp^3 hybridization of the carbon bonded to lithium [16, 17]. Furthermore, it seems that the excess charge density on this carbon is small, implying again that the carbon-lithium bond of these compounds is basically covalent. For example, excess charge density on the carbon in methyllithium over that in methane was estimated to be of the order of 0.1 electron [16] only.

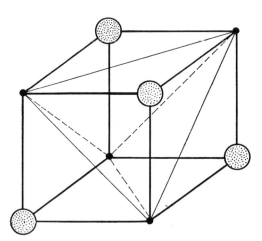

Figure 5.   Cubic structure for the methyllithium tetramer. Large strippled spheres represent the carbon atoms of the methyl groups, and small spheres represent lithium atoms [16]. This is a special case of the general structure shown in Fig. 1.

Surprisingly, the results summarized in Table 3 show that the value of $J_{^{13}C-^1H}$ for methyllithium and butyllithium is appreciably smaller than that reported for the corresponding sp^3 hybridized hydrocarbons. Apparently, due to reasons not yet understood, negative charge decreases $J_{^{13}C-^1H}$ values [25, 26].

Recent theoretical studies by Peyton and Glaze [27] and Cowley and White [28] provide additional support for the low degree of polar character of alkyl carbon-lithium bonds. The energy level diagram for the alkyllithium

Table 3   Nuclear Magnetic Resonance of Alkyllithiums

Compound	Solvent	$\delta(^{13}C)^a$ (ppm)	$J_{^{13}C-^1H}$ (Hz)	$J_{^{13}C-^7Li}$ (Hz)
MeLi	THF	+209	98	15
MeLi	Ether	+206	99	15
MeLi	Triethylamine	+205	97	15
Methane	Neat	+196^b	125	—
n-BuLi	Hexane	+182	100	—
n-BuLi	Ether	+182	98	14
n-Butane	Neat	+181^c	125	—
t-BuLi	Cyclohexane	+182	—	11
t-BuLi	Toluene	—	—	~10
Isobutane	Neat	+169^b	—	—

^{a} Relative to external carbon disulfide.
^{b} H. Spiesecke and W. G. Schneider, *J. Chem. Phys.*, **35**, 722 (1962).
^{c} E. G. Paul and D. M. Grant, *J. Am. Chem. Soc.*, **85**, 1701 (1963).

tetramer is given in Fig. 6. Application of the CNDO/2 method [27] as well as the LCAO-SCF approximation [28] lead to the conclusion that the formation of the polymer (tetramer) results in transfer of excess negative charge from carbon to lithium atoms. Indeed, Peyton and Glaze [27] suggest that the electron-deficient network in the polymer acts as an electron sink and the resulting delocalization of charge decreases the polar nature of the carbon-lithium bond. In agreement with the ^{13}C NMR data, these calculations indicate that the excess negative charge at the carbon atom in the methyllithium tetramer is low. According to the CNDO-SCF calculations [28] the carbon atoms in the tetramer are negatively charged but there is an appreciable positive overlap integral between carbon and lithium indicating the covalency of this linkage.

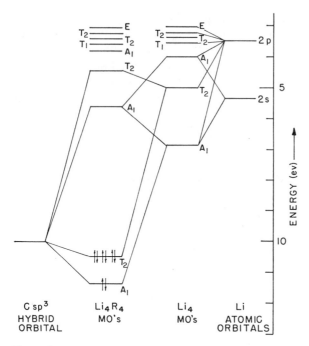

Figure 6.   Energy level diagram for alkyllithium tetramer [3].

In conclusion, it appears that chemical reactivity, NMR spectral studies, and theoretical calculations are in general accord, all indicating a predominant polar covalent character of carbon-lithium bonds in alkyllithiums.

2.2. Aryllithiums

Extensive NMR studies of phenyllithium have been reported [29–31]. In contrast to alkyllithiums, whose protons associated with carbons alpha to the lithium resonate at high field, the *ortho* protons in phenyllithium exhibit a low field resonance. Since only one low energy valence bond structure exists for the phenyl anion, protons *ortho* to the lithium are expected to experience a relatively large diamagnetic shift arising from the inductose effect. The same effect should operate to a lesser degree for the *meta-* and *para*-protons. The resonance of the *meta-* and *para*-protons is indeed upfield from those of benzene but, as noted, the *ortho*-protons resonate at lower field. This de-shielding effect has been accounted for in terms of the paramagnetic contribution to the chemical shift [29–31]. As mentioned earlier, this contribution arises from mixing of the ground state with low-lying excited states under the influence of the external magnetic field. In accord with the expected partial covalent character of the carbon-lithium bond, the excited state has been tentatively identified as π^* [29–31].

The ^{13}C chemical shifts of the *ortho* carbons, like those of the *ortho* protons, are downfield with respect to benzene [30]. Similarly, the resonance of the carbon alpha to the lithium is also shifted downfield. On the basis of charge polarization the opposite is anticipated; that is, the *ortho* carbon as well as the α-carbon should resonate at high field with respect to benzene. Thus, as in the proton NMR data [29, 31], the observed ^{13}C shifts have to be explained in terms of a paramagnetic contribution, requiring that the ground state be mixed with an excited state.

Interpretation of the NMR data for phenyllithium, like that of other organolithiums, is difficult because these species are apparently aggregated [32] in solution. Since the structure of the dissolved aggregates is unknown, it is difficult to unequivocally assess the effects on the chemical shifts. Even more perplexing is Ladd and Parker's recent suggestion that phenyllithium is monomeric in diethylether [66]. Based on the temperature dependence of the ^{7}Li NMR spectra of equimolar mixture of phenyl and p-tolyllithium, they argued that monomeric rather than dimeric aryllithium exists in the solution. However, by analogy with the spectral parameters of pyridine [29, 30], it was concluded that the carbon-lithium bond in phenyllithium is largely ionic in character; that is, the relevant spectra were attributed to the phenyl carbanion. Moreover, the ionic character of the carbon-lithium bond seems to decrease in the *meta* or *para* substituted phenyllithium [31].

2.3. Delocalized Organolithiums

In contrast to alkyllithiums, the bonding in the "delocalized" or resonance-stabilized organolithiums varies substantially with the nature of the organic moiety and the solvent. For example, the carbon-lithium bond in fluorenyllithium is essentially ionic in donor solvents, whereas the analogous bond is basically covalent in benzyllithium dissolved in hydrocarbons [33]. Many of the delocalized organolithiums form highly colored and stable species of relatively low reactivity [34].

The nature of the organic moiety R is expected to significantly influence reactivity as well as the nature of the carbon-lithium interaction. The polarity of a $R^{\delta-}$-$Li^{\delta+}$ bond depends on the stability of R^-, which is reflected in the pK_a values listed in Table 4. The lower the pK_a, the more stable the carbanion and the more "ionic" the bond.

Proton NMR spectra of three "delocalized" organolithiums, viz. triphenylmethyl, diphenylmethyl, and benzyllithium in tetrahydrofuran solution were reported by Sandel and Freedman [36]. The spectrum of triphenyllithium, shown in Fig. 7, is interpreted by first-order analysis. The three

Table 4 Relative Carbanion
Stabilities as Determined by
pK_a [35]

Compound	pK_a
Cumene	37
Benzene	37
Propene	35.5
Toluene	35
Diphenylmethane	35
Triphenylmethane	33
Fluorene	23
Indene	19
Cyclopentadiene	15
Fluoradene	11

multiplets have the predicted intensity ratios of 2:2:1. The spectra of these compounds in strong donor solvents such as dimethylsulfoxide and hexamethylphosphoramide are identical with that observed in THF, suggesting that in all these solvents the carbon-lithium bond in triphenylmethyllithium is virtually ionized. The charge distributions are summarized in Table 5.

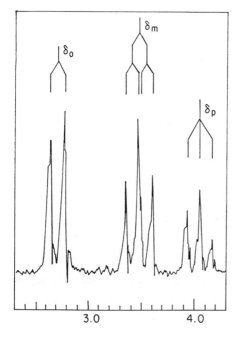

Figure 7. Proton NMR spectrum of triphenylmethyllithium in THF [36].

Table 5 Charge Distribution in Phenyl-Substituted Methyl Carbanions [36]

Carbanion	NMR Method[a]		SCF MO[a]	LCAO MO[a]
$(C_6H_5)_3C^-Li^+$	ortho	0.00	−0.05	−0.08
	meta	−0.08	−0.06	0.00
	para	−0.13	−0.10	−0.08
$(C_6H_5)_2CH^-Li^+$	ortho	−0.08	−0.08[b]	−0.10
	meta	−0.07	−0.06[b]	0.00
	para	−0.16	−0.23	−0.10
$C_6H_5CH_2^-Li^+$	ortho	−0.12	−0.14	−0.14
	meta	−0.10	−0.07	0.00
	para	−0.18	−0.23	−0.14

[a] In units of the absolute value of the charge of an electron.
[b] These are average values since the two ortho and meta positions in each ring are not equivalent.

They were estimated from the NMR data by using the conversion constant, 10 ppm/electron [7]. These values are compared with those derived from theoretical calculations and a particularly good qualitative agreement is found for those determined by the self-consistent field method. Recent theoretical studies by Hoffmann, Bissell, and Farnum [37] indicate that the stability of carbanions such as triphenylmethyl arises from a delicate balance of steric and conjugative effects.

Carbon-13 and lithium-7 NMR studies of these arylmethyllithiums were reported by Waack and co-workers [25]; their results are summarized in Table 6. In each case, the α-^{13}C resonance is shifted downfield from that of the parent sp^3 hybridized hydrocarbon. In the absence of rehybridization. δ_{13C} in aromatic molecules is proportional to the local charge density [8], For example, in cyclopentadienyllithium, δ_{13C} is shifted upfield by 30 ppm from that of cyclopentadiene. Hence, on the basis of the established relationships between charge density, hybridization, and ^{13}C chemical shifts, Waack and co-workers [25] have interpreted their data as a direct evidence for sp^2 hybridization of the α-carbon atoms in the arylmethyllithiums. However, the small ^{13}C shift observed in benzyllithium suggests that the carbon-lithium bond has more s character than in diphenyl- or triphenylmethyllithium, indicating a greater degree of covalency in benzyllithium than in the other two compounds.

Studies of the effect of solvent on the ^{13}C NMR shifts of benzyllithium revealed that the anion-cation interaction increases with decreasing solvent polarity [33]. Proton, carbon, and lithium-7 NMR data for benzyllithium in a variety of solvents are summarized in Table 7. As has been established

Table 6 ^{13}C and ^{7}Li NMR of Phenylmethyllithium Compounds [25]

Compound	Solvent	$\delta_{13_C}{}^a$	$J_{13_{C\text{-}H}}$ (cps)	$\delta_{Li}{}^b$	$\rho_{\alpha\text{-}C}{}^c$
$(C_6H_5)_3CLi$	THF	+102 (singlet)	—	+1.07 (sharp)	0.32 (0.13)
$(C_6H_5)_3CH$	CDCl$_3$	+132^d	107	—	—
$(C_6H_5)_2CHLi$	THF	+114 (doublet)	142	+1.16 (broad)	0.40 (0.08)
$(C_6H_5)_2CH_2$	CDCl$_3$	+157	126	—	—
$C_6H_5CH_2Li$	THF	+163 (triplet)	133	+0.12 (broad)	0.57 (0.38)
$C_6H_5CH_3$	Neat	+172^e	126 (THF)	—	—

^{a} In ppm relative to CS_2.

^{b} Relative to aqueous $LiNO_3$.

^{c} Hückel charge densities for anion calculated by standard LCAO MO methods. These values are given only to indicate relative magnitudes. The distribution of charge in these molecules no doubt differs considerably from that calculated according to Hückel theory owing to the influence of the negative charge on the effective electronegativity of the carbon cores and to anion-cation interaction. Values in parentheses are the sum of electron densities at the ring protons, as determined by the NMR method of ref. 2, subtracted from unity. Charge densities calculated by a self-consistent molecular orbital treatment are reported by A. Brickstock and J. A. Pople, *Trans. Faraday Soc.*, 50, 901 (1954).

^{d} It is interesting to compare these $\delta^{13}C$ with the values calculated using the bond parameters of G. B. Savitsky and K. Namikawa, *J. Phys. Chem.*, 68, 1956 (1964). The calculated values are 144 (($C_6H_5)_3CH$), 160 (($C_6H_5)_2CH_2$), and 176 ($C_6H_5CH_3$), and although uniformly high, are in reasonable agreement with the measured values.

^{e} P. C. Lauterbur, *Ann N.Y. Acad. Sci.*, 70, 841 (1958).

Table 7 NMR of Benzyllithium

Solvent	$\delta_{13_C}{}^a$ (ppm)	$J_{13_{C}\text{-}^1H}$ (Hz)	$\delta_{^1H}$ (ppm)^b para	α-CH$_2$	$\delta_{^7Li}{}^c$ (ppm)
THF	+163	132	+1.8	+0.79	+1.06
Et$_2$O/1.5 THF:BzLi	+167.6	129	—	—	
Et$_2$O	+168.5	135	+1.3	+0.70	+1.47
Benzene	+174.5	116	~+0.7	+0.21	+2.07
Toluene^d	+172				

^{a} Relative to external CS_2 (neat).

^{b} Relative to internal toluene (all chemical shifts upfield of toluene).

^{c} Relative to internal butyllithium (measured independently).

^{d} Chemical shift is for toluene in THF.

by previous proton NMR studies [36], negative charge is delocalized throughout the phenyl ring of benzyllithium. The extent of delocalization, reflected in the proton NMR shifts, is moderated by anion-cation interaction. Polar solvents interact with the lithium cation, moderate its Lewis acid strength, and facilitate charge separation [38]. This leads to greater shielding of the protons. On the other hand, solvation of the lithium cation is negligible in hydrocarbon solvents. Consequently, lithium interacts strongly with the negatively polarized organic moiety and thus the shielding of the protons is reduced.

Studies of the solvent dependency of the electronic absorption spectra of 1,1-diphenyl-*n*-hexyllithium established that the λ_{max} shifts from 410 mμ in hexane to 496 mμ in THF [39]. This apparently reflects the increased charge delocalization in the anion caused by the increased solvation of the lithium cation in more polar solvents. The proton NMR spectrum of diethylzinc-1,1-diphenyl-*n*-hexyllithium mixtures showed that interaction of the organolithium with the Lewis acid, diethylzinc, decreases the charge delocalization [40]. In this system, the equilibrium between complexed and uncomplexed organolithium is observed. The change in fractional charge density occurring at each of the *ortho*, *meta*, and *para* protons varies with the organolithium/diethylzinc ratio, and it is probable that a corresponding change occurs at the α-carbon. Thus the result of anion-Lewis acid interaction, considering the incipient lithium cation to be a Lewis acid of strength varying with the extent of solvent interaction, is to transfer some of the negative charge density from the anion to the acceptor.

Although the charge density in the ring can be evaluated from the proton NMR data [36], the evaluation of the charge density on the α-carbon is less certain. The negative charge density on the organic moiety cannot exceed unity, this extreme case probably being attained in the most polar solvents. The maximum of charge density on the α-carbon may be evaluated by difference using the change of the chemical shift of the *para* proton as a diagnostic tool. The evidence based on changes observed in chemical shifts of *all* the protons is less desirable, because the chemical shifts of the *ortho* and *meta* protons are affected by changes in molecular geometry and by magnetic anisotropy [41], whereas the chemical shift of the *para* proton is influenced to a lesser degree by these factors.

As noted earlier, ^{13}C chemical shifts vary substantially with the state of hybridization and charge density. The conversion constant of 160–200 ppm/electron, established for sp^2-carbon, apparently is also applicable for sp^3-carbon [42–45]. The ^{13}C-^{1}H coupling constants, like ^{13}C chemical shifts, are also useful in assessing the state of carbon hybridization, although the $J_{^{13}C^{-1}H}$ seems to be by several hertz smaller for species bearing greater negative charge density than those of the corresponding neutral molecule [25].

The proton NMR data (cf. Table 7) indicate that charge density in the ring of benzyllithium in THF solution (relative to benzene) is $\sim$0.6 electron [36]. Thus only $\sim$0.4 of electron charge remains on the α-carbon (assuming unit charge density on the benzyl moiety). In benzene, the proton chemical shifts indicate that the excess charge density on the ring decreases to $\sim$0.2 electron. On the assumption that the α-carbon charge density is reduced in the same proportion as that of the ring, the maximum density on the α-carbon of benzyllithium in benzene is calculated to be $\sim$0.15 electron. Similar changes in the proton NMR shifts (and corresponding charge densities) in benzene and the THF solutions have been observed by Bywater and Worsfold [67], although the magnitudes of the shifts are not precisely the same as in Table 7.

Assuming sp^2 hybridization of benzyllithium in THF solution, it is concluded that $\sim$0.8 excess electron density is needed on the α-carbon to account for the experimentally observed ^{13}C chemical shift. This is unlikely in view of the high ring charge densities estimated from ^1H NMR data. Hence the interpretation of the ^{13}C chemical shift of benzyllithium in THF can only be reconciled with the other observations if a substantial degree of sp^3 character is attributed to the α-carbon. The degree of sp^3 character increases as polarity of the solvent decreases, this being indicated by the upfield shift of δ_{13C}. Similar conclusions become apparent by considering the change occurring in $J_{13C\text{-}1H}$ of the α-carbon of benzyllithium caused by solvent, that is, $J_{13C\text{-}1H}$ decreases from 132 Hz in THF to 116 Hz in benzene, reflecting increased sp^3 character in benzene.

Thus both chemical shifts and coupling constants can be utilized to show how solvent influences the degree of coordination between anion and cation and consequently how the state of α-carbon hybridization in compounds such as benzyllithium varies with the solvent's nature.

In conclusion, the dependency of the NMR data on solvent is interpreted as evidence that the α-carbon of benzyllithium has a considerable sp^3 character which increases with decreasing solvent polarity. Similar NMR studies of 1,1-diphenyl-n-hexyllithium lead McKeever and Waack [68] to conclude that the larger π-system of the diphenyl-n-hexyl moiety moderated the dependency of anion-cation interaction on solvent and resulted in greater (than in benzyllithium) sp^2 character to the α-carbon. In contrast, the proton NMR spectra of styryllithium in benzene and THF are very similar, suggesting that styryllithium is largely ionic in both polar and nonpolar solvents [67]. In their NMR study of 1,1-diphenyl-n-butyllithium (DPB), Okamoto and Yuki [71] concluded that DPB exists as solvent separated ion pairs in THF but is dimeric in benzene solvent.

The nature of the carbon-lithium bond in allylic type organolithiums is of great interest. West and co-workers [46] interpreted the proton NMR spectrum of allyllithium in tetrahydrofuran and diethylether as indicative of a predominantly delocalized organolithium species. The proton spectrum was examined over a temperature range -100 to $+60°$. The AA'BB'C pattern observed at lower temperatures changes reversibly into an AB_4 pattern at higher temperatures:

$$\left[\begin{array}{c} H_C \\ H_{B'} \qquad H_B \\ {}^*H_{A'} \quad {}^*H_A \end{array} \right] Li^+ \rightleftharpoons \left[\begin{array}{c} H_C \\ {}^*H_{B'} \qquad {}^*H_B \\ H_{A'} \quad H_A \end{array} \right] Li^+$$

All four terminal protons are equivalent at $37°$, but at $-87°$ two distinct types of terminal hydrogen are observed. The ^7Li resonance remains essentially unchanged in both THF and ether solution over the whole temperature range -80 to $25°$ [47]. Similarly, the line-width remains essentially constant, being only slightly broader at $-80°$, presumably due to an increase in solvent viscosity.

Interpretation of the NMR spectral data is complemented by ultraviolet and infrared spectral data. Examination of UV spectra over the temperature range -100 to $25°$ reveals an absorption peak within the range 310–320 mμ, in agreement with the theoretically estimated absorption maximum for the allyl carbanion of 326 mμ [48]. Thus all the spectral data confirm the predominantly "delocalized" ionic structure of the organic moiety of allyllithium in ethers.

Nuclear magnetic resonance techniques are useful in studying the nature of the chain ends of "living" polydienes [49–55] such as polybutadienyllithium and polyisoprenyllithium. Addition of sec-butyl or tert-butyllithium to butadiene or isoprene yields polymers of low degree of polymerization, and consequently it becomes feasible to examine the proton spectrum of a "living" polymer [56].

Glaze and Jones [54] concluded from their NMR spectral data on cyclohexane solutions of 1:1 adducts of s- and t-butyllithium and 1,3-butadiene that in this nonpolar solvent there are two organolithium species, I and II, both σ-bonded allyl type. They are formed in a 3:1 ratio and apparently do not interconvert on the NMR time scale.

Ultraviolet and infrared spectral data complement the NMR results. It is notable that the addition of a donor solvent, THF, changes the spectrum significantly; it becomes consistent with that of a delocalized species, III.

$$\left\{ \begin{array}{c} \text{R—CH}_2 \overset{\text{H}}{\underset{\text{H}}{\text{C}}} = \overset{\text{H}}{\underset{\text{H}}{\text{C}}} \overset{\text{H}}{\underset{\text{H}}{\text{C}}} \end{array} \right\}^{-} ; \text{Li}^+$$

III

Similar conclusions were drawn by Naylor, Hsieh, and Randall [55], who examined the ¹H-NMR spectrum of *s*-butyllithium/1,3-butadiene adducts in benzene-d_6.

The proton spectra of "living" polyisoprene oligomers were studied by Schué, Worsfold, and Bywater [51–53] and are shown in Fig. 8. The NMR data for *s*- or *t*-butyl adducts of isoprene in benzene solution indicate that *trans* and *cis* isomers are formed.

$$\underset{trans}{\overset{\text{Li}^+}{ \underset{\text{R—CH}_2 \qquad \text{CH}_3}{\overset{\text{H}_{\delta^-} \qquad \text{CH}_2 \delta^-}{\text{C}=\text{C}}} }} \qquad \underset{cis}{ \underset{\text{R—CH}_2 \qquad \underset{\text{Li}^+ \delta^-}{\text{CH}_2}}{\overset{\text{H}_{\delta^-} \qquad \text{CH}_3}{\text{C}=\text{C}}} }$$

The high-field resonance of the terminal methylene was explained by its close proximity to the lithium. Since the olefinic carbon-3 also exhibited a high-field resonance, some charge was apparently delocalized to this position.

The polyisoprenyllithium oligomer retains its structure, characteristic of benzene solution, when transferred from that solvent to THF ($-80°$) although the two olefinic triplets (—CH=) moved appreciably upfield indicating redistribution of charge to carbon-3. Upon warming to $-40°$, a complete *cis/trans* isomerization occurred in about an hour. This observation shows again how sensitive the carbon-lithium interaction is to the solvent environment.

An equilibrium between covalent and delocalized oligomers of living polyisoprene was suggested by Morton and co-workers [49, 50], who studied their NMR spectra. In their later study [50], ethyllithium-d_5 was added to 1,3-butadiene-d_6, which was then "terminated" with 1,3-butadiene. Thus it was assumed that only the proton resonance of the "living" chain end was observed.

Nuclear magnetic resonance studies also provide an interesting insight into the nature of the propagation reaction in diene polymerization. The polymer microstructure is predominantly 1,4 [57] when the polymerization proceeds in hydrocarbon solvent. The occurrence of 1,4- addition has been accounted for in terms of formation of a complex between the living polymer

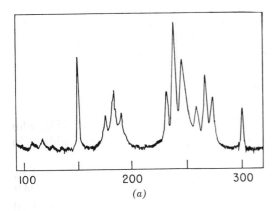

(a)

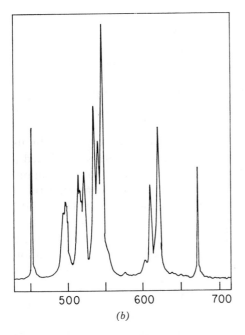

(b)

Figure 8. Proton spectrum of polyisoprenyllithium in benzene [53]. (*a*) 100 Mc NMR spectrum of polyisoprenyllithium (DP ca. 1.2) in benzene at 40°C. Olefinic proton region. Side bands at 150 and 300 cycles upfield from benzene. (*b*) 100 Mc NMR spectrum of the same polymer in benzene at 60°C in the high-field region. Side bands at 452 and 672 cycles upfield from benzene. Deuterated tert-butylltihium used as initiator to remove interference from butyl group signals.

281

and the monomer [58]. However, Morton and co-workers suggest on the basis of their NMR results that the microstructure is determined after the addition of the next monomer unit [50]. Addition of monomer to the living polymer is exclusively 1,4- (4,1- in the case of isoprene). Vinyl 1,2-type microstructure is suggested to result from the addition of a delocalized tautomeric species in a 1,4-manner. Thus the amount of 1,2-adduct is determined by the nature of the living polymer, rather than specific complexation with monomer.

These results are supported by the studies of Randall and co-workers who also concluded that the configuration of the ultimate unit is determined at the moment of further addition of butadiene. In their proposed second-order Markov process, the final configuration would depend not only on the manner in which the reacting unit approaches, but also on the configuration sequence of the two preceding units [69]. Thus all the evidence available now indicates that the configuration of the butadiene terminal unit is not "fixed" until after the addition of a next monomer unit [69–70].

Sandel, Freedman, and co-workers [59, 60] have examined by proton NMR techniques the rotational barriers in phenylallyllithium derivatives.

They generated the 1,3-diphenylallyllithium anion from *cis* and *trans* 1,3-diphenylpropene [59]. The NMR data imply that this largely delocalized anionic species has a low activation energy for rotation around the partial double bonds. On the basis of the chemical shift of the β proton, the AB_2 type spectrum was interpreted to conform with the *trans, trans* structure although the reaction could lead to other geometrical isomers. Only the *trans, trans* isomer can attain a completely coplanar, strain-free conformation. Similar results were reported by Heiszwolf and Kloosterzeil [61].

Sandel, McKinley, and Freedman [60] examined the temperature dependence of the proton spectra of phenylallyllithium, -sodium, and -potassium. The proton NMR spectra for phenylallyllithium and pentadeuteriophenylallyllithium are shown in Fig. 9. The downfield shift of the central (δ_C) proton and high-field resonance of the *para* phenyl proton seem to indicate a *trans* geometry at the 2,3 carbon-carbon bond. It is also suggestive of the ionic nature of the species. Rotational barriers were determined to be $\Delta H^{\ddagger} = 19.8$ kcal/mole for the rotation about the C_1-C_2 bond and $\Delta H^{\ddagger} = 13.9$ kcal/

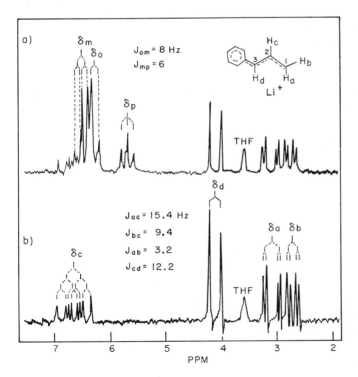

Figure 9. Nuclear magnetic resonance spectra, 60 MHz, at 5° in THF-d_8 of (*a*) phenylallyllithium and (*b*) pentadeuteriophenylallyllithium [60].

mole for rotation around the phenyl-carbon-3 bond. The respective π-bond orders, predicted from Hückel molecular orbital theory, are in good qualitative agreement with such rotational barriers. Although the authors favored a delocalized species [59, 60], they recognize that their results are also consistent with an equilibrium between a covalent and ionic species:

In this interpretation, the data would provide the temperature dependence of k_1 and k_2, the rate of collapse of the ion pair, and the rate of ionization of the covalent species. Differentiation between the two alternative interpretations could be possible if a system involved more than two coexisting rotamers. The rotational barrier is lower in ether than in THF. Similarly, the proton resonance is shifted downfield in ether from that in THF. Both these observations may indicate decreased ionic character of the carbon-lithium bond in the less polar ether solvent.

Bates, Gosselink, and Kaczynski [62] examined rotational barriers of various pentadienyllithiums and interpreted their data in terms of an equilibrium between a covalent and ionic organolithium species. Similar explanations have been invoked to explain proton NMR data of other allyl carbanions [63, 64].

In conclusion, examination of the temperature dependency of reversible changes observed in the proton magnetic resonance spectra of the substituted allyllithiums [46; 59–63] suggests that these species exist predominantly as delocalized ionic moieties. Furthermore, exchange of the terminal hydrogens is rapid at higher temperatures either as a result of rotation around partial double bonds in the allylic anion or via transient covalent bond formation with the lithium cation to give the simple covalent organolithium compounds with a single carbon-carbon bond which is available for rapid rotation.

Klein and Brenner [65] examined proton NMR spectra of propargylic dianions. The data indicated that delocalization of charge in the pentenylic anion is less than that found in allylic systems [59, 60] or pentadienyl systems [62]. Invoking the principle of maximum π-orbital overlap, the authors proposed that the carbanionic species has the linear acetylene or a sesqui acetylene structure:

$$R-C\!\equiv\!C\!\equiv\!C-R'^{2-}$$

Unequivocal interpretation of the NMR data for organolithiums is difficult because these species are highly aggregated in solution [32, 46]. For example, as noted by West and co-workers [46], allyllithium at a concentration of $1.5M$ in ether has an apparent degree of aggregation of ~10, whereas in tetrahydrofuran solution, the degree of aggregation is about 1.4 at $0.8M$. Thus consideration of these complex organolithiums as discrete ions, ion pairs, or solvent-separated ion pairs is an oversimplification.

Although completely unequivocal NMR studies of structure and bonding in organolithiums is not yet possible, this assessment of the current status of this rapidly expanding field suggests that it will continue to be a fruitful area of investigation.

REFERENCES

1. H. Gilman, in *Organic Chemistry*, Vol. I, H. Gilman, Ed., Wiley, New York, 1943, pp. 520–24.

2. R. Waack and M. A. Doran, *J. Am. Chem. Soc.*, **91**, 2456 (1969); P. West and R. Waack, *J. Am. Chem. Soc.*, **92**, 840 (1970).

3. T. L. Brown, in *Advances in Organometallic Chemistry*, F. A. G. Steon and R. West, Eds., Academic Press, New York, 1965, pp. 365–395.

4. T. L. Brown, *Accts. Chem. Res.*, **1**, 23 (1968).

5. G. E. Coates, M. L. H. Green, and K. Wade, *Organometallic Compounds*, Vol. I, Methuen, London, 1967, pp. 1–42.

6. J. A. Pople, *Proc. Roy. Soc., London*, **A239**, 550 (1957).

7. G. Fraenkel, R. E. Carter, A. McLachlan, and J. H. Richards, *J. Am. Chem. Soc.*, **82**, 5846 (1960).

8. H. Spiesecke and W. G. Schneider, *Tetrahedron Lett.*, **14**, 468 (1961).

9. P. C. Lauterbur, *Tetrahedron Lett.*, **14**, 274 (1961).

10. J. W. Emsley, J. Feeney, and L. H. Sutcliffe, *High Resolution Nuclear Magnetic Resonance Spectroscopy*, Vol. 2, Pergamon Press, Oxford, 1966, pp. 988–1031.

11. F. A. Cotton and G. Wilkinson, *Advanced Inorganic Chemistry*, Interscience, New York, 1962, pp. 157–159.

12. E. Weiss and E. A. C. Lucken, *J. Organomet. Chem., Amsterdam*, **2**, 197 (1964).

13. L. M. Seitz and T. L. Brown, *J. Am. Chem. Soc.*, **88**, 2174 (1966).

14. T. L. Brown, D. W. Dickerhoof, and D. A. Bafus, *J. Am. Chem. Soc.*, **84**, 1371 (1962).

15. L. D. McKeever, R. Waack, M. A. Doran, and E. B. Baker, *J. Am. Chem. Soc.*, **90**, 3244 (1968).

16. L. D. McKeever, R. Waack, M. A. Doran, and E. B. Baker, *J. Am. Chem. Soc.*, **91**, 1057 (1969).

17. L. D. McKeever and R. Waack, *Chem. Comm.*, **1969**, 750.

18. D. Margerison and J. P. Newport, *Trans. Faraday Soc.*, **59**, 2058 (1963).

19. H. L. Lewis and T. L. Brown, private communication.

20. T. L. Brown, R. L. Gerteis, D. A. Bafus, and J. A. Ladd, *J. Am. Chem. Soc.*, **86**, 2135 (1964).

21. I. Craubner, *Z. Physik. Chem.*, **51**, 225 (1967).

22. T. L. Brown, L. M. Seitz, and B. Y. Kimura, *J. Am. Chem. Soc.*, **90**, 3245 (1968).

23. M. Weiner, G. Vogel, and R. West, *Inorg. Chem.*, **1**, 654 (1962).

24. G. Fraenkel, D. G. Adams, and J. Williams, *Tetrahedron Lett.*, 767 (1963).

25. R. Waack, M. A. Doran, E. B. Baker, and G. A. Olah, *J. Am. Chem. Soc.*, **88**, 1272 (1966).

26. R. M. Hammaker, *J. Mol. Spectrosc.*, **15**, 506 (1965).

27. G. R. Peyton and W. H. Glaze, *Theoret. Chim. Acta, Berl.*, **13**, 259 (1969).

28. A. H. Cowley and W. D. White, *J. Am. Chem. Soc.*, **91**, 34 (1969).

29. G. Fraenkel, D. G. Adams, and R. R. Dean, *J. Phys. Chem.*, **72**, 944 (1968).

30. A. J. Jones, D. M. Grant, J. G. Russell, and G. Fraenkel, *J. Phys. Chem.*, **73**, 1624 (1969).

31. J. A. Ladd, *Spectrochim. Acta*, **22**, 1157 (1966); J. A. Parker and J. A. Ladd, *J. Organomet. Chem.*, **19**, 1 (1969).

32. P. West and R. Waack, *J. Am. Chem. Soc.*, **89**, 4395 (1967).

33. R. Waack, L. D. McKeever, and M. A. Doran, *Chem. Comm.*, **1969**, 117.

34. R. Waack and M. A. Doran, *J. Am. Chem. Soc.*, **85**, 1651 (1963).

35. D. J. Cram, *Fundamentals of Carbanion Chemistry*, Academic Press, New York, 1965, Chapter 1.

36. V. R. Sandel and H. H. Freedman, *J. Am. Chem. Soc.*, **85**, 2328 (1963).

37. R. Hoffmann, R. Bissell, and D. G. Farnum, *J. Phys. Chem.*, **73**, 1789 (1969).

38. R. Waack, M. A. Doran, and P. E. Stevenson, *J. Am. Chem. Soc.*, **88**, 2109 (1966).

39. R. Waack and M. A. Doran, *J. Phys. Chem.*, **67**, 148 (1963).

40. R. Waack and M. A. Doran, *J. Am. Chem. Soc.*, **85**, 4042 (1963).

41. T. K. Wu and B. P. Dailey, *J. Chem. Phys.*, **41**, 2796 (1964).

42. T. Yonczawa, I. Morishima, and H. Kato, *Bull. Chem. Soc., Japan*, **39**, 1398 (1966).

43. B. V. Cheney and D. M. Grant, *J. Am. Chem. Soc.*, **89**, 5319 (1967).

44. T. D. Alger, D. M. Grant, and E. G. Paul, *J. Am. Chem. Soc.*, **88**, 5397 (1966).

45. W. J. Horsley and H. Sternlicht, *J. Am. Chem. Soc.*, **90**, 3738 (1968).

46. P. West, J. I. Purmont, and S. V. McKinley, *J. Am. Chem. Soc.*, **90**, 797 (1968).

47. P. West and L. D. McKeever, unpublished results.

48. K. Kuwata, *Bull. Chem. Soc., Japan*, **33**, 1091 (1960).

49. R. Sakata, R. D. Sanderson, and M. Morton, ACS Central Regional Meeting, Akron, Ohio, May, 1968.

50. R. D. Sanderson and M. Morton, Gordon Research Conference on Elastomers, New London, New Hampshire, July, 1969. M. Morton, R. D. Sanderson, and R. Sakata, *Polym. Lett.*, **9**, 61 (1971).

51. S. Bywater, Gordon Research Conference on Elastomers, New London, New Hampshire, July, 1969.

52. F. Schué, D. J. Worsfold, and S. Bywater, Fifteenth Canadian High Polymer Forum, Kingston, Ontario, September, 1969.

53. F. Schué, D. J. Worsfold, and S. Bywater, *Polym. Lett.*, **7**, 821 (1969); *Macromolec.*, **3**, 509 (1970).

54. W. H. Glaze and P. C. Jones, *Chem. Comm.*, **1969**, 1434.

55. F. E. Naylor, H. L. Hsieh, and J. C. Randall, *Polymer Preprints*, **II**, 90 (1970); *Macromolec.*, **3**, 486 (1970).

56. M. Szwarc, *Carbanions, Living Polymers, and Electron Transfer Processes*, Interscience, New York, 1968.

57. W. Cooper and G. Vaughan, *Recent Developments in the Polymerization of Conjugated Dienes*, in *Progress in Polymer Science*, Vol. 1, 1967.

58. M. Szwarc, *Carbanions, Living Polymers, and Electron Transfer Processes*, Interscience, New York, 1968, p. 436.

59. H. H. Freedman, V. R. Sandel, and B. P. Thill, *J. Am. Chem. Soc.*, **89**, 1762 (1967).

60. V. R. Sandel, S. V. McKinley, and H. H. Freedman, *J. Am. Chem. Soc.*, **90**, 495 (1968).

61. G. J. Heiszwolf and H. Kloosterziel, *Rec. Trav. Chim.*, **86**, 1345 (1967).

62. R. B. Bates, D. W. Gosselink, and J. A. Kaczynski, *Tetrahedron Lett.*, 205 (1967).

63. D. Seyferth and T. F. Jula, *J. Organomet. Chem.*, *Amsterdam*, **8**, P13 (1967).

64. E. Grovenstein, S. Chondra, C. E. Collum, and W. E. Davis, Jr., *J. Am. Chem. Soc.*, **88**, 1275 (1966).

65. J. Klein and S. Brenner, *J. Am. Chem. Soc.*, **91**, 3904 (1969).

66. J. A. Ladd and J. Parker, *J. Organometal. Chem.*, *Amsterdam*, **28**, 1 (1971).

67. S. Bywater and D. J. Worsfold, private communication.

68. L. D. McKeever and R. Waack, *J. Organometal. Chem.*, *Amsterdam*, **28**, 145 (1971).

69. J. C. Randall and R. S. Silas, *Macromolec.*, **3**, 491 (1970).

70. J. C. Randall, F. E. Naylor, and H. L. Hsieh, *Macromolec.*, **3**, 497 (1970).

71. Y. Okamoto and H. Yuki, *J. Organometal. Chem.*, *Amsterdam*, **32**, 1 (1971).

7

Nuclear Magnetic Resonance Studies of Alkali Radical Ion Pairs

EGBERT DE BOER and JAN L. SOMMERDIJK*

Department of Physical Chemistry, University of Nijmegen, Nijmegen, The Netherlands

* Present address: Philips Research Laboratories Eindhoven, The Netherlands.

1. INTRODUCTION

Paramagnetic systems can be investigated by nuclear magnetic resonance [1–3] as well as by electron spin resonance. Since each group of equivalent nuclei in a radical ion pair is characterized by only one single NMR signal, the NMR spectrum of a radical ion pair is, in most cases, more easily interpreted than its corresponding ESR spectrum. A special advantage of the NMR method arises from the fact that the sign as well as the magnitude of the hyperfine splitting constant (HFSC) of a given nucleus can be inferred directly from the sign and magnitude of the relevant contact shift observed in the NMR spectrum of the investigated species, while only the magnitude, but not sign, is given by ESR. Small HFSCs, falling below the range of resolution of an ESR spectrometer, can be measured by NMR spectroscopy, and this advantage still does not preclude determination of large HFSCs up to 5.0 gauss. Since every nucleus endowed with a magnetic moment can be studied by NMR, this technique could be used directly to investigate the state of the alkali nuclei in radical ion pairs; resonance experiments can be performed on the aromatic moiety of the ion pair as well as on the cation. The study of the line-widths of the resonance lines may provide information about intramolecular relaxation processes and this in turn may give information about the structure of the ion pair.

Hitherto the NMR technique has not been used extensively in studies of alkali radical ion pairs. The high radical concentration (about $0.1–1M$) necessary to obtain sufficiently short correlation times and a plausible intensity of the signal complicates such investigations. Nevertheless, a few systems have been studied recently; these include the alkali biphenyl [4–8] and the alkali naphthalene systems [9, 10].

2. GENERAL CONSIDERATIONS

2.1. Contact and Paramagnetic Shifts

NMR spectra of paramagnetic particles in solution are modified in two ways with respect to the spectra of the corresponding diamagnetic species. First, the resonance lines are broadened; second, the lines are shifted by the Fermi contact interaction. The physical principles giving rise to these effects have been described in a review article by de Boer and van Willigen [2].

For radicals in the doublet ground state, the Fermi contact shift δ_c^0, expressed in gauss, is given by [2, 11]

$$\delta_c{}^0 = -\frac{(g\beta_e)^2}{g_N\beta_N}\frac{aH}{4kT} \tag{1}$$

where g is the isotropic g-value of the radical (usually that of the free electron), a is the HFSC (expressed in gauss), and H is the magnetic field at which resonance occurs; the other symbols in Eq. 1 have their usual meaning.

Since

$$a = \frac{8\pi}{3} g_N\beta_N\rho_N \tag{2}$$

the contact shift $\delta_c{}^0$ is independent of the magnetic moment of the nucleus. From the direction of the shift the sign of the spin density can be inferred; a high field shift indicates a negative ρ_N, a low field shift indicates a positive ρ_N.

In a partly reduced solution of negatively charged paramagnetic ions the observed contact shift δ_c is given by [11–13]

$$\delta_c = f_P \delta_c{}^0 \tag{3}$$

where f_P is the fraction of negative particles. By measuring δ_c as a function of the degree of the reduction f_P, we should be able to construct a linear plot of δ_c against f_P. From the slope of this straight line the HFSC can be determined.

The determination of f_P involves some problems. In principle f_P can be determined from the paramagnetic shift of the solvent resonance signal δ_s caused by the paramagnetic particles. In the absence of special effects, δ_s should be linearly related to the susceptibility change of the solution when paramagnetic ions are introduced. For cylindrical sample tubes δ_s is given by the Langevin formula [14],

$$\delta_s(\text{in ppm}) = -\tfrac{2}{3}\pi 10^3 \frac{N\beta_e{}^2}{kT} c \tag{4}$$

N being Avogadro's number, β_e the Bohr magneton, and c the concentration of radicals in moles/liter. This expression has often been utilized to determine the concentration of paramagnetic particles from the observed solvent shift and, in the absence of scalar interactions with the solvent molecules, the procedure appears to be reliable [11, 12]. However, recent experiments indicate that the nuclear shift of ethereal solutions of certain radical anion salts is considerably smaller than the theoretically predicted value [7, 15].

de Boer et al. [16] found that the molar shift for a solution of sodium triphenylene in tetrahydrofuran is 20% lower than the theoretical value of 2.57 ppm at 33°C and that a 0.1M solution of the green sodium coronene salt at 33°C showed a solvent shift of only 3–4 Hz with respect to pure tetrahydrofuran instead of the theoretically expected value of 15 Hz. Addition of equimolar amounts of glymes to these solutions (see Chapter 3) shifted the THF peaks to their theoretically predicted position, whereas the CH_2 peaks of the glymes were shifted by more than 4 ppm to low field in the case of sodium triphenylene, and by 6–10 ppm for the sodium coronene system [17]. These results demonstrate the existence of a Fermi contact interaction between the unpaired electron and the protons of those solvent molecules, which are coordinated with the ion pairs. This interaction makes a concentration determination based on the theoretical value of the molar shift unreliable, although δ_s is still linearly related to the concentration of radicals [16]. Therefore the fraction of paramagnetic particles f_P should be obtained from the relation

$$f_P = \frac{\delta_s}{\delta_{s,\max}} \tag{5}$$

where $\delta_{s,\max}$ is the maximum solvent shift observed on completion of the reduction.

2.2. Broadening Mechanisms

The line-width of a resonance peak depends on intermolecular and intramolecular interactions. For a resonating nucleus in a paramagnetic species the intramolecular interactions are the most important, and the line-width parameter T_2 is governed by a sum of three intramolecular contributions,

$$T_2^{-1} = (T_2^{-1})_{Fc} + (T_2^{-1})_D + (T_2^{-1})_Q, \tag{6}$$

the three terms representing the contributions due to the Fermi contact interaction, the anisotropic electron-nucleus dipolar interaction, and the quadrupole interaction, respectively. For a radical undergoing a rapid Brownian motion in solution, the following expressions have been derived for the effect of the three different types of interactions upon T_2^{-1} [1, 18–20]:

$$(T_2^{-1})_{Fc} = \frac{1}{3} a^2 \left(\frac{g\beta_e}{\hbar}\right)^2 S(S+1)\left\{\tau_e + \frac{\tau_e}{1+\omega_e^2\tau_e^2}\right\} \tag{7}$$

$$(T_2^{-1})_D = \frac{1}{15} \frac{(g\beta_e g_N\beta_N)^2}{\hbar^2 r^6} S(S+1)\left\{7\tau_d + \frac{13\tau_d}{1+\omega_e^2\tau_d^2}\right\} \tag{8}$$

$$(T_2^{-1})_Q = \frac{3}{40} f(I)\left(\frac{e^2Qq}{\hbar}\right)^2\left(1+\frac{\eta^2}{3}\right)\tau_r \tag{9}$$

where τ_e is the electron correlation time and τ_d the dipolar correlation time defined by

$$\tau_d^{-1} = \tau_e^{-1} + \tau_r^{-1} \tag{10}$$

in which τ_r is the rotational correlation time. The factor $f(I)$ is given by

$$f(I) = \frac{2I + 3}{I^2(2I - 1)} \tag{11}$$

eQ is the quadrupole moment and eq the electric field gradient. The other symbols in Eqs. 7, 8, and 9 have their usual meaning.

Equation 8 is not applicable to radical-ions. It has been derived for the interaction involving magnetic point dipoles separated by a distance r [18]; however, the unpaired electron delocalized over the whole radical-ion must be described by a molecular orbital and not by a magnetic point dipole. Extending Solomon's theory for point dipoles [18], it can be shown [21] that for $S = \frac{1}{2}$, Eq. 8 must be replaced by

$$(T_2^{-1})_D = \frac{(D:D)}{120\hbar^2}\left[7\tau_d + \frac{13\tau_d}{1 + \omega_e^2\tau_d^2}\right] \tag{12}$$

where

$$(D:D) = \sum_{\alpha,\beta}(D_{\alpha\beta})^2 \tag{13}$$

in which α and β are components of r (the distance between the unpaired electron and nucleus N) referred to the molecular coordinate frame. The elements of the D tensor are given by

$$D_{\alpha\beta} = \langle\varphi_0|\frac{r^2\,\delta_{\alpha\beta} - 3\alpha\beta}{r^5}|\varphi_0\rangle g\beta_e g_N\beta_N \tag{14}$$

in which φ_0, the MO containing the unpaired electron, can be expressed as linear combination of the carbon $2p_z$ functions χ_i

$$\varphi_0 = \sum_i C_{0i}\chi_i \tag{15}$$

Defining

$$\rho_{ij} = C_{0i}C_{0j}^* \tag{16}$$

and

$$D_{ij} = \langle\chi_i|\frac{r^2\,\delta_{\alpha\beta} - 3\alpha\beta}{r^5}|\chi_j\rangle g\beta_e g_N\beta_N \tag{17}$$

we find for the elements of the D tensor

$$D_{\alpha\beta} = \sum_{i,j}\rho_{ji}D_{ij} = T_r(\rho D) \tag{18}$$

where T_r stands for the sum of the diagonal elements of the product matrix

ρD. The elements D_{ij} can be calculated by using the formulas of McConnell and Strathdee [22] and Derbyshire [23].

In solutions of paramagnetic ion pairs investigated so far, the correlation times are within the range 10^{-10} to 10^{-11} sec [4, 7], whereas the ESR frequency ω_e is about 3×10^{11} rad/sec. Hence only the secular parts of the various relaxation mechanisms contribute to T_2^{-1} and (for doublet systems) Eqs. 7, 8, and 9 can be simplified to

$$(T_2^{-1})_{Fc} = \frac{1}{4}\left(\frac{g\beta_e}{\hbar}\right)^2 a^2 \tau_e \tag{19}$$

$$(T_2^{-1})_D = \frac{7}{120}\frac{(D:D)}{\hbar^2}\tau_d \tag{20}$$

$$(T_2^{-1})_Q = \frac{3}{40}f(I)\left(\frac{e^2 Qq}{\hbar}\right)^2 \tau_r \qquad (\eta = 0) \tag{21}$$

The line-widths should therefore be proportional to a particular correlation time. The correlation times depend on the radical concentration c, the absolute temperature T, and the viscosity η. According to the Debye-Einstein model the dependence of τ_r on η and T is given by [1–3]

$$\tau_r \sim \frac{\eta}{T} \tag{22}$$

while according to the model of Pake and Tuttle [24] the dependence of τ_e on c, T, and η is given by

$$\tau_e \sim \frac{\eta}{T} \cdot \frac{1}{c} \tag{23}$$

From these relations it follows that the total line-width is proportional to η/T, provided Eqs. 19, 20, and 21 are valid.

3. APPLICATIONS

3.1. Alkali Naphthalene Ion Pairs

Thorough NMR investigations of the alkali naphthalene ion pairs were reported by Hendriks et al. [10]. Both proton and alkali resonance were studied using 1,2-dimethoxyethane (DME) as solvent, although the sodium naphthalene system (Na-Nl) was also studied in tetrahydrofuran (THF). Beside proton resonance, deuterium resonance studies were carried out on completely deuterated samples. Some of the results are discussed in the following sections.

3.1.1. ¹H *and* ²D *Resonance*

Figure 1 gives the NMR spectra resulting from the ¹H and ²D resonance. According to Eqs. 2, 7, and 8, the magnetic contributions to T_2^{-1} of the

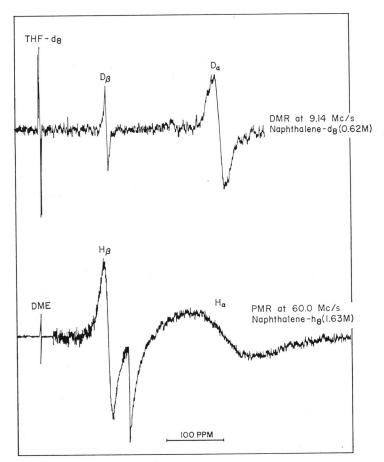

Figure 1. ¹H and ²D NMR spectra of a solution containing $1.61M$ NI-h_8 and $0.62M$ NI-d_8 in DME completely reduced with Na. As internal reference about 5 volume % THF-d_8 was added. The peculiar shape of the reference signal THF-d_8 was caused by modulation effects. The peaks are recorded with different spectrometer settings.

²D resonance peaks are by a factor $(\gamma_H/\gamma_D)^2 \approx 40$ smaller than those contributing to T_2^{-1} of the ¹H peaks, whereas Eqs. 1 and 2 predict the contact shift to remain the same, if the experiments are performed at the same

magnetic field as the ^{1}H experiments. This results in a better resolution of the ^{2}D spectrum, as is clearly manifested in Fig. 1. The analysis of the ^{1}H and ^{2}D spectra is given in Table 1. The proton NMR HFSCs agree with those

Table 1 Analysis of ^{1}H and ^{2}D NMR Spectra of Naphthalene-h8 and -d8 Negative Ion

Method	Parameter	α-Position		β-Position
ESR	a_H	4.925 ± 0.003		1.820 ± 0.002
^{1}H NMR	a_H	-4.87 ± 0.03	$[6.93 \pm 0.20]$	-1.85 ± 0.01
	T_2^{-1} (sec^{-1})	$(3.30 \pm 0.06) \times 10^4$	$[6.85 \pm 0.26]$	$(0.482 \pm 0.009) \times 10^4$
^{2}D NMR	$a_D \cdot \left(\dfrac{\gamma_H}{\gamma_D}\right)$	-4.80 ± 0.02	$[6.81 \pm 0.23]$	-1.84 ± 0.01
	T_2^{-1} (sec^{-1})	$(8.5 \pm 0.2) \times 10^2$	$[5.3 \pm 0.3]$	$(1.60 \pm 0.05) \times 10^2$

derived from ESR experiments [25]. The small differences between the NMR and the ESR HFSC of various protons are not significant and may be expected in view of the different conditions under which the experiments were performed, dilute solutions ($\sim 10^{-4}M$) in the ESR experiments and concentrated solutions ($\sim 1M$) in the NMR studies. The negative signs of the HFSCs are accounted for by spin-polarization mechanisms [3].

It can be shown [4, 7] that for large HFSCs, encountered in the naphthalene negative ion, the only important contribution to the line-width arises from the Fermi contact interaction; the line-width should then be proportional to a^2. Therefore we have included between brackets in Table 1 the ratio of the line-widths and the ratio of the squares of the NMR HFSCs, all normalized to the values referring to the β-proton peak. Within the accuracy of the measurements the ^{1}H line-widths indeed are found to be proportional to a^2. By using Eq. 7 and the value of T_2^{-1} reported in Table 1, a value of $(1.8 \pm 0.1) \times 10^{-11}$ sec is calculated for τ_e.

The predominance of the Fermi contact interaction over all the other types of interactions is confirmed by the results of studies of the line-width as a function of radical-ions concentration. It was found that T_2^{-1} is linearly related to $1/c$, in accordance with Eqs. 19 and 23. However, as shown by the data given in Table 1, the ^{2}D line-width is not proportional to a^2. Hendriks et al. [10] ascribed this deviation to the quadrupole contribution to the line width, stemming from the field gradient in the C—D bond and probably from the field gradient caused by the counterion and solvent molecules.

2.1.2. Alkali Resonance

A survey of the measured metal HFSCs is given in Fig. 2. The ESR data are taken from the work of Dodson and Reddoch [26]. The figure shows that

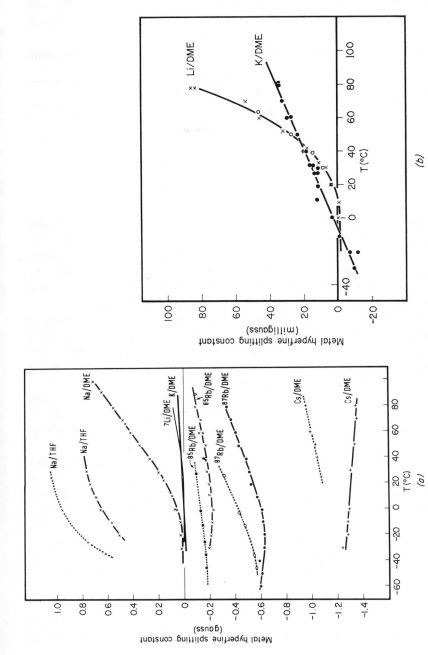

Figure 2. Temperature dependence of the alkali metal HFSC observed for the alkali naphthalene ion pairs in THF and DME. Solid lines refer to NMR data, dashed lines to ESR results of Dobson and Reddoch [26].

297

the Li and Na HFSCs are positive, whereas those of Rb and Cs are negative. The sign of the HFSC of K depends on temperature, changing from negative to positive.

The plot of the Na HFSC versus temperature for the system Na-Nl-DME (Fig. 2) is curved. It suggests that loose, solvent-separated ion pairs, existing at low temperatures, are changed into tight, contact ion pairs at high temperatures. This view is supported by the data obtained in THF. In this solvent, of lower solvating power than DME, the Na HFSC varies only slightly with temperature and eventually it approaches its limiting value. This behavior can be explained by assuming the existence of contact or tight ion pairs in the entire range of investigated temperatures.*

The reversal of sign observed for the K HFSC can best be rationalized by assuming the existence of contact ion pairs in the entire temperature range. For solvent-separated ion pairs we would not expect a sign reversal of the spin density at the alkali nucleus.

The temperature dependence of the Rb HFSC implies that Rb-Nl forms contact ion pairs at high temperatures (above about $-20°C$) and solvent-separated ion pairs become abundant at low temperatures.

The almost constant value of the Cs HFSC is typical for contact ion pairs, whose structural conformation remains unaffected by temperature.

Two general features of these systems emerge from the data. First, the formation of contact ion pairs is facilitated by increasing effective radius of the alkali ions due to decreasing solvation. Second, the HFSC tends to become negative for the larger cations. In Section 4 of Chapter 8 various spin transfer mechanisms are discussed and explanations are presented for the observed negative alkali spin densities.

In Fig. 3 the measured line-width is plotted as a function of η_0/T. The viscosity of the pure solvent has been used in calculation. This is not entirely correct, since the viscosity depends on the concentration of radicals. However, if the concentration-dependent terms vary only slightly with the temperature (evidence for this is found in [27] and [28]), the use of η_0 instead of η affects the slope of the plots by a constant factor only.

Hendriks et al. [10] analyzed the line-width in terms of the various relaxation mechanisms (Eq. 6). The results are summarized in the following sections.

Li-Nl-DME. Comparison of the line-width of the two Li isotopes led the authors to conclude that the quadrupole contributions to T_2^{-1} are very small and that the anisotropic intramolecular dipolar interaction provides

---

* These conclusions are at variance with the ESR observations of Hirota [40] and the results of Szwarc et al. [41] based on studies of electron-transfer equilibria. Both groups find tight sodium pairs in THF at 20°C and virtually only loose pairs at −70°C (editor).

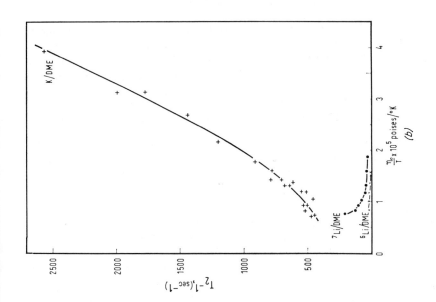

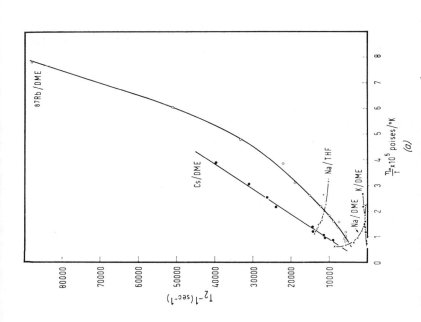

Figure 3. Alkali metal NMR line-widths (T_2^{-1}) versus η_0/T.

299

the most important relaxation mechanism. This interaction is expected to increase with temperature as the distance between Nl$^-$ and Li$^+$ becomes smaller due to the decreased degree of cation's solvation. This explanation is consistent with the increase of the Li HFSC at higher temperatures (Fig. 2).

Na-Nl-DME, THF.   The results of line-width analysis for two temperatures is given in Table 2. The main part of the line-width arises from dipolar and

Table 2   Analysis of the Sodium Line-Width for Sodium Naphthalene

Solvent	T (°C)	a_{Na} (gauss)	τ_e (sec)	T_2^{-1} (sec^{-1})	$(T_2^{-1})_{Fc}$ (sec^{-1})	$(T_2^{-1})_D + (T_2^{-1})_Q$ (sec^{-1})
DME	+30	0.200	7×10^{-11}	1,200	200	1,000
	+90	0.634	3×10^{-11}	5,900	900	5,000
THF	+30	0.764	—	12,900	—	—

quadrupolar interaction. The large increase of the line-width when DME is replaced by THF is accompanied by a large increase of the HFSC, indicating the formation of contact ion pair in THF. The Na$^+$ ion is then closer to the aromatic plane, and this enhances the magnetic contributions to the line-width. The increase of the HFSC with temperature (Fig. 2) parallels the increase of the line-width (Fig. 3).

K-Nl-DME.   In view of the small value of the HFSC, it is unlikely that magnetic interactions contribute appreciably to the line-width. This is also indicated by the plot of T_2^{-1} versus η_0/T, which does not show a singularity at the point where the HFSC is zero. The total line-width has been ascribed to the quadrupole interactions and on this basis values ranging from 1.3 and 4.1 MHz have been derived for e^2Qq/h.

Rb-Nl-DME.   A complete analysis of the line-widths was possible because the studies could be performed on two Rb isotopes, ^{85}Rb and ^{87}Rb. Table 3 summarizes the results. The anisotropic dipolar interaction appeared

Table 3   Analysis of the Rubidium Line-Width at   +30°C for Rubidium Naphthalene

Isotope	T_2^{-1} (sec^{-1})	$(T_2^{-1})_Q$ (sec^{-1})	$(T_2^{-1})_{D+Fc}$ (sec^{-1})	$(T_2^{-1})_{Fc}$ (sec^{-1})
^{85}Rb	$7,780 + 150$	$7,670 \pm 190$	106 ± 35	112 ± 28
^{87}Rb	$8,700 \pm 330$	$7,480 \pm 180$	$1,210 \pm 400$	$1,010 \pm 260$

to be unimportant and the quadrupole relaxation seemed to contribute almost 100% to the ^{85}Rb line-width and to account for about 85% of the line-width of ^{87}Rb.

Cs-Nl-DME. Figure 3 shows that the Cs line-width varies linearly with η_0/T, in agreement with the relative insensitivity of the Cs HFSC to temperature variation (see Fig. 2). The total line-width could be ascribed to the Fermi contact interaction and an upper limit of 1.0×10^{-10} sec is calculated for the relevant τ_e. The lack of dependence of the HFSC on temperature and the proportional relation between T_2^{-1} and η_0/T seem to be characteristic for tight or contact ion pairs.

Takeshita and Hirota [9] reported NMR determinations of sign of alkali HFSCs in some alkali radical ion pairs. Their results for Na and Cs naphthalene obtained at one temperature agree qualitatively with the results of Hendriks et al. [10].

3.2. Alkali Biphenyl Ion Pairs

The alkali biphenyl ion pairs have been extensively studied by Canters et al. [4, 7, 8], who reported NMR resonance investigations of all commonly encountered alkali isotopes. Tetrahydropyran (THP), 2-methyltetrahydrofuran (MTHF), THF, DME, diglyme (Dg), triglyme (Tg), and tetraglyme (Ttg) were used as solvents. The results are discussed in the following sections.

3.2.1. ^{1}H and ^{2}D Resonance

Figure 4 shows the ^{2}D spectrum of deuterated sodium biphenylide (Bp$^-$) dissolved in DME. The spectrum is well resolved and again illustrates the potentialities of the method. The meta-deuterium peak is shifted to low field, pointing to a positive spin density on the *meta*-deuterium atoms and hence to a negative spin density on the adjacent carbon atoms. The agreement between the ESR and NMR HFSCs was excellent [4, 7, 8].

The ^{1}H and ^{2}D line-width of the *ortho* and *para* resonance lines were found to be proportional to the square of the HFSC [4, 8]. However, the ^{1}H and ^{2}D *meta* line-width deviated from the expected relation, the deviations being attributed to significant dipolar contributions to the line-width caused by the large spin densities on the neighboring *ortho* and *para* carbon atoms [4, 8]. On the basis of Eq. 19 τ_e was found to be equal 2×10^{-11} sec for the system Na-Bp-DME (conc. $1M$) at $+30°$C.

3.2.2. Alkali Resonance

Resonance experiments on all commonly encountered alkali isotopes— ^{6}Li, ^{7}Li, ^{23}Na, ^{39}K, ^{85}Rb, ^{87}Rb, and ^{133}Cs—were carried out. These studies were performed in a variety of solvents differing in their solvating power.

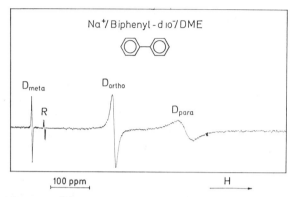

Figure 4. ^{2}D NMR spectrum of a completely reduced solution of NaBp-d_{10} in 1,2-dimethoxyethane (DME) at $T = +30°C$. The radical concentration was kept at approximately $1M$. The unusual appearance of the two peaks of the internal standard, tetrahydrofuran-d_8, which are indicated by R, is caused by overmodulation. Spectra were measured on a Varian DP-60 spectrometer equipped with a V 4210 transmitter, which was stabilized by a crystal-stabilizer at 9.1 Mc/sec. Different peaks were recorded different spectrometer settings.

Positive values of the Li HFSC were found in all the investigated solvents. Figure 5 summarizes the results and illustrates the influence of the solvent on the metal HFSC. The Li HFSC decreases as the solvating power of the medium increases. In Dg and in Tg, the Li-Bp ion pairs seem to form solvent-separated ion pairs; hence the HFSC should be virtually zero in both solvents. The observed small shift might be caused by ring current or by some specific chelating effects (solutions of alkali halides were used as reference).

Sodium HFSC in Na-Bp could be positive or negative, depending on solvent and temperature [7]. In MTHF sign reversal is observed.

The K HFSC for the K-Bp solutions increases from negative value to zero when DME is substituted by Tg. This can be explained again by an increased solvation of the potassium ions. Extrapolating these results, we could expect that the K HFSC would be practically zero in Ttg in contrast with the experimental findings. It is assumed that in Ttg specific solvation of the K$^+$ ion occurs, affecting the resonance position of the metal nucleus. It is interesting to note that in this solvent K-Bp forms single crystals having the composition KBpTtg$_3$ [29], making it clear that the solvent plays an important role in the formation of the crystals and probably ascertains their thermodynamic stability.

Finally, as seen from Fig. 6, the Rb and Cs HFSCs appear to be negative for their biphenyl salts even in glymes. The Cs signal is shifted 370 ppm to high field with respect to the signal of the reference compound (CsCl in

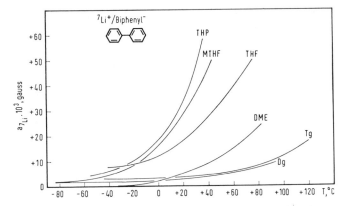

Figure 5. Temperature dependence of the Li HFSC in Li-Bp measured in different solvents.

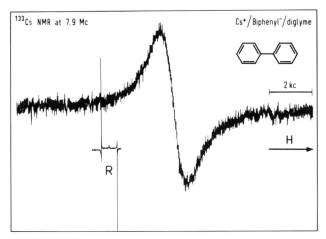

Figure 6. Cs NMR spectrum at +50°C of a 0.4M solution of Cs$^+$-biphenyl$^-$ in diglyme. The trace designated by R is the spectrum of a 1.5M reference solution of CsCl in H$_2$O recorded in the sideband mode of operation.

H$_2$O). The cause of the negative alkali spin density is discussed in Section 4 of Chapter 8.

The pertinent data permit us to assess the relative importance of the different intramolecular interactions in determining the rate of relaxation of the alkali nuclei in radical ion pairs [10].

Li.   The relaxation of ^{6}Li as well as of ^{7}Li usually is determined by the

Fermi contact interaction. However, if this interaction is very weak, the anisotropic dipolar relaxation determines the relaxation.

Na.   The relaxation of ^{23}Na is determined by the Fermi contact interaction provided the Na HFSC is large ($\geq$ 1 gauss). Otherwise the quadrupole interaction determines the relaxation and, in some cases, the anisotropic dipolàr interaction may become equally important.

K.   The relaxation of ^{39}K nuclei is determined to a great extent (80–100%) by the quadrupolar interaction.

Rb.   The same applies for the relaxation of ^{85}Rb, whereas the relaxation of ^{87}Rb, although determined to a large extent by the quadrupole interaction, is affected also by the Fermi contact interaction.

Cs.   The relaxation of ^{133}Cs is always determined by the Fermi contact interaction.

3.2.3. Dynamic versus Static Model

The behavior of cesium biphenyl ion pair in diglyme studied by Canters et al. [6] show some interesting features. The plot of Cs HFSC versus temperature is appreciably curved, as seen in Fig. 7. Two physically different models

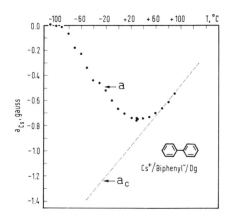

Figure 7. Cs HFSC as a function of temperature in the system Cs$^+$-biphenyl$^-$-diglyme determined from NMR measurements. The dotted line presents the presumed variation of a_c, the HFSC of the contact ion pairs, with temperature.

may account for this behavior. In one model a dynamic equilibrium between rapidly interconverting solvent-separated and contact ion pairs [30–33] is assumed. The observed HFSC would be then equal to

$$a = f_c a_c + (1 - f_c)a_s \qquad (24)$$

in which f_c and $(1 - f_c)$ are the fractions of contact and solvent-separated ion pairs, respectively and a_c and a_s are the corresponding Cs HFSCs. In an alternate model, which is essentially static in nature [32, 34, 35], the potential energy well of the ion pair gradually changes with temperature and thus the ion pair is a contact pair at high temperature and it becomes a solvent-separated ion pair at low temperatures. At intermediate temperatures the pair acquires some intermediate properties.

The temperature behavior of the Cs HFSC does not allow us to differentiate between these two models. However, the variation of Cs line-width with temperature permits the differentiation. Canters has shown [7] that the Cs line-width is determined merely by the Fermi contact interaction. The static model implies that only one kind of ion-pair species is present at each temperature. In this case the Cs line-width should be proportional to

$$T_2^{-1} \sim a^2 \frac{\eta}{T} \tag{25}$$

(cf. Eqs. 19 and 23). On the other hand, if the dynamic model applies, two different species would be present in solution in rapid equilibrium and then the line-width should be given by

$$T_2^{-1} = (1 - f_c)\left(\frac{1}{T_2}\right)_s + f_c\left(\frac{1}{T_2}\right)_c \tag{26}$$

where

$$\left(\frac{1}{T_2}\right)_s \sim a_s^2 \frac{\eta}{T} \quad \text{(solvent-separated ion pair)} \tag{27}$$

and

$$\left(\frac{1}{T_2}\right)_c \sim a_c^2 \frac{\eta}{T} \quad \text{(contact ion pair)} \tag{28}$$

As seen from Fig. 7, the HFSC approaches zero at low temperatures and hence $a_s = 0$. Substituting this into Eqs. 24 and 26, we find for the dynamic model

$$T_2^{-1} \sim a a_c \frac{\eta}{T} \tag{29}$$

These two different relations may be checked provided the value of a_c is known at each temperature. To obtain these values the assumption was made that at high temperatures the observed HFSC a is equal to a_c, whereas at other temperatures a_c varies linearly with the temperature, as found for other systems, for example, Na-Bp-MTHF and Rb-Bp-DMe. In this way a_c can be found by extrapolation (dotted line in Fig. 7).

The result of the analysis is given in Fig. 8. The linear correlation of T_2^{-1} with $a^2\eta/T$ seems to be much better than the correlation of T_2^{-1} with $a a_c \eta/T$.

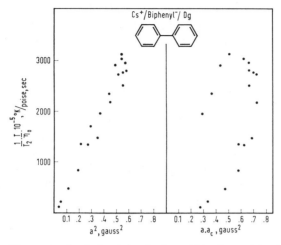

Figure 8. Analysis of the Cs line-width observed in the system Cs^+-biphenyl$^-$-diglyme in the temperature range of -70 to $+90°C$. For η the viscosity of the pure solvent has been used [7], while the values for a^2 and $a \cdot a_c$ have been derived from Fig. 7.

Hence the static model describes better the Cs-biphenyl ion pairs than the dynamic model. It is interesting that Williams et al. [36] came to the same conclusion on the basis of their study of g-value of the Cs-naphthalene ion pairs. Further discussion of the two models of ion pairs is given elsewhere [42].

3.3. Alkali 2,2′-Dipyridyl Ion Pairs

An interesting phenomenon has been observed by Takeshita and Hirota [9] for the alkali 2,2′-dipyridyl ion pairs. The structure of the ion pairs is assumed [37] to be

The cation is located in the nodal plane of the MO ψ_0 containing the unpaired electron and attached to the two N-atoms. The fact that the Li and Na HFSC in the ESR spectra do not change with temperature and solvent provides a supporting evidence for this model. As shown in Table 4, the magnitudes of the ^7Li splittings measured by NMR agree well with those determined by ESR, its sign being negative, as expected for a cation located in the nodal plane of ψ_0. The negative spin density is brought about by the exchange

Table 4 ^{7}Li and ^{23}Na HFSC (in gauss) for the Ion Pairs of 2,2′-Dipyridyl at 300°K

Solvent	^{7}Li		^{23}Na	
	NMR	ESR	NMR	ESR
DMF	−0.64		−0.36 (0.5M)	
DME	−0.66	0.71	−0.21 (0.5M)	
			−0.001 (1.5M)	0.57
THF	−0.67	0.71	+2.23	0.57
MTHF	−0.71	0.71	+3.04	

mechanisms (see Chapter 7). The sodium 2,2′-dipyridyl system is strikingly different from the lithium dipyridyl. In dimethylformamide (DMF) the sign of the sodium splitting is negative, as expected for ion pairs, but in DME the HFSC is concentration dependent. The splitting is negative at low concentrations, but it gradually increases and approaches zero at high concentrations. Very large positive splittings were obtained in THF and MTHF.

It is known that alkali 2,2′-dipyridyl forms aggregates in solvents such as THF and MTHF [38, 39]. These aggregates undoubtedly are present in the concentrated solutions used in NMR studies. Thus the NMR HFSC refers to the sodium ions in an aggregate and its large positive value indicates that the Na$^+$ ion does not reside in the nodal plane of ψ_0. In such a case, the direct and the overlap mechanisms produce positive spin density at the Na nucleus (see Chapter 8). Similarly, the concentration dependence of the alkali splitting in DME can be rationalized in terms of an equilibrium between ion pairs and ion aggregates.

4. CONCLUSION

The available results of NMR studies of ion pairs are promising. The method proves to be applicable for studies of alkali radical salts in solution in the concentration range of 0.1–1.0M. The spectra are easily interpreted and provide information on magnitude and sign of the HFSCs. The study of the line-widths furnishes information on correlation times and on the intramolecular relaxation mechanisms. Moreover, important information about the structure of the ion pairs can be obtained.

The required high concentrations of ion pairs is a drawback of the method. This may cause complications arising from the formation of aggregated ion systems. The choice of a proper diamagnetic reference solution may constitute an additional problem. The difference between the shielding constants of a given nucleus in the reference and in the sample solution may affect the value of the contact shifts to an unknown extent. However, if the contact shifts are large, as they usually are, this leads to only a minor error in the value of the HFSCs.

The interpretation of the alkali line-widths presents the most difficult problem due to the fact that the position of the counterion in the complex is unknown for most ion pairs. The formation of single crystals of alkali biphenyl salts [29] may be very helpful in this respect. It is hoped that X-ray studies of these crystals may reveal the position of the alkali ion. Provided the structures of the ion pairs in solution are similar to that in the crystals, the electronic properties of these ion pairs could be calculated on that basis.

REFERENCES

1. A. Abragam, *The Principles of Nuclear Magnetism*, Oxford University Press, London, 1961.
2. E. de Boer and H. van Willigen, *Progress in Nuclear Magnetic Resonance Spectroscopy*, Vol. 2, J. W. Emsley, J. Feeney, and L. M. Sutcliffe, Eds., Pergamon Press, Oxford, 1967, Chapter 3.
3. A. Carrington and A. D. McLachlan, *Introduction to Magnetic Resonance*, Harper and Row, New York, 1967.
4. G. W. Canters and E. de Boer, *Mol. Phys.*, **13**, 395 (1967).
5. G. W. Canters, E. de Boer, B. M. P. Hendriks, and H. van Willigen, *Chem. Phys. Lett.*, **1**, 627 (1968).
6. G. W. Canters, E. de Boer, B. M. P. Hendriks, and A. A. K. Klaassen, *Proc. Coll. Ampère XV*, 242 (1969).
7. G. W. Canters, Thesis, Nijmegen, The Netherlands (1969).
8. G. W. Canters, B. M. P. Hendriks, and E. de Boer, *J. Chem. Phys.*, **53**, 445 (1970).
9. T. Takeshita and N. Hirota, *Chem. Phys. Lett.*, **4**, 369 (1969).
10. B. M. P. Hendriks, G. W. Canters, C. Corvaja, J. W. M. de Boer, and E. de Boer, *Mol. Phys.*, **20**, 193 (1971).
11. E. de Boer and C. MacLean, *Mol. Phys.*, **9**, 191 (1965).
12. E. de Boer and C. MacLean, *J. Chem. Phys.*, **44**, 1334 (1966).
13. E. de Boer and J. P. Colpa, *J. Phys. Chem.*, **71**, 21 (1967).
14. J. A. Pople, W. G. Schneider, and H. J. Bernstein, *High-Resolution Nuclear Magnetic Resonance*, McGraw-Hill, New York, 1959.
15. R. E. Jesse, Thesis, Amsterdam, The Netherlands (1966).
16. E. de Boer, A. M. Grotens, and J. Smid, *J. Am. Chem. Soc.*, **92**, 4742 (1970).
17. E. de Boer, A. M. Grotens, and J. Smid, *Chem. Comm.*, **1970**, 1035.
18. I. Solomon, *Phys. Rev.*, **99**, 559 (1955).
19. N. Bloembergen, *J. Chem. Phys.*, **27**, 572 (1957).
20. R. A. Bernheim, T. H. Brown, H. S. Gutowsky, and D. E. Woessner, *J. Chem. Phys.*, **30**, 950 (1959).

21. J. A. M. van Broekhoven, B. M. P. Hendriks, and E. de Boer, *J. Chem. Phys.*, **54**, 1988 (1971).

22. H. M. McConnell and J. Strathdee, *Mol. Phys.*, **2**, 129 (1959).

23. W. Derbyshire, *Mol. Phys.*, **5**, 225 (1962).

24. G. E. Pake and T. R. Tuttle, Jr., *Phys. Rev. Lett.*, **3**, 423 (1959).

25. B. G. Segal, A. Reymond, and G. K. Fraenkel, *J. Chem. Phys.*, **51**, 1336 (1969).

26. C. L. Dodson and A. H. Reddoch, *J. Chem. Phys.*, **48**, 3226 (1968).

27. R. L. Kay, T. Vituccio, L. Zawoyski, and D. F. Evans, *J. Phys. Chem.*, **70**, 2336 (1966).

28. F. J. Millero, *J. Phys. Chem.*, **72**, 3209 (1968).

29. G. W. Canters, A. A. K. Klaassen, and E. de Boer, *J. Phys. Chem.*, **74**, 3299 (1970).

30. S. Winstein and G. C. Robinson, *J. Am. Chem. Soc.*, **80**, 169 (1958).

31. E. Grunwald, *Anal. Chem.*, **26**, 1696 (1954).

32. N. Hirota and R. Kreilick, *J. Am. Chem. Soc.*, **88**, 614 (1966).

33. T. E. Hogen-Esch and J. Smid, *J. Am. Chem. Soc.*, **87**, 669 (1965).

34. N. M. Atherton and S. I. Weissman, *J. Am. Chem. Soc.*, **83**, 1330 (1961).

35. P. Chang, R. V. Slates, and M. Szwarc, *J. Phys. Chem.*, **70**, 3180 (1966).

36. W. G. Williams, R. J. Pritchett, and G. K. Fraenkel, *J. Chem. Phys.*, **52**, 5584 (1970).

37. A. Zahlen, F. W. Heineken, M. Bruin, and F. Bruin, *J. Chem. Phys.*, **37**, 683 (1962).

38. I. M. Brown, S. I. Weissman, and L. C. Snyder, *J. Chem. Phys.*, **42**, 1105 (1965).

39. J. D. W. van Voorst, W. G. Zijlstra, and R. Sitters, *Chem. Phys. Lett.*, **1**, 321 (1967).

40. N. Hirota, R. Carraway, and W. Schook, *J. Am. Chem. Soc.*, **90**, 3611 (1968).

41. Y. Karasawa, G. Levin, and M. Szwarc, *J. Am. Chem. Soc.*, **93**, 4614 (1971).

42. M. Szwarc, *Account Chem. Res.*, **2**, 87 (1969).

Appendix

Nuclear Magnetic Resonance Studies of Solvation of Ions and Ion Pairs

MICHAEL SZWARC

Polymer Research Center, State University College of Forestry, Syracuse, New York

Problems pertaining to the structure and behavior of ions and ion pairs which were elucidated by NMR techniques, but not reviewed in Chapters 6 and 7, are briefly discussed in this appendix.

The presence of free ions affects the NMR spectrum of the solvent in a variety of ways. The electric field generated by ions perturbs the charge distribution in the adjacent solvent molecules, hence it modifies the electronic shielding of their nuclei. This, in turn, changes the position of their NMR lines.

Two extreme cases should be distinguished. When solvent molecules surrounding the ions are retained in their solvation shells for a time substantially longer than $1/\Delta\nu$, $\Delta\nu$ being the difference in the resonance frequencies of a nucleus in question in an unperturbed and perturbed solvent molecule, then the respective NMR lines split into two—one set characterizing solvent molecules in the bulk of the liquid and the other referring to the molecules forming the solvation shell. Under such conditions individual chemical shifts of both kinds of solvent molecules can be determined and the integration of the respective NMR peaks gives the ratio of their concentrations. Since the concentration of the ions is known, the solvation number can be obtained through this approach. In the other extreme, when the residence time of solvent molecules in the solvation shell is much shorter than $1/\Delta\nu$, the original spectrum of the solvent is shifted but not split. At low concentration of salt, the shift is proportional to its concentration.

The first successful application of the technique utilizing the splitting of NMR lines was reported by Taube [1] who investigated the proton NMR spectrum of aqueous methanol solution of Mg^{2+}, $(ClO_4^-)_2$ at $-75°C$. The results gave the number of water and methanol molecules solvating each Mg^{2+} ion (total about 6) and, in principle, this work could also provide information on the relative abilities of these two solvents to compete for the sites in the solvation shell. The latter problem has been recently studied in the gaseous phase; the relative abilities of methanol and water to solvate gaseous ions has been determined (see Chapter 2, Section 7).

The approach of Taube was subsequently utilized by other workers, for example, Matwiyoff [2] determined the solvation number for Co^{2+} and Ni^{2+} ions in DMF.

Extensive proton NMR studies of salt solutions were reported by Fratiello and his associates. NMR spectra of DMF, DMSO, or N-methylformamide in aqueous solutions of $AlCl_3$ showed two sets of proton lines at room temperature [3]. The set of lines shifted downfield was attributed to the molecules of the organic component in the solvation shell of Al^{3+}, the other was assigned to the molecules present in the bulk of the solution. Integration demonstrated that under these conditions only one organic molecule resides in the solvation shell, its lifetime being at least 10^{-2} sec; however, water produced only one NMR line indicating that its molecules (or protons) are rapidly exchanged between the solvation shells and the bulk of the liquid. These studies were extended to solutions of other salts [4] and to some binary salt mixtures [7].

Proton exchange becomes sufficiently slow at low temperatures ($-47°C$) and thus two NMR peaks of water could be discerned in concentrated $AlCl_3$ solution [9]. The line shifted by 252 cps downfield was attributed to the water in the solvation shell, and the integration led to a value of 6 for the number of H_2O's coordinated with Al^{3+}. Extension of this work to binary solvents (H_2O-DMSO and H_2O-DMF) showed again [10] that one can distinguish whether the components of the solution are present in the solvation shell or in the bulk. The composition of the solvation shell and the coordination numbers were determined. Similar studies were reported in subsequent papers [11, 12].

Thomas and Reynolds [5] observed separate peaks for the bulk DMSO and DMSO in the solvation of Al^{3+} when $AlCl_3$ was dissolved in anhydrous dimethyl sulfoxide at $20°C$. The integration led to the solvation number of about 6; the intensity of the peak due to protons coupled to ^{13}C served as the reference. The broadening of the lines observed at higher temperatures led to the rate constant of 3.6 sec^{-1} for the dissociation of the complex at $65°C$.

The enthalpy of dissociation was calculated to be about 20 kcal/mole. Similarly, separate proton NMR peaks of acetonitrile in the bulk and in the solvation shell were observed for acetonitrile solutions of $Al(ClO_4)_3$ containing a trace amount of water [8].

In many systems only one water NMR line appears in the spectrum. The question arises whether this result is caused by a rapid exchange of whole water molecules or the exchange of their protons. The answer was provided by studies of ^{17}O NMR first reported by Taube [6]. Even at 20°C, when the exchange of protons is fast, the ^{17}O NMR spectrum of ^{17}O enriched water showed two well-separated peaks when the solutions of perchlorates of Al^{3+}, Ga^{3+}, and Be^{2+} were investigated. However, only one peak was recorded for the solutions of salts of Mg^{2+}, Sn^{2+}, Ba^{2+}, and Hg^{2+}. These results demonstrate, therefore, that water molecules reside in the solvation shells of Al^{3+}, Ga^{3+}, and Be^{2+} for a time longer than 10^{-4} sec, although their protons exchange rapidly.

The ^{17}O technique was quickly improved. Thus the coordination number of Be^{2+} and Al^{3+} ions in aqueous solutions were obtained by Connick and Fiat [13, 16], of Ga^{3+} by Swift et al. [14], and of Co^{2+} by Matwiyoff and Darley [17]. A very thorough study of hydration shells of paramagnetic Ni^{2+} salts [15] led to the rate constant for water exchange of 3×10^4 sec^{-1} and enthalpy of activation of 10.8 ± 0.5 kcal/mole.

Studies of hydration shells by means of ^{17}O NMR were facilitated by an interesting device [6, 18]. Addition of paramagnetic ions to a solution substantially shifts its NMR lines (this is the Knight or contact shift described in Chapter 5, Section 6.2.2), provided the exchange between the solvation shell of the paramagnetic ion and the bulk is rapid ($1/\tau \gg 10^4$ sec^{-1}). The magnitude of the Knight shift depends on the ratio of water molecules in the hydration shells of the paramagnetic ions and those in the bulk. The addition of diamagnetic ions, strongly retaining water in their hydration shells ($\tau > 10^{-4}$ sec), increases the Knight shift. This effect results from a decrease in the number of water molecules in the bulk which were available for the exchange with the solvation shells of the paramagnetic ions. Consequently, this phenomenon serves to magnify the difference in the chemical shift of the bulk solvent and that embedded in the solvation shell of a diamagnetic ion [6], in addition, the difference in the Knight shifts observed in the absence and presence of diamagnetic cations provides information from which the solvation number for diamagnetic ions can be deduced [18]. Taube used the paramagnetic Co^{2+} ions in his pioneering studies [6], but even stronger shifts result from the addition of salts of Dy^{3+} [19].

NMR studies may also reveal the point of attachment of a solvating agent to the solvated ion. For example, the addition of $SbCl_5$ to DMF shifts the resonance of the —CHO protons by 56 cps downfield while the resonance of protons in the two CH_3 groups is shifted only by 32 and 43 cps, respectively [3]. This indicates that the DMF molecule is attached to the antimony ion through its oxygen atom rather than through nitrogen. Interestingly, the two lines, corresponding to the protons of the two methyl groups in the bulk of the liquid, coalesce when the sample is heated to 100°C, whereas the relevant lines from the complexed DMF remain unchanged [3]. Apparently the double-bond character of the C—N bond increases when the oxygen becomes complexed to $SbCl_5$ and this prevents the onset of free rotation around C—$N(CH_3)_2$, even at 100°C.

In pyridine-water solution of $BeCl_2$ or $MgCl_2$ the resonance of α protons of pyridine is displaced further downfield than that of the β or γ protons [20]. This shows that the cation is bounded to the nitrogen of the pyridine molecule. Similar observations were reported for the solutions of $LiClO_4$, $NaClO_4$, and $AgBF_4$ in THF, MeTHF, and THP [21]. Again the presence of the salt shifted the resonance of α protons more than of β, proving that the cations are linked to the oxygen atoms of the ethers.

In the studies discussed above, "tight" solvation shells are formed around cations. The interaction of solvent molecules with anions is relatively weak and therefore their time of residence in solvation shells of anions is very short. Only one example was reported in the literature [22] when the solvation of anion seemed to be observed, namely, solvation of perchlorate anion by dioxane. Of course, the presence of anions contributes to the shift of NMR lines of the solvent, and their contribution can be calculated whenever the contribution of cations is known [23].

Solvation of an ion affects its chemical shift if the nucleus of the ion has a spin. In the absence of cation-anion interaction the shift and shape of the line are expected to be independent of salt concentration. Hence, if any concentration effects are observed they can be attributed to collisions between cations and anions and, in the extreme case, to the formation of ion pairs.

Such effects were studied by Deverell and Richards [25] who investigated the resonance lines of $^{23}Na^+$, $^{39}K^+$, $^{87}Rb^+$, and $^{133}Cs^+$ in aqueous solutions. These relaxation effects were investigated also by observing the resonance of halogen's nuclei in LiCl and LiBr [26–28]. Extensive studies of relaxation phenomena induced by ion pairing and manifested by the shape of resonance line of alkali ions were reported by Eisenstadt and Friedman [29, 30].

Studies of chemical shift of ^7Li, ^{23}Na, and other alkali ions in nonaqueous solutions are scanty. The first studies of this type were reported by Maciel et al. [31] for Li salts and by Bloor and Kidd [32] who determined the ^{23}Na chemical shift of NaI (extrapolated to infinite dilution) dissolved in 14 different oxygen and nitrogen donor organic solvents. The latter workers correlated the shifts with the solvent basicity and thus established a relation between the screening effect and the strength of the Na^+–O) (or Na^+–N) bond. Recently this work was extended to other salts, for example, sodium thiocyanate, perchlorate, and tetraphenylboride [33–35]. Perchlorate and tetraphenyl-boride show different behavior from iodide [35, 36]. Apparently iodide forms tight ion pairs in the investigated solvents whereas perchlorate and tetra-phenylborides form loose pairs. An excellent linear relation was observed between ^{23}Na chemical shift and solvent's donicity (see Chapter 1, Table 2 for solvent's donicity). Extension of these studies to mixed solvents has been reported recently [37].

Interesting information about solvent-ion interaction in nonaqueous media were derived from studies of NaAl(Butyl)$_4$. This salt is soluble in hydrocarbons, hence the interaction with the added solvating agents could be investigated with a minimum interference from the medium. Schaschel and Day [43] determined the proton resonance of tetrahydrofuran added to a solution of NaAl(Butyl)$_4$ in cyclohexane for different THF/NaAl(Butyl)$_4$ ratios. The chemical shift was virtually constant (about 0.22 ppm in respect to THF-cyclohexane solution) until equimolar amount of THF was added and then it decreased. Thus the 1:1 stoichiometry of the complex Na$^+$(THF), Al(Butyl)$_4{}^-$ is established. A distinct break in the curve giving the chemical shift versus (THF)/NaAl(Bu)$_4$ ratio was noted for the THF/NaAl(Butyl)$_4$ = 4, implying eventual formation of Na(THF)$_4$, Al(Butyl)$_4$ complex. Similar studies, reported by Wuepper and Popov [34] for the solvation of NaAl-(Butyl)$_4$ by DMSO in dioxane, indicate the formation of a solvation complex involving six molecules of DMSO.

Studies of the proton resonance of Al in lithium and sodium tetramethyl aluminate dissolved in ether, THF, or DME were reported by Gore and Gutowski [44] who focused their attention on the shape of NMR line. Coupling of ^{17}Al (I = $\frac{5}{2}$) with the protons of CH$_3$ groups is visible when the relaxation of ^{27}Al is slow. This is the case for the loose separated ion pairs in which the Al(CH$_3$)$_4{}^-$ ion is placed in a symmetrical environment, being surrounded by solvent molecules only. On the other hand, the electric field gradient at Al nucleus is sufficiently large in a tight pair to lead to a rapid relaxation of Al and consequently the resonance of CH$_3$ protons appears as a singlet. The results permitted the evaluation of the equilibrium constants of the conversion of tight pairs into loose ones for the sodium and lithium salts in several solvents. The relevant thermodynamic parameters, ΔH and ΔS, were also reported.

Similar studies were carried out by de Boer et al. [45] who investigated the relaxation of ^{23}Na nucleus in sodium tetraphenylboride. The width of the ^{23}Na line was found to be different in THF solution and in THF-glyme (or crown ether) mixture. The results led to the determination of $1/T_2$ as a func-tion η/T (η is the viscosity of the medium), hence to the estimate of the rota-tional correlation time. The latter depends on the electric gradient at sodium nucleus which changes with the structure of the ion pair.

Studies of solvation and structure of ion pairs derived from carbanions were initiated by the work of Dixon [38]. Proton NMR lines of THF and DME showed substantial upfield shifts when the ethers were added to a solution of fluorenyl lithium in deuterated benzene. These observations, coupled with the shifts found for the ^{7}Li resonance, led him to conclude that the cation is located directly above the plane of the aromatic anion. Hence the ether molecules, which are coordinated with the cation, become magnetically shielded by the field generated by the diamagnetic current induced in the aromatic system. Similar effects are observed on the addition of glymes or macrocyclic crown ethers [39, 40]. Detailed investigation of the individual proton shifts indicates that the crown ether (and probably also the glyme) surrounds the cation while occupying a plane parallel to the aromatic ring [40].

Fluorenyl salts form tight as well as loose pairs (see Chapter 3). The presence of solvent molecules in the loose pairs was demonstrated through studies of NMR spectra of the solvent in solutions of fluorenyl salt in the presence and absence of glyme. For example, in THF solution of lithium fluorenyl (loose ion pair) the solvent NMR peaks were shifted upfield by about 8 to 10 cps. When an equimolar amount of glyme was added the THF peaks returned to the position observed in the pure solvent whereas the peaks of the glyme were shifted by 30 cps upfield relative to their resonance in pure THF [39]. This observation manifests that the glyme replaced THF molecules in the solvation shell of the loose pair converting it into a loose glyme-separated pair.

Although the binding of glymes is sufficiently weak to permit a rapid exchange between the solvation shell and the bulk of the liquid, the binding of macrocyclic ether is so much stronger that the rate of exchange could be investigated by NMR technique [40]. At a sufficiently low temperature ($-40°C$) separate lines were observed from the bounded and free crown ether present in 1:1 proportion. The lines of the polyether ring protons coalesce at $2 \pm 1°C$. Hence the rate of exchange could be calculated at this temperature, that is, $k = \frac{1}{2}\tau$ [free ether] where $\tau = \sqrt{2}/2\pi \, \Delta\nu$ and $\Delta\nu$ denote the difference (56 cps) of the relevant resonance frequencies of the free and bound ether, respectively. Similarly, the coalescence was observed at $-18°C$ ($\Delta\nu = 10$ cps) for the lines of the aromatic protons and at $-27°C$ for the lines of the CH$_3$ protons ($\Delta\nu = 4$ cps). Thus the rate constants of exchange are 220, 550, and 3200 M^{-1} sec^{-1} at $-27°$, $-18°$, and $+2°C$, respectively [40, 41], and the activation energy is 12.5 kcal/mole. The actual exchange may be more complex than implied, for example, it may involve the dissociation of the ion pair into free ions followed by exchange of the solvating agent in the free cation.

The rate of exchange is much faster for the potassium fluorenyl than for the lithium salt—no separation of peaks is observed even at $-60°C$. This proves that the complexation of the investigated crown ether with Li$^+$ is much stronger than with K$^+$.

The NMR pattern of glyme-5 complexed to sodium fluorenyl is temperature dependent [40]. Apparently the average position of the glyme with respect to the pair changes with temperature, suggesting that two different loose pairs may be formed in this system. This finding supports the similar conclusion drawn from the ESR studies of glyme-separated sodium naphthalenide [42]. Investigation of electron exchange between this salt and the parent hydrocarbon demonstrated that two distinct loose pairs are present in its solution, one rapidly and the other slowly transferring its electron to the naphthalene molecule.

Interesting NMR patterns have been observed in solutions of radical anions. The resonance of solvent's protons is shifted upfield, but frequently to a lesser extent than predicted by the theory of solutions of paramagnetic compounds. Apparently the solvent molecules imbedded in the solvation shells acquire some spin density due to their proximity to the unpaired electron. This leads to a downfield shift that partially balances the upfield shift caused by the bulk paramagnetic susceptibility. When glyme or crown ether is added to such solutions the theoretically predicted shift is observed [46]. The powerful solvating agent replaces the solvent molecules in the solvation shells and thus annihilates the contact shift experienced by the protons of the solvent.

The observed deviations of the proton resonance caused by the presence of paramagnetic salts have been described previously when we discussed the paramagnetic transition metal ions. These deviations are attributed to the contact or pseudo-contact shift. The anisotropy of the g-tensor of the transition metal ions favors the operation of the pseudocontact shift (a direct dipole-dipole interaction). However, the g-tensor of the large aromatic radical anions is virtually isotropic; hence the effects observed in their solutions have to be ascribed to a Fermi contact shift.

Conversion of a tight carbanion pair into a loose one should affect the resonance of its protons. In several solvents and for a variety of carbanion salts the temperature dependence of such chemical shifts was found to be given by sigmoid curves having two plateaus, one at high and the other at low temperatures [47]. These results were interpreted in terms of equilibria between the two types of ion pairs. The plateaus apparently give the chemical shifts of each kind of ion pairs. On this basis, the relevant equilibrium constants and the respective ΔH's and ΔS's were calculated. This approach was previously utilized by Hirota [48] in his ESR studies of equilibrium between tight and loose ion pairs. The assumption invoked by him and by Jackman (constancy of a or of chemical shift of each kind of pair in the investigated range of temperatures) may be questioned [42]. This may slightly affect the numerical values of the reported equilibrium constants but a serious error can be introduced into the calculated values of ΔH and ΔS.

REFERENCES

1. J. H. Swinehart and H. Taube, *J. Chem. Phys.*, **37**, 1579 (1962). J. H. Swinehart, T. E. Rogers, and H. Taube, *J. Chem. Phys.*, **38**, 398 (1963).
2. N. A. Matwiyoff, *Inorg. Chem.*, **5**, 788 (1966).
3. A. Fratiello and D. P. Miller, *Molec. Phys.*, **11**, 37 (1966).
4. A. Fratiello, D. P. Miller, and R. Schuster, *Molec. Phys.*, **12**, 111 (1967).
5. S. Thomas and W. L. Reynolds, *J. Chem. Phys.*, **44**, 3148 (1966).
6. J. A. Jackson, J. F. Lemons, and H. Taube, *J. Chem. Phys.*, **32**, 553 (1960).
7. A. Fratiello, R. E. Lee, D. P. Miller, and V. M. Nishida, *Molec. Phys.*, **13**, 349 (1967).
8. L. D. Supran and N. Sheppard, *Chem. Comm.*, 832 (1967).
9. R. E. Schuster and A. Fratiello, *J. Chem. Phys.*, **47**, 1554 (1967).
10. A. Fratiello, R. E. Lee, V. M. Nishida, and R. E. Schuster, *J. Chem. Phys.*, **47**, 4951 (1967).
11. A. Fratiello, V. M. Nishida, R. E. Lee, and R. E. Schuster, *J. Chem. Phys.*, **50**, 3624 (1969).
12. A. Fratiello, R. E. Lee, and R. E. Schuster, *Molec. Phys.*, **18**, 191 (1970).
13. R. E. Connick and D. N. Fiat, *J. Chem. Phys.*, **39**, 1349 (1963).
14. T. J. Swift, O. G. Fritz, and T. A. Stephenson, *J. Chem. Phys.*, **46**, 406 (1967).
15. R. E. Connick and D. Fiat, *J. Chem. Phys.*, **44**, 4103 (1966).
16. D. Fiat and R. E. Connick, *J. Am. Chem. Soc.*, **88**, 4754 (1966); **90**, 608 (1968).
17. N. A. Matwiyoff and P. E. Darley, *J. Phys. Chem.*, **72**, 2659 (1968).
18. M. Alei and J. A. Jackson, *J. Chem. Phys.*, **41**, 3402 (1964).
19. W. B. Lewis, J. A. Jackson, J. F. Lemons, and H. Taube, *J. Chem. Phys.*, **36**, 694 (1962).
20. A. Fratiello and E. G. Christie, *Trans. Faraday Soc.*, **61**, 306 (1965).
21. D. Nicholls and M. Szwarc, *J. Phys. Chem.*, **71**, 2727 (1967).
22. J. F. Hinton, L. S. McDowell, and E. S. Amis, *Chem. Comm.*, 776 (1966).
23. R. N. Butler, E. A. Phillpott, and M. C. R. Symons, *Chem. Comm.*, 371 (1968).
24. Z. Luz and S. Meiboom, *J. Chem. Phys.*, **40**, 1058 (1964).
25. C. Deverell and R. E. Richards, *Mol. Phys.*, **10**, 551 (1966).
26. C. Hall, G. L. Heller, and R. E. Richards, *Mol. Phys.*, **16**, 337 (1969).
27. C. Deverell and R. E. Richards, *Mol. Phys.*, **16**, 421 (1969).
28. C. Hall, R. E. Richards, G. N. Schulz, and R. R. Sharp, *Mol. Phys.*, **16**, 528 (1969).
29. M. Eisenstadt and H. L. Friedman, *J. Chem. Phys.*, **44**, 1407 (1966).
30. M. Eisenstadt and H. L. Friedman, *J. Chem. Phys.*, **46**, 2182 (1967).
31. G. E. Maciel, J. K. Hancock, L. F. Lafferty, P. A. Muller, and W. K. Musker, *Inorg. Chem.*, **5**, 554 (1966).
32. E. G. Bloor and R. G. Kidd, *Can. J. Chem.*, **46**, 3425 (1968).
33. B. W. Maxey and A. I. Popov, *J. Am. Chem. Soc.*, **90**, 4470 (1968).
34. J. L. Whuepper and A. I. Popov, *J. Am. Chem. Soc.*, **92**, 1493 (1970).
35. R. H. Erlich, E. Roach, and A. I. Popov, *J. Am. Chem. Soc.*, **92**, 4989 (1970).
36. R. H. Erlich and A. I. Popov, *J. Am. Chem. Soc.*, **93**, 5620 (1971).

37. R. H. Erlich and A. I. Popov, *J. Am. Chem. Soc.*, **93** (1972).

38. J. A. Dixon, P. A. Gwinner, and D. C. Lini, *J. Am. Chem. Soc.*, **87**, 1379 (1965).

39. L. L. Chan and J. Smid, *J. Am. Chem. Soc.*, **89**, 4547 (1967).

40. K. H. Wong, G. Konizer, and J. Smid, *J. Am. Chem. Soc.*, **92**, 666 (1970).

41. K. H. Wong, Ph.D. Thesis, Syracuse, N.Y. (1971).

42. K. Höfelmann, J. Jagur-Grodzinski, and M. Szwarc, *J. Am. Chem. Soc.*, **91**, 4645 (1969).

43. E. Schaschel and M. C. Day, *J. Am. Chem. Soc.*, **90**, 503 (1968).
    C. N. Hammonds, T. D. Westmoreland, and M. C. Day, *J. Phys. Chem.*, **73**, 4374 (1969).

44. E. S. Gore and H. S. Gutowski, *J. Phys. Chem.*, **73**, 2515 (1969).

45. E. de Boer, A. M. Grotens, and J. Smid, *Chem. Comm.*, 759 (1971).

46. E. de Boer, A. M. Grotens, and J. Smid, *J. Am. Chem. Soc.*, **92**, 4742 (1970).

47. J. Grutzner, J. Lawlor, and L. Jackman, *J. Am. Chem. Soc.*, **93** (1972).

48. N. Hirota, *J. Am. Chem. Soc.*, **90**, 3603 (1968).

8

Electron Spin and Nuclear Magnetic Resonance Studies of Ion Pairs—Quantitative Approach

JAN L. SOMMERDIJK* and EGBERT DE BOER

Department of Physical Chemistry, University of Nijmegen, Nijmegen, The Netherlands

1. INTRODUCTION

Alkali radical-anion pairs can be studied by ESR and NMR spectroscopy. The experimental results of ESR studies have been reviewed in Chapter 5 and those derived from NMR investigations in Chapters 6 and 7. Here we treat quantitatively the theoretical aspects of problems encountered in studies of ion pairs through magnetic resonance techniques.

The main theories of magnetic resonance are briefly described in Section 2, and these can be utilized in the interpretation of magnetic resonance spectra. In Section 3 the quantitative treatment of perturbation of anion by cation is outlined; the influence of radical anion on the cation paired with it is discussed in Section 4. Also in this section much consideration is given to the mechanism of spin transfer and to the occurrence of negative spin density at the alkali nucleus.

Dynamic effects observed in spectra of ion pairs caught the attention and interest of many students of ESR and NMR. Useful information concerning the nature of ion pairs can be abstracted from the resonance spectra exhibiting these effects. This is demonstrated in Section 5, where the discussion is focused on the alkali pyracene ion pairs. This choice was dictated not only by historical reasons but also by the fact that the ESR spectra of these ion pairs illustrate all the effects of ion pairing, such as polarization, g-value shift, negative alkali spin density, cation exchange, and cation oscillation. Studies of these effects have led to a complete determination of the position of the alkali ion in the ion pair.

Finally, in Section 6 cation effects on triplet systems are discussed. Two classes of triplet systems can be distinguished, biradicals and triplet dianions. Cations may strongly perturb the electronic properties of the second type of triplet system, as has become evident from the relevant ESR studies.

* Present address: Philips Research Laboratories, Eindhoven, The Netherlands.

2. THEORY OF ESR AND NMR LINE SHAPES

2.1. Bloch Equations

The behavior of the bulk magnetization $\mathbf{M}$ in an ESR or an NMR experiment can be described by a phenomenological set of differential equations [1–5]. These were first proposed by Bloch [6] and are commonly referred to as "Bloch equations." When a steady magnetic field $\mathbf{H_0}$ is applied along the z-axis on an assembly of electronic or nuclear spins, the Bloch equations are given by

$$\frac{d\mathbf{M}}{dt} = -\frac{\mathbf{k}(M_z - M_0)}{T_1} - \frac{(\mathbf{i}M_x + \mathbf{j}M_y)}{T_2} + \gamma(\mathbf{M} \times \mathbf{H_0}) \qquad (1)$$

where $\mathbf{i}$, $\mathbf{j}$, and $\mathbf{k}$ are unit vectors along the x-, y-, and z-axes and γ is the gyromagnetic ratio of the electronic or nuclear spins.

According to Eq. 1 the spin vectors perform a precession (Larmor precession) about the steady magnetic field $\mathbf{H_0}$ with an angular velocity of $\omega_0 \, (= \gamma H_0)$. The rotating transverse components of $\mathbf{M}$ decay to zero with characteristic time T_2, while the longitudinal component M_z relaxes to M_0 with characteristic time T_1. The value M_0 is proportional to the static magnetic susceptibility

$$M_0 = \chi_0 H_0 \qquad (2)$$

On applying a radio-frequency (RF) field $\mathbf{H_1}$, which rotates in the xy-plane in the same sense as the Larmor precession, resonance may occur and the Bloch equations become

$$\frac{d\mathbf{M}}{dt} = -\frac{\mathbf{k}(M_z - M_0)}{T_1} - \frac{(\mathbf{i}M_x + \mathbf{j}M_y)}{T_2} + \gamma(\mathbf{M} \times \mathbf{H_0}) + \gamma(\mathbf{M} \times \mathbf{H_1}) \qquad (3)$$

where

$$\mathbf{H_1} = H_1(\mathbf{i} \cos \omega t - \mathbf{j} \sin \omega t) \qquad (4)$$

It is convenient to choose a new coordinate system which rotates with the RF field and in which the new x-axis is taken along the $\mathbf{H_1}$ direction, whereas the z-axis remains unaltered. In this rotating frame the Bloch equations are

$$\frac{d\mathbf{M}}{dt} = -\frac{\mathbf{k}(M_z - M_0)}{T_1} - \frac{(\mathbf{i}M_x + \mathbf{j}M_y)}{T_2} + \mathbf{M} \times [\mathbf{k}(\omega_0 - \omega) + \mathbf{i}\gamma H_1] \qquad (5)$$

where all the unit axes and vector components refer to the system of rotating coordinates. Defining the complex transverse magnetization G as

$$G = M_x + iM_y \qquad (6)$$

we can derive from Eq. 5

$$\frac{dG}{dt} + \left[\frac{1}{T_2} + i(\omega_0 - \omega)\right]G = i\gamma H_1 M_z \tag{7}$$

If the microwave power is low enough to avoid saturation, M_z may be replaced by M_0. By introducing

$$\alpha = \frac{1}{T_2} + i(\omega_0 - \omega) \tag{8}$$

Eq. 7 is reduced to

$$\frac{dG}{dt} + \alpha G = i\gamma H_1 M_0 \tag{9}$$

Equation 9 is valid for a system which contains only one type of species, but not for systems involving, say, N different species, each characterized by its own relaxation time and γ. Provided conversion of one species into another is not permitted, the macroscopic magnetization of each kind of species is independent of the others, and then the total magnetization is described by a set of N independent equations:

$$\frac{dG_n}{dt} + \alpha_n G_n = i\gamma H_1 M_{0,n} \qquad (n = 1, \ldots, N) \tag{10}$$

where α_n is defined as

$$\alpha_n = T_{2,n}^{-1} + i(\omega_n - \omega) \tag{11}$$

in which ω_n is the resonance frequency of species n. These equations have to be modified if exchange occurs between the different species [7]. To account for the exchange phenomena McConnell [8] proposed the following modified Bloch equations:

$$\frac{dG_n}{dt} + \alpha_n G_n = i\gamma H_1 M_{0,n} - \frac{G_n}{t_n} + \sum_{m \neq n}^{N} P_{mn} \frac{G_m}{t_m} \tag{12}$$

where t_n and t_m are the mean lifetimes of the species n and m and P_{mn} the probability that species m is converted into n. The additional terms appearing in Eq. 12 but not in 10 describe the change in magnetization due to the exchange process. The mathematical form of the exchange terms is analogous to the concentration-dependent terms in kinetic equations.

Under slow passage conditions dG_n/dt can be taken equal to zero. The resonance line shape is then obtained by solving the resulting linear equations and by taking the imaginary part of the total complex magnetization given by

$$G = \sum_n G_n \tag{13}$$

For the simple case of two exchanging species A and B it is convenient to define

$$p_A = \frac{\tau_A}{\tau_A + \tau_B}, \qquad p_B = \frac{\tau_B}{\tau_A + \tau_B} \tag{14}$$

Then the separate macroscopic moments may be written as

$$M_{0A} = p_A M_0, \qquad M_{0B} = p_B M_0 \tag{15}$$

The total complex magnetization is found to be equal to

$$G = G_A + G_B$$

$$= i\gamma H_1 M_0 \frac{\tau_A + \tau_B + \tau_A \tau_B (\alpha_A p_A + \alpha_B p_B)}{(1 + \alpha_A \tau_A)(1 + \alpha_B \tau_B) - 1} \tag{16}$$

In Section 5 we discuss some applications of Eq. 16 for relaxation processes due to cation movements in radical ion pairs.

2.2. Density Matrix Formalism

ESR and NMR line shapes can also be described with the density matrix formalism [4, 9] developed by Kaplan [10] and Alexander [11]. The density matrix ρ_{ij} is defined as

$$\rho_{ij} = c_i c_j^* \tag{17}$$

where c_i and c_j are expansion coefficients of the wave function ψ in terms of a complete set of orthogonal and normalized basis functions φ_i, φ_j, . . . , thus

$$\psi = \sum_i c_i \varphi_i \tag{18}$$

The ensemble average of the expectation value of an operator O for a system with wave function ψ is given by

$$\langle \bar{O} \rangle = \int \psi^* O \psi \, dV \tag{19}$$

Defining

$$O_{ij} = \int \varphi_i^* O \varphi_j \, dV \tag{20}$$

and using Eqs. 17, 18, and 19, we find

$$\langle \bar{O} \rangle = \sum_{i,j} \rho_{ij} O_{ji} = \mathrm{Tr}\,(\rho O) \tag{21}$$

where Tr stands for the sum of the diagonal elements of the product matrix ρO. The density matrix formalism can be used for calculating $\langle S^+ \rangle$ or $\langle I^+ \rangle$

for an ensemble of electronic or nuclear spins. These quantities are proportional to the complex magnetization in the xy-plane, the imaginary part of which determines the microwave absorption, as we have seen in the preceding section. Thus, for example, in an ESR experiment the line shape is determined by the elements of the density matrix ρ and of the S^+ matrix with respect to the solutions of the time-independent part of the spin Hamiltonian,

$$\overline{\langle S^+ \rangle} = \mathrm{Tr}\,(\rho S^+) \tag{22}$$

If the S^+ matrix is known, only the ρ matrix has to be calculated in order to obtain $\overline{\langle S^+ \rangle}$. This can be done with the aid of the quantum mechanical form of Liouville's theorem [4]:

$$\frac{d\rho}{dt} = \frac{i}{\hbar}\,[\rho, \mathscr{H}] \tag{23}$$

where $\mathscr{H}$ is the total spin Hamiltonian of the system.

In ESR experiments involving relaxation and exchange processes, $\mathscr{H}$ is given by

$$\mathscr{H} = \mathscr{H}_0 + \mathscr{H}_1 + \mathscr{H}_{\mathrm{rel}} + \mathscr{H}_{\mathrm{exc}} \tag{24}$$

in which $\mathscr{H}_0$ is the time-independent part of the spin Hamiltonian, while $\mathscr{H}_1$, $\mathscr{H}_{\mathrm{rel}}$, and $\mathscr{H}_{\mathrm{exc}}$ are the contributions of the RF field, the relaxation mechanisms, and the exchange processes, respectively. Eq. (23) can also be written as

$$\frac{\hbar}{i}\frac{d\rho}{dt} = [\rho, \mathscr{H}_0 + \mathscr{H}_1] + \left(\frac{d\rho}{dt}\right)_{\mathrm{rel}} + \left(\frac{d\rho}{dt}\right)_{\mathrm{exc}} \tag{25}$$

Consider now a rotating coordinate system with the z-axis along the magnetic field $\mathbf{H}_0$ and with the x-axis along the $\mathbf{H}_1$ direction of the oscillating RF field; the effective magnetic field is given by

$$\mathbf{H} = \left(H_0 - \frac{\omega}{\gamma_e}\right)\mathbf{k} + H_1\mathbf{i} \tag{26}$$

Then for a system of one electron spin $\mathbf{S}$ and n nuclear spins $\mathbf{I}_i$ the time-independent part of Eq. 24 becomes

$$\begin{aligned}\mathscr{H}_0' &= \mathscr{H}_0 + \mathscr{H}_1 \\ &= \gamma_e \hbar \mathbf{H}\cdot\mathbf{S} + \sum_i \gamma_i \hbar \mathbf{H}\cdot\mathbf{I}_i + \sum_i a_i \mathbf{S}\cdot\mathbf{I}_i\end{aligned} \tag{27}$$

where the first two terms represent the Zeeman interactions and the last term the Fermi contact interaction.

Let us consider now the second term of Eq. 25. The diagonal elements ρ_{ii}, which represent population densities, are assumed to decay exponentially to the thermal equilibrium value $(\rho_0)_{ii}$ with decay time T_1. Therefore we may write for the diagonal elements

$$\left(\frac{d\rho}{dt}\right)_{\text{rel, D}} = \frac{1}{T_1}(\rho_0 - \rho)_{\text{D}} \qquad (28)$$

According to Alexander [11] the nondiagonal elements decay exponentially to zero with decay time T_2. Thus the nondiagonal part of the second term of Eq. 25 is given by

$$\left(\frac{d\rho}{dt}\right)_{\text{rel, ND}} = \frac{1}{T_2}(-\rho)_{\text{ND}} \qquad (29)$$

To describe the exchange processes it is convenient to introduce an exchange operator P [10]. Working on the product spin functions P effects an interchange of the exchanging spins, e.g., $\cdots \alpha(1)\beta(2) \cdots$ becomes $\cdots \alpha(2)\beta(1) \cdots$. The exchange effect on the density matrix is then given by

$$\Delta\rho = P^{-1}\rho P - \rho \qquad (30)$$

The result of two interchanges restores the original situation, so that $P^{-1} = P$. Hence

$$\left(\frac{d\rho}{dt}\right)_{\text{exc}} = \frac{P\rho P - \rho}{\tau} \qquad (31)$$

By substituting Eqs. 27, 28, 29, and 31 into Eq. 25 and by putting $d\rho/dt$ equal to zero (slow passage condition), the elements of ρ can be calculated from the remaining set of linear equations. Then the line shapes are easily obtained from $\overline{\langle S^+ \rangle}$, which can be expressed in the elements of ρ according to Eq. 22.

The density matrix formalism can be applied to many exchange processes. It can be shown [12] that in the limits of slow and rapid exchange the results are the same as those obtained from the modified Bloch equations.

2.3. Relaxation Matrix Method

In the foregoing section we saw that ESR and NMR line shapes could be described in terms of the density matrix elements ρ_{ij} (see Eq. 22). In the formalism of Kaplan and Alexander the elements ρ_{ij} are determined by using Eq. 23. Another method for obtaining these elements has been developed by Ayant [13], Bloch [14], and Redfield [15, 16] and has been investigated extensively by Freed and Fraenkel [17–19]. It has been shown that for rapid fluctuations the elements of the density matrix obey a set of linear differential equations of the following form:

$$\frac{d\rho_{ij}}{dt} = \sum_{k,l} R_{ij,kl} e^{i(\omega_{ij}-\omega_{kl})t} \rho_{kl} \tag{32}$$

where

$$\omega_{ij} = \frac{E_i - E_j}{\hbar} \tag{33}$$

The coefficients $R_{ij,kl}$ are called relaxation coefficients and are given by:

$$R_{ij,kl} = J_{ik,jl}(\omega_{jl}) + J_{ik,jl}(\omega_{ik}) - \delta_{jl} \sum_n J_{nk,ni}(\omega_{nk}) - \delta_{ik} \sum_n J_{nj,nl}(\omega_{nl}) \tag{34}$$

where $J_{ij,kl}$ are the spectral densities defined by

$$J_{ij,kl}(\omega) = \frac{1}{2\hbar^2} \int_{-\infty}^{+\infty} \overline{\langle i| \mathscr{H}_1(\tau) |j\rangle\langle k| \mathscr{H}_1^*(t + \tau) |l\rangle} e^{-i\omega t} dt \tag{35}$$

in which the $\mathscr{H}_1(t)$ is the time-dependent part of the Hamiltonian. The diagonal elements of the relaxation matrix are related to the transverse relaxation time $T_{2,k}$ [17]:

$$T_{2,k}^{-1} = -R_{ij,ij} \tag{36}$$

where k refers to one of the components of the resonance line.

Since Eq. 32 is valid only for rapid fluctuations of $\mathscr{H}_1(t)$, this theory is not applicable to systems in which exchange processes are slow or proceed with intermediate rates. In this respect this approach is inferior to the other two since they can be applied, in principle, to any rate of exchange.

An advantage of the relaxation matrix method is its capability handling nonsecular relaxation processes. These processes are caused by the non-secular terms of the time-dependent Hamiltonian such as $S^\pm I^\mp$. These terms are not considered in the Bloch equations or in the density matrix method. However, for most systems studied so far, these effects have been found unimportant.

All three methods lead, in first approximation, to the same results [17] for rapid exchanges between two species.

3. CATION PERTURBATIONS OF RADICAL ANIONS

3.1. Changes in Proton Hyperfine Splitting Constants

In the early years of ESR studies of aromatic radical anions the proton hyperfine splitting constants (HFSCs) were considered to be characteristic properties of a radical [20, 21], and within the experimental accuracy they were indeed found to be independent of solvent, counterion, and temperature. Therefore the proton splittings were compared with theoretical values predicted for the isolated radical [22–24].

However, studies of electrical conductivity [25, 71] and of optical spectra [26] revealed that the radical anions and the counterions are often paired in solution. The reality of ion pairing was firmly confirmed by ESR studies since alkali hyperfine splittings were observed in the ESR spectra of alkali salts of aromatic radical anions, first by Atherton and Weissman [27] for the sodium naphthalenide system in tetrahydrofuran. The sodium HFSC is greatly affected by solvent and temperature. On the other hand, the proton HFSCs of the naphthalene anion were approximately independent of solvent and temperature and were virtually the same as those found for the free anions [27].

On raising the precision of the ESR experiments, several effects of cation and solvent upon proton hyperfine splitting constants were reported. Variable proton HFSCs were observed for many radical anions of aromatic ketyls [28–30] and nitroaromatics [31, 32] and later for some aromatic hydrocarbon anions [33–35]. Experimental data showing these variations have been summarized in recent review articles [36, 37] (see Chapter 5).

Not many attempts have been undertaken to account quantitatively for the observed effects. This may be due to the fact that most of the observed changes are relatively small and that the details of ion-pair structure are far from being known. In the following sections we discuss some approximations which either have been or may be used for calculation of proton hyperfine splitting constants of ion pairs of radical-ions.

3.1.1. *Modified Hückel Method*

The Hückel molecular orbital (MO) theory for aromatic systems has been proved to be remarkably successful in accounting for the observed proton HFSCs [5, 38–40]. According to this theory, the unpaired electron of an aromatic radical anion occupies an MO given by a linear combination of atomic orbitals (LCAO):

$$\varphi_0 = \sum_i C_{0i}\chi_i \tag{37}$$

where χ_i is the $2p_z$ carbon orbital centered on C_i. The integrated spin density ρ_i at this atom is given then by

$$\rho_i = C_{0i}{}^2 \tag{38}$$

The HFSC a_i of the proton H_i bonded to C_i can be obtained from the well-known McConnell-Weissman relation [41, 42]:

$$a_i = Q\rho_i \tag{39}$$

where Q is a proportionality constant which can be determined by comparing the theoretical and experimental HFSCs.

Since the Hückel theory and its applications for aromatic radicals have been described in many books [38, 39, 43, 44], it is unnecessary to discuss it further. Only its modifications for treating cation effects shall be considered. McClelland [45] suggested that the cation effects can be accounted for by using the following effective Hamiltonian:

$$\mathscr{H}'_{\text{eff}} = \mathscr{H}_{\text{eff}} - \frac{e^2}{r} \tag{40}$$

where $\mathscr{H}_{\text{eff}}$ is the usual effective Hückel Hamiltonian and $-e^2/r$ is the electrostatic attraction of the aromatic anion and the cation, the latter being represented by a point charge. The matrix element of $\mathscr{H}'_{\text{eff}}$ between two atomic functions χ_i and χ_j is given by

$$(\mathscr{H}'_{\text{eff}})_{ij} = (\mathscr{H}_{\text{eff}})_{ij} + \langle \chi_i | - \frac{e^2}{r} | \chi_j \rangle \tag{41}$$

According to McClelland the second term of Eq. 41 can be approximated by

$$\langle \chi_i | - \frac{e^2}{r} | \chi_j \rangle = - \frac{2e^2}{r_i + r_j} \langle \chi_i | \chi_j \rangle \tag{42}$$

where r_i and r_j are the distances between the cation and the carbon atoms C_i and C_j, respectively. The overlap integral may be taken as [39, 43]

$$\langle \chi_i | \chi_j \rangle = \begin{cases} 1 & \text{if } i = j \\ 0.25 & \text{if } C_i, C_j \text{ are adjacent} \\ 0 & \text{if } C_i, C_j \text{ are nonadjacent.} \end{cases} \tag{43}$$

From the modified Hückel matrix the eigenvalues and eigenfunctions (MOs) can be constructed in the same way as from the unmodified Hückel matrix. The spin densities are then obtained from the modified MO φ'_0 containing the unpaired electron:

$$\rho_i = C'^2_{0i} \tag{44}$$

This method was shown to give a fairly good account for the observed cation effects on the proton HFSCs of anthracene$^-$, azulene$^-$, and acenaphthene$^-$ when reasonable cation positions were chosen [46, 47]. The results obtained for the radical anion of pyracene (Fig. 1) will be discussed in greater detail.

de Boer and Mackor [33, 48] found that upon ion-pair formation the eight methylene protons of pyracene, which were equivalent in the free ion, became unequivalent, forming two groups of four equivalent protons. Moreover, an alternating line-width effect was observed in some cases (see Section 5). To

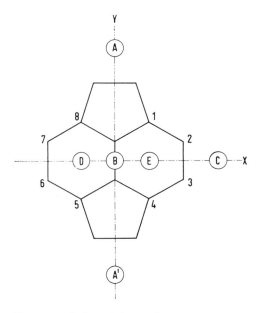

Figure 1. Carbon skeleton of pyracene.

explain these phenomena it was suggested that the cation is localized in position A or A′ (see Fig. 1) and that jumping of the cation between these two equivalent sites is responsible for the alternating line-widths.

However, the proposed structure of pyracene ion pair implies that the aromatic protons 2 and 7 should not be equivalent to protons 3 and 6. Indeed, the calculation of Reddoch [46], who utilized the method of McClelland, showed that the HFSCs of those aromatic protons should be markedly different, contrary to experimental observations, since the spin densities on the adjacent carbon atoms (2, 3, 6, and 7) are drastically changed by the perturbing effect of the cation (see Fig. 2). Reddoch therefore suggested that the cation is placed at a reasonable distance of 3.5 Å above the plane of pyracene and oscillates along the X axis between positions D and E (see again Fig. 1). The results of his calculations, summarized in Fig. 3, showed that for this model the spin densities on the carbon atoms 2, 3, 6, and 7 remain nearly identical, in agreement with experimental results. Recent ESR studies of partially deuterated pyracene confirmed the validity of this model [49].

The modified Hückel method may also be applied to radical anions containing oxygen and nitrogen atoms. The presence of the heteroatoms can be accounted for by adjusting the Coulomb and resonance integrals [38, 43].

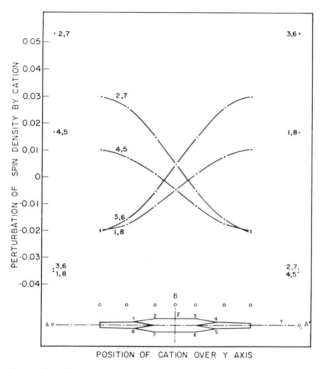

Figure 2. Proton spin density perturbation in pyracene⁻ by a cation located over the *y*-axis. Reproduced by permission from [46].

For instance, for oxygen,

$$\alpha_O = \alpha + h\beta \qquad (45)$$

$$\beta_{CO} = k\beta \qquad (46)$$

Here α_O and α are the Coulomb integrals for oxygen and for an aromatic carbon atom, while β_{CO} and β are the resonance integrals of the carbonyl C—O bond and of the aromatic C—C bond, respectively. Rieger and Fraenkel [50] found that the best values of the parameters h and k are around 1.6 and 1.5, respectively.

It would be desirable to apply the modified Hückel method to systems treated qualitatively in Chapter 5 by Sharp and Symons.

3.1.2. Perturbation Treatment

The second term of Eq. 40 may be considered as a perturbation of the first term. According to first-order perturbation theory the perturbed MO $\varphi_0^{(1)}$

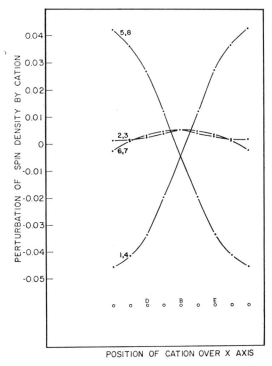

Figure 3. Proton spin density perturbation in pyracene⁻ by a cation located over the x-axis. Reproduced by permission from [46].

is given by

$$\varphi_0^{(1)} = \varphi_0 + \sum_x \frac{\langle \varphi \mid -e^2/r \mid \varphi_x \rangle}{E_0 - E_x} \varphi_x \tag{47}$$

where φ_0 and φ_x are eigenfunctions of the effective unperturbed Hückel Hamiltonian with eigenvalues E_0 and E_x, respectively. By expressing φ_0 and φ_x as a linear combination of atomic orbitals and using Eq. 42, the matrix element in Eq. 47 becomes

$$\langle \varphi_0 \mid -\frac{e^2}{r} \mid \varphi_x \rangle = -2e^2 \sum_{i,j} C_{0i} C_{xj} \frac{\langle \chi_i \mid \chi_j \rangle}{r_i + r_j} \tag{48}$$

where $\langle \chi_i \mid \chi_j \rangle$ again may be approximated by Eq. 43. The spin densities are now given by

$$\rho_i = [C_{0i}^{(1)}]^2. \tag{49}$$

Reddoch used this theory to explain the changes in the proton HFSCs of anthracene⁻ [51]. It was shown that the splittings were functions of solvent, cation, and temperature. The extreme values of the splittings are given in Table 1. Species A refers to the sodium ion pair in tetrahydrofuran; species B

Table 1   Extreme Values (in gauss) of the Proton HFSCs in Anthracene⁻ [51]

Species	a_1	a_2	a_9
A	2.757	1.498	5.294
B	2.545	1.551	5.484

refers to the potassium ion pair in tetramethylethylenediamine. In contrast to A, species B shows an alkali hyperfine splitting. Hence the cases A and B reflect the minimum and maximum cation perturbations, respectively.

In Table 2 the calculated proton HFSCs are given for the free anion and

Table 2   Calculated Proton HFSCs (in gauss) of Anthracene⁻ [51]

Species	a_1	a_2	a_9
Free ion	2.688	1.344	5.432
Perturbed ion	2.615	1.350	5.521

for the perturbed anion calculated by the first-order perturbation theory on the assumption that the cation is located 3.5 Å above the center of the anthracene skeleton. Comparison between Tables 1 and 2 shows that the sign and the approximate magnitude of the experimental variations are very well predicted by the theory. However, similar calculations on ion pairs of azulene⁻ failed in predicting the observed effects [51]. At present little can be said about the general reliability of this method because only a few attempts to utilize it have been reported in the literature.

3.1.3. Self-Consistent Field and Configuration Interaction

The methods just described have two serious shortcomings. First, they do not account for the electronic repulsion in the π-electron system, and second, only one MO is considered in the spin-density calculations. Studies of unperturbed aromatic systems demonstrated that π-electronic repulsion and excited configurations often play an important role in determining the electronic properties [39, 44]. Adequate theories accounting for these factors are the self-consistent field (SCF) theory supplemented by the configuration interaction (CI) theory.

In SCF calculations on aromatic systems, the zero-differential-overlap (ZDO)-LCAO approximation of Pariser, Parr, and Pople (PPP) has been used most frequently [52, 53]. Generally, the Hückel MOs are utilized as a basic set for the iterative SCF procedure. It seems that the SCF approach can be applied equally well to the perturbed radical anions. The effect of the perturbing cation may be accounted for by using the modified Hückel MOs as a basic set (see Section 3.1.1). The calculation then proceeds in the same way as for unperturbed systems. The spin densities are obtained from the coefficients of the SCF MO containing the unpaired electron.

To our knowledge the SCF calculations of spin densities of perturbed radical ions have been reported only for one case—the triphenylene radical monoanion. Unfortunately, this approach, like the modified Hückel theory, was found inadequate to account for the observed large shifts in the proton HFSCs [54]. This is not surprising in view of some theoretical difficulties due to the degeneracy of the lowest antibonding MOs, which is lifted when ion pairs are formed, provided the cation is not localized in a central position. Removing of the degeneracy gives rise to relatively strong perturbations of the spin densities.

For unperturbed radicals the Hückel theory usually gives a better agreement with experiment than does the SCF theory. This may be due to cancelling of errors in the Hückel method [39]. Hence the ZDO-LCAO SCF theory is not as useful for calculating spin densities of perturbed radical anions as has been expected. For free radical-ions the results of calculations are appreciably improved if the SCF procedure is combined with the CI theory. A slight modification of the CI approach permits calculation of spin densities for ion pairs.

In the Hückel and restricted SCF methods the ground state of a radical is described by a single Slater determinant,

$$\psi_0 = |\varphi_1 \bar{\varphi}_1 \cdots \varphi_x \bar{\varphi}_x \cdots \varphi_0| \tag{50}$$

where φ_x denotes an MO occupied by an electron with α-spin ($m_s = +\frac{1}{2}$) and $\bar{\varphi}_x$ represents an MO occupied by an electron with β-spin ($m_s = -\frac{1}{2}$). The CI method also takes excited configurations into account. Since the spin-density operator is a one-electron operator, only singly excited configurations with respect to the ground configuration have to be considered. Important entities in the CI calculations are the matrix elements of the total Hamiltonian between two configurations. The CI calculations on systems perturbed by a cation differ from those on unperturbed systems by virtue of the extra term $-e^2/r$ that appears in their Hamiltonian. The matrix element of $-e^2/r$ between two configurations can be written in terms of the matrix element between two MOs. For instance, if an excited configuration is given by

$$\psi_k = |\varphi_1 \bar{\varphi}_1 \cdots \varphi_x \bar{\varphi}_x \cdots \varphi_k| \tag{51}$$

the respective matrix element of $-e^2/r$ is

$$\langle \psi_0| - \frac{e^2}{r} |\psi_k\rangle = \langle \varphi_0| - \frac{e^2}{r} |\varphi_k\rangle \tag{52}$$

The evaluation of this matrix element follows that of Eq. 48. The final CI spin-density calculation is analogous to that for the unperturbed systems [55–57]. We could use, for instance, the Hückel MOs of the unperturbed systems as the basis MOs for the calculations; however, improved results are to be expected if the modified Hückel MOs or, still better, the modified SCF MOs of the perturbed system are used.

As far as we know, no such detailed spin-density calculations on ion pairs have been reported yet in the literature, but we expect that this method will provide a better understanding of the observed perturbation effects on the proton hyperfine splitting constants.

3.2. Changes in Hyperfine Splitting Constants of Other Nuclei

The hyperfine coupling of an unpaired electron with a ^{13}C, ^{14}N, or ^{17}O nucleus is determined not only by the spin density on the nucleus itself but also by the spin densities on the neighboring nuclei. For instance, the ^{13}C HFSC is given by [58]

$$a(^{13}C) = Q_1\rho_1 + \sum_i \rho_i Q_i \tag{53}$$

where ρ_1 is the spin density at the ^{13}C nucleus and the summation extends over the neighboring nuclei with spin density ρ_i.

Since the second term in Eq. 53 is of the same order of magnitude as the first, the effect of cationic perturbation on the splitting constants of these nuclei is usually much greater than on proton hyperfine splittings [36, 37]. In addition, in ion pairs of ketyls and nitroaromatics the cation is located near the functional group; consequently, the perturbation of the spin density on the oxygen or nitrogen nuclei should be large, causing an appreciable change of the ^{17}O or ^{14}N coupling constants.

The experimental data are surveyed in review articles by Hirota [36] and in Chapter 5 by Sharp and Symons. The latter authors also give some qualitative explanations accounting for the various cation effects. The calculation methods described in the previous sections may lead to more quantitative explanations.

3.3. g-Value Shifts

In aromatic free radicals the orbital angular momentum is largely quenched. Therefore the g-values of such radicals are expected to be very close to the free electron value g_0, which is equal to 2.0023192 [59]. As has been shown by Stone [60], deviations from g_0 can be attributed to spin-orbit interaction via coupling with the electrons in localized σ-orbitals of the C—C and C—H bonds. It could be demonstrated that as a result of this mechanism the g-values of aromatic radicals are linearly related to the Hückel orbital energy of the MO occupied by the unpaired electron. Expressed in the Coulomb and resonance integrals, α and β, this energy is given by

$$E_0 = \alpha + \lambda_0\beta \tag{54}$$

The shift in the g-value is then equal to

$$g - g_0 = B + \lambda_0 C \tag{55}$$

where B and C are constants independent of the radical studied. This linear relationship was confirmed experimentally by Blois et al. [61] and by Segal et al. [62] for numerous positive and negative radical ions, and the g-values were found to be independent of solvent, counterion, and temperature.

Exceptions to this relationship were found only for some radicals having a twofold orbitally degenerate electronic ground state [62, 63]. Such deviations have been attributed to Jahn-Teller distortions causing enhanced spin-orbit interactions [64–66].

Recently, more accurate g-value data [67–69] have revealed that the g-value of a radical anion may change upon ion-pair formation. In an ion pair the unpaired electron may be delocalized onto the cation, and this increases the spin-orbit interaction, causing, as will be shown later, a negative shift of the g-value.

Delocalization into the valence s-orbital of the metal does not affect the g-value. However, if a small amount of the metal p_z-orbital is admixed in the MO containing the unpaired electron, coupling with the metal p_x- and p_y-orbitals results in a shift of the g-value. The shift of the diagonal elements of the g-tensor due to the spin-orbit interaction is then given by

$$\Delta g_{ii} = f\zeta \sum_{j=x,y} \frac{\langle p_z | l_i | p_j \rangle \langle p_j | l_i | p_z \rangle}{E_z - E_j} \tag{56}$$

in which f is a positive proportionality factor depending on the amount of mixing, ζ is the spin-orbit coupling parameter for the relevant cation, and l_i is a component of the orbital momentum operator; E_z and $E_{x,y}$ are the energies of the perturbed metallic p_z-orbital and of the degenerate orbitals

p_x and p_y, respectively. In Eq. 56 the contributions of higher metallic orbitals have been neglected.

Evaluating the matrix elements of Eq. 56, we get

$$\Delta g_{xx} = \Delta g_{yy} = \frac{f\zeta}{E_z - E_x} \tag{57}$$

$$\Delta g_{zz} = 0$$

and thus the shift in the g-value is given by

$$\Delta g = \frac{\frac{2}{3}f\zeta}{E_z - E_x} \tag{58}$$

The parameter ζ can be shown to be always positive. Since we have assumed that the np_z-orbital is perturbed by the radical anions, whereas the np_x- and np_y-orbitals are not, we have

$$E_z < E_x = E_y \tag{59}$$

Therefore Eq. 58 predicts a negative g-shift proportional to the spin-orbit coupling parameter. The mixing factor f is not constant, but it depends on the nature of the cation. Nevertheless, since ζ substantially increases with atomic number, the negative shift is expected to increase along the series Li $<$ Na $<$ K $<$ Rb $<$ Cs.

The g-value measurements by de Boer [67] on alkali pyracene systems verified these predictions. Although the g-value of the lithium and the sodium ion pairs is almost the same as for the free pyracene anion, its value for the cesium pyracene ion pair is considerably lower. Dodson and Reddoch [68] found the g-values of the cesium and the rubidium naphthalenides to be appreciably lower than of the free anion, the largest effect was observed for

Table 3   g-Value Dependence on the Alkali Spin Orbit Parameter ζ for Alkali Naphthalene Ion Pairs [69]

Cation	$\Delta g \times 10^5$*		ζ (cm^{-1})
	DME	THF	
Li$^+$	−0.07	+0.07	0.29
Na$^+$	−1.13	−0.56	11.46
K$^+$	−1.19	−0.32	38.48
Rb$^+$	−4.84	−4.73	158.40
Cs$^+$	−18.66	−20.69	369.41

* $\Delta g = g$(ion pair) $- g$(free ion) at -20C.

the cesium salt. Although they did not observe g-shifts for other alkali naphthalenides, the more accurate measurements by Williams et al. [69] detected a small g-shift in the predicted direction for such systems.

The g-value shifts determined in dimethoxyethane (DME) and tetrahydrofuran (THF) are compared in Table 3 with the spin-orbit coupling parameter ζ. The predicted trends are clearly shown, a finding supporting the proposed mechanism of the g-value shift.

3.4. Temperature Dependence

Several factors must be considered to explain the temperature dependence of hyperfine splittings and g-values in alkali radical ion pairs. The structure of the ion pair may change with temperature, resulting in a large change of the cation perturbation of the π-system. This type of temperature dependence corresponds to the static models of ion pairs proposed by Atherton and Weissman [27] and by Szwarc et al. [70–71].

The investigated solution may contain different types of ion pair [72, 73], or mixtures of ion pairs and free ions in dynamic equilibrium each with the other. For rapid interconversion between the species, the ESR parameters become weighted averages of those of the individual species. As the temperature varies, the equilibrium is shifted and this, in turn, changes the mean values of the ESR parameters. The dynamic model of ion pairs has been elaborated by Hirota and Kreilick [74] and by Hogen-Esch and Smid [75].

The proton hyperfine splitting constants may vary with temperature even in free anions [76], possibly due to changes in the out-of-plane C—H vibrations.

From the foregoing analysis, it follows that it is difficult to predict quantitatively the temperature dependences of HFSCs. In most studies the measured temperature trends in hyperfine splitting and g-value shifts have been associated with the measured alkali splitting observed in the same system. Hirota [77] found a linear correlation between the proton and the alkali hyperfine splittings in the alkali anthracenide ion pairs and suggested that the presence of two distinct ion pairs having the same geometrical structures but different cation-anion separation may account for this observation (dynamic model). However, such a systematic correlation was not observed for the sodium naphthalenide system. This was explained by assuming an equilibrium between two geometrically different ion pairs. In such a case the cation perturbation is no longer a unique function of the cation-anion distance, and consequently the correlation between the alkali and proton splittings is lost.

For two distinct species A and B in rapid equilibrium with each other, the measured alkali HFSC is given by

$$a_M = \frac{a_A + Ka_B}{1 + K} \tag{60}$$

where K is the equilibrium constant $[A]/[B]$, and a_A and a_B are the alkali HFSCs of A and B, respectively.* Similarly, for the g-value shift we find

$$\Delta g = \frac{\Delta g_A + K\Delta g_B}{1 + K} \tag{61}$$

Elimination of K yields

$$\Delta g = \frac{\Delta g_B - \Delta g_A}{a_B - a_A} a_M + \Delta g_A - \frac{\Delta g_B - \Delta g_A}{a_B - a_A} a_A, \tag{62}$$

that is, a linear relationship is expected between Δg and a_M, as indeed has been found for the sodium naphthalenide system by Williams et al. [69], who measured these ESR parameters at various temperatures. However, the plot of Δg against a_M is far from being linear for cesium naphthalenide in DME. Apparently the dynamic model is inadequate for describing the temperature dependence in this system, and it has also been proved invalid for other ion pairs of cesium. This was demonstrated for the ion pairs of the tetraphenyl-boron anion by electrical conductivity studies [78] and for the biphenyl anion by a detailed analysis of the Cs NMR line-width as a function of temperature [79]. Probably in these systems the cesium cation forms a tight ion pair with the radical anion, and the structure of the pair is modified by temperature (static model). The temperature dependence of a_M and Δg are ascribed then to variations in the relative populations of the various vibrational levels of the ion pair. Further theoretical and experimental work are necessary to elaborate this point.

4. ALKALI HYPERFINE SPLITTING CONSTANTS IN RADICAL ION PAIRS

4.1. Introduction

Anion-cation interactions in aromatic radical ion pairs are often revealed through the appearance of alkali hyperfine splittings in the ESR spectra. Such a splitting was observed first for sodium naphthalenide in THF [27].

* Here it is implicitly assumed that a_A and a_B are temperature independent. Most probably this is not the case. [Editor.]

Each proton hyperfine line was split into four lines, their separation being sensitive to changes in solvent and temperature. Analysis of the spectrum led to the conclusion that this additional splitting was caused by the hyperfine interaction of the unpaired electron of naphthalene$^-$ with the spin ($I = \frac{3}{2}$) of the sodium nucleus. Such an interaction results from a nonzero electron spin density at that nucleus. Subsequently, numerous observations of alkali splittings were reported [36, 80] see also Chapter 5.

The sign of the alkali coupling constant was not investigated in the early studies. Its direction, although not given by the ESR spectra, was believed to be positive, implying a positive spin density at the alkali nucleus. However, the anomalous temperature dependence of the cesium splitting observed by de Boer for the cesium pyracene ion pair [67] led to the reconsideration of this assumption. When the temperature was lowered, the cesium splitting decreased to zero and thereafter increased upon further cooling. This phenomenon was eventually explained by postulating a change of sign of the spin density at the alkali nucleus as shown in Fig. 4.

The assumption that spin density at the alkali nucleus may be positive as well as negative was confirmed by NMR studies of alkali ion pairs [79, 81] and this subject is further discussed in Chapter 7. The sign of the alkali

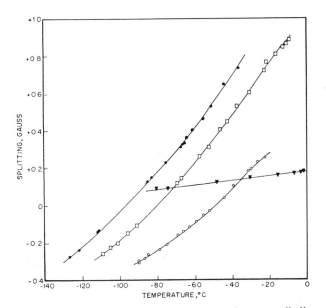

Figure 4. Alkali HFSC versus temperature for some alkali pyracene ion pairs. $\cdot$, a_{C_s} in 2-methyltetrahydrofuran. $\square$, a_{C_s} in tetrahydrofuran. $\bigcirc$, a_{C_s} in 1,2-dimethoxyethane. $\blacktriangledown$, a_{Na} in tetrahydrofuran. Reproduced by permission from [67].

coupling constant, a_M, can be derived from the observed contact NMR shift (δ_c, in gauss),

$$a_M = - \frac{g_M \beta_M}{(g\beta_e)^2} \frac{4kT}{H} \delta_c \qquad (63)$$

all the symbols having their usual meaning (cf. Eq. 1, Chapter 7). Both negative and positive contact shifts were observed and for some systems a sign reversal at some temperature was reported [79, 81].

Several serious difficulties have to be overcome before alkali HFSCs in ion pairs can be treated quantitatively. The results of calculations depend on the assumed structure of the investigated ion pair and, unfortunately, this information is lacking. Moreover, the splittings are relatively small when compared with the alkali splittings of the relevant free atoms. Hence a theory still must be developed to quantitatively account for a small effect of a relatively large radical anion on a small cation located in an unknown position. Not surprisingly, such a general theory of alkali HFSCs in ion pairs has not yet been developed and only a few attempts have been made to calculate quantitatively the observed splittings for some simple systems. Most of the discussions of this subject have dealt with the interpretation of the order of magnitude, the temperature dependence, or the sign of the alkali HFSCs.

4.2. Spin Transfer Mechanisms

4.2.1. Direct Mechanism

The simplest mechanism of spin transfer from the radical anion to the alkali cation assumes a direct contact interaction between the unpaired electron of the radical anion and the alkali nucleus. If $\varphi_0(r_M)$ is the amplitude of the MO, φ_0, containing the unpaired electron at the position occupied by the alkali nucleus, the alkali spin spin density is given by

$$\rho_M = \varphi_0^2(r_M) \qquad (64)$$

It has been shown [27] that this mechanism accounts for only a small fraction of the observed spin density since the amplitude of φ_0 nearly vanishes at positions accessible to the cation. The resulting spin density calculated by this method is always positive, and therefore this mechanism cannot explain a negative alkali spin density. Hence some other mechanisms must contribute to the spin transfer.

4.2.2. Overlap Mechanism

Mixing of the pure aromatic MO φ_0 with the alkali valence orbital χ_{ns} leads to a nonzero spin density at the alkali nucleus and gives some covalent character to the bonding between the ions. The hybridized molecular orbital is

$$\varphi_0' = \varphi_0 + \lambda \chi_{ns} \tag{65}$$

and the spin density at the alkali nucleus is then

$$\rho_M = \varphi_0^2(r_M) + 2\lambda\varphi_0(r_M)\chi_{ns}(r_M) + \lambda^2\chi_{ns}^2(r_M) \tag{66}$$

The first-order perturbation gives the mixing coefficient [82]

$$\lambda = \frac{\langle \varphi_0 | \, e^2/r \, | \chi_{ns} \rangle}{E_0 - E_{ns}} \tag{67}$$

where E_0 and E_{ns} are the energies of the MO φ_0 and of the metallic orbital χ_{ns}, respectively. Combining Eq. 37 with Eq. 67, we find

$$\lambda = \sum_i C_{0i} \frac{\langle \chi_i | \, e^2/r_i \, | \chi_{ns} \rangle}{E_0 - E_{ns}} \tag{68}$$

in which r_i is the distance between the cation and the carbon atom C_i of the radical. The matrix elements of this equation have been computed by Aono and Oohashi [82] for several reasonable structures of the alkali naphthalenide ion pair. Empirical evaluation of these matrix elements has been reported by Goldberg and Bolton [83], who used a modified Wolfsberg-Helmholtz approximation [84]:

$$\langle \chi_i | \frac{e^2}{r_i} | \chi_{ns} \rangle = (I_i + I_M)\langle \chi_i | \chi_{ns} \rangle \tag{69}$$

where I_i and I_M are the ionization potentials of the carbon atom C_i and of the metal atom M. Substitution of Eq. 69 into Eq. 68 gives

$$\lambda = \sum_i \frac{C_{0i}(I_i + I_M)}{E_0 - E_{ns}} \langle \chi_i | \chi_{ns} \rangle \tag{70}$$

This equation shows that the mixing coefficient is proportional to the overlap integrals $\langle \chi_i | \chi_{ns} \rangle$, which were tabulated by Aono and Oohashi [82]. Having λ, we calculate the spin density at the alkali nucleus from Eq. 66. For any reasonable cation-anion distances the first two terms of Eq. 66 may be neglected [82], and this leads to a reasonable approximation:

$$\rho_M = \lambda^2 \chi_{ns}^2(r_M) \tag{71}$$

The theoretical alkali hyperfine splitting constant derived from this treatment therefore is

$$a_M = Q_M \lambda^2 \tag{72}$$

where Q_M denotes the hyperfine splitting constant of the free alkali atom in the ^2S ground state.

Calculations based on this mechanism gave in many cases the right order of magnitude for the alkali HFSCs. For example, Aono and Oohasi [82] calculated for alkali naphthalenide the alkali splittings of 0.7 and 0.8 gauss in good agreement with experiments [27]. In their calculations the cation was localized 3.2 or 4.2 Å above the center of a benzene ring. Similar and even more extensive calculations on alkali ion pairs of the naphthalene, anthracene, and biphenylene radical anions were reported by Goldberg and Bolton [83]. The spin density at the alkali nucleus has been calculated for several plausible cation positions and the ion-pair association energies were obtained, using the modified Hückel approximation discussed in Section 3.1.1. This method has been employed previously by other workers to compute potential energy surfaces of alkali ion pairs of pyrazine [85] and acenaphthene [47]. The association energy E_a of an ion pair is approximated as

$$E_a = \sum_x \nu_x (E_x' - E_x) - \sum_i \frac{e^2}{r_i} \tag{73}$$

where E_x and E_x' are the energies of the Hückel MO φ_x of the free anion and of the modified Hückel MO φ_x' of the ion pair, respectively, and ν_x is the occupation number of the relevant MOs. The last term represents the repulsion between the cation and the effective positive carbon cores. The cation was assumed to be located above the nuclear plane of the radical ion, at a distance given by the sum of the effective height of the π-system (1.9 Å), and the ionic radius of the alkali cation.

The alkali spin densities for the sodium naphthalenide ion pair with the sodium at 3 Å above the plane are given in Fig. 5. It is important to note that the predicted sodium splitting is very small when the cation is located near the center of the molecule, whereas the maximum spin density is anticipated for the positions near the center of one of the two benzene rings (cf. [82]). The experimental sodium splittings are 1–2 gauss corresponding to a spin density of about $3–6 \times 10^{-3}$—a value accurately predicted provided the cation is placed near the center of a benzene ring. Calculations of the association energies expected for the various placements of the cation showed that this position is also energetically the most stable.

Similar results were obtained for lithium naphthalenide by Pedersen and Griffin [86], who applied the INDO (intermediate neglect of differential overlap) technique introduced by Pople et al. [87]. The placement of lithium over the center of a benzene ring is again preferred to that over the center of the naphthalene skeleton. However, the energy differences between the various placements are small and perhaps not too meaningful in view of the involved approximations.

The spin density map for the sodium anthracenide ion pair is shown in Fig. 6. Here the largest density is found for the cation positions near the

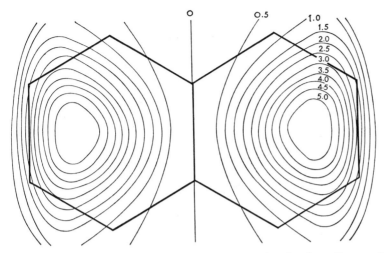

Figure 5. Calculated spin density at the sodium nucleus for the sodium naphthalene ion pair. The cation is 3 Å above the aromatic plane. The spin density is given in units of 10^{-3}. Reproduced by permission from [83].

center and these placements are also preferred energetically. Again the order of magnitude of the calculated spin densities agrees very well with the experimental results. Comparison of Fig. 5 and Fig. 6 shows that the sodium splitting is larger for the anthracenide than for the naphthalenide pair, in spite of the fact that the cation is more loosely bound in the former pair than in the latter [71]. This prediction is confirmed by the experimental data [74, 77]. It therefore seems that cation is indeed located above the central ring of anthracenide, contradicting the calculations based on optical spectra [88–89] but in accord with the calculations of the proton splitting perturbations [51].

A different situation is encountered in the alkali biphenylenide system for which the positions of maximum association energy corresponds to minimum spin densities. This may account for the fact that the alkali splittings are much smaller for biphenylenide ion pairs than for the corresponding naphthalenide or anthracenide ion pairs [83].

In conclusion, the charge transfer mechanism accounts reasonably well for the observed alkali splittings. Such calculations give only the positive contributions to the HFSCs if the approximations just outlined are used in the calculations (see Eqs. 66, 71, and 72). However, more refined calculations may lead to negative contributions also, and in fact Goldberg and Bolton [83] showed that such a result may be obtained if the method of different orbitals for different spins (DODS) [39] is applied. In the DODS method

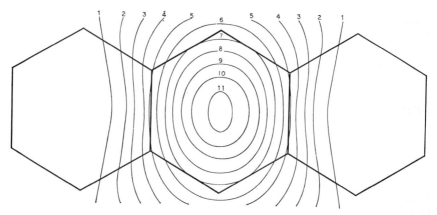

Figure 6. Calculated spin density at the sodium nucleus for the sodium anthracene ion pair. The cation is 3 Å above the aromatic plane. The spin density is given in units of 10^{-3}. Reproduced by permission from [83].

one takes into account the polarization of the closed shell MOs by the unpaired electron. Consequently, an α-spin MO is not equivalent to a β-spin MO, and therefore the closed-shell MOs may also contribute to the alkali HFSCs. The polarization or exchange effects are also incorporated in the INDO method [87], and its application led to negative spin densities for some conformations of lithium naphthalenide [86].

The DODS and the INDO methods are tedious and suffer from many computational problems [86]. It is therefore desirable to develop another approximation to explain the negative HFSCs. It will be shown in the following section that explicit consideration of the exchange effects leads directly to negative spin densities on alkali nucleus.

4.2.3. Exchange Mechanisms

The modified Hückel treatment cannot account for the negative alkali HFSCs, just as the conventional Hückel method failed to explain negative carbon spin densities observed in some hydrocarbon radicals [24]. It was shown that exchange effects involving the unpaired electron and the electrons of the closed shells are responsible for the negative spin densities. Adequate description of these effects is achieved by the CI calculations (Section 3.1.3), which often predict, in agreement with experiment, negative spin densities, for sites at which the Hückel zero-order spin densities are small [90, 91]. Since the zero-order Hückel spin density at the alkali nucleus of an ion pair is also small, we may expect that CI calculations will predict negative spin densities at these nuclei.

The zero-order electronic ground state of an ion pair may be described by the following Slater determinant:

$$\psi_0 = |\overline{\varphi_a'\varphi_a'} \cdots \overline{\varphi_x'\varphi_x'} \cdots \overline{\varphi_0'\chi_{1s}'\chi_{1s}'} \cdots \overline{\chi_\mu'\chi_\mu'}| \tag{74}$$

All functions in this determinant are considered as MOs belonging to the entire ion-pair system

$$\varphi_x' = \varphi_x + \sum_\mu C_{x,\mu}\chi_\mu \tag{75}$$

$$\chi_\mu' = \chi_\mu + \sum_x C_{\mu,x}\varphi_x \tag{76}$$

The coefficients $C_{x,\mu}$ and $C_{\mu,x}$ are supposed to be very small, implying that the MOs φ_x' are mainly aromatic, whereas the MOs χ_μ' are mainly metallic.

A better description of the ground state is obtained if configuration interaction is taken into account. For spin-density calculations only singly excited configurations must be considered. From this class of excited configurations only the configurations $\psi_{x,y}$ where an electron has been excited from a doubly occupied MO, say X, to an empty MO, say Y, can give rise to important first-order corrections to the spin density. If the ground state is described by

$$\psi_0^{(1)} = \psi_0 + \sum_{x,y} \lambda_{x,y}\psi_{x,y} \tag{77}$$

where the coefficients $\lambda_{x,y}$ are determined by first-order CI, the first-order spin density at the alkali nucleus is equal to

$$\rho_{\text{M}}^{(1)} = 2\sum_{x,y} \frac{\langle \varphi_0'X \mid Y\varphi_0'\rangle}{E_y - E_x} X(r_{\text{M}})Y(r_{\text{M}}) \tag{78}$$

where

$$\langle \varphi_0'X \mid Y\varphi_0'\rangle = \iint \varphi_0'(1)\, X(2)\, \frac{e^2}{r_{12}}\, Y(1)\varphi_0'(2)\, dV_1\, dV_2 \tag{79}$$

Since in Eq. 78 each X and Y may be either aromatic (φ_x') or metallic (χ_v') MO, four different types of excitations can be discerned.

X, Y Metallic. By applying Eq. 78 to all metal-to-metal excitations, the following first-order CI contribution to the spin density is obtained:

$$\rho_{\text{M,M}}^{(1)} = 2\sum_{\mu,v} \frac{\langle \varphi_0'\chi_\mu' \mid \chi_v'\varphi_0'\rangle}{E_v - E_\mu} \chi_\mu'(r_{\text{M}})\chi_v'(r_{\text{M}}) \tag{80}$$

If we consider only the admixture of valence metal orbitals in φ_0, $\rho_{\text{M,M}}^{(1)}$ [92]

is in good approximation equal to

$$\rho_{M,M}^{(1)} = C_{0,ns}^2 \left[2 \sum_{\mu,\nu} \frac{\langle X_{ns}\chi_\nu \mid \chi_\mu\chi_{ns}\rangle}{E_\nu - E_\mu} \chi_\mu(r_M)\chi_\nu(r_M) \right]$$

$$+ C_{0,np}^2 \left[2 \sum_{\mu,\nu} \frac{\langle \chi_{np}\chi_\nu \mid \chi_\mu\chi_{np}\rangle}{E_\nu - E_\mu} X_\mu(r_M)X_\nu(r_M) \right] \qquad (81)$$

where $C_{0,ns}$ and $C_{0,np}$ are the mixing coefficients for the metal orbitals χ_{ns} and χ_{np}, respectively, with the MO φ_0.

The first sum in Eq. 81 can be compared with the spin density at the alkali nucleus for the alkali atom in the n^2S state, brought about by the unpaired electron in the ns valence orbital through core polarization. This part gives rise to positive spin densities. The second sum can be considered as a first-order CI contribution to the spin density at the alkali nucleus for the alkali atom being in the n^2P state. Hence it can be set equal to [92]

$$C_{0,np}^2 \rho_M(n^2P) \qquad (82)$$

The contribution to the alkali HFSC is then given by

$$a_M = C_{0,np}^2 a_M(n^2P) \qquad (83)$$

where $a_M(n^2P)$ is the alkali HFSC for the free atom in the n^2P state.

According to Eq. 83 the sign of the alkali HFSC in an ion pair due to polarization of the core by an np electron is equal to the sign of $a_M(n^2P)$. The experimental values of $a_M(n^2P)$ are given in Table 4 together with the

Table 4   Alkali HFSCs (in MHz) of the Free Atoms in the n^2S and n^2P States [92]

Atom	$a_M(n^2S)$	$a_M(n^2P)$
^{6}Li	152.1	−12.0
^{7}Li	401.8	−31.6
^{23}Na	886	−0.5[a]
^{39}K	231	−0.2[a]
^{85}Rb	1,012	2
^{87}Rb	3,417	8
^{133}Cs	2,298	—

[a] Sign uncertain.

experimental $a_M(n^2S)$ values. These values were calculated from the spectral terms in the n^2P and n^2S states of the free alkali atoms [93, 94]. It appears that $a_M(n^2P)$ is only a very small fraction of $a_M(n^2S)$. Therefore, for a given alkali metal, the range of the measured positive HFSCs should be much larger than the range of the negative HFSCs, contradicting the experimental facts. Furthermore, it appears that $a_M(n^2P)$ is distinctly negative only for lithium. Thus we would expect that the strongest tendency to have a negative HFSC would be exhibited by Li ion pairs but not by the Rb and Cs pairs, contrary to experimental results. For example, the alkali HFSC is negative for Rb and the Cs biphenylides and naphthalenides [79, 81, 95], whereas it is positive in the corresponding Li ion pairs [95]. Therefore an MO description involving only the excitations of the metallic part of the ion pair is not adequate to explain the negative spin densities at the alkali nucleus.

X, Y Aromatic. For local aromatic excitations Eq. 78 is reduced to

$$\rho_{Ar,Ar}^{(1)} = 2 \sum_{x,y} \frac{\langle \varphi_0' \varphi_x' \mid \varphi_y' \varphi_0' \rangle}{E_y - E_x} \varphi_x'(r_M) \, \varphi_y'(r_M) \tag{84}$$

which can be approximated by

$$\rho_{Ar,Ar}^{(1)} = 2 \sum_{x,y} \frac{\langle \varphi_0 \varphi_x \mid \varphi_y \varphi_0 \rangle}{E_y - E_x} C_{x,ns} C_{y,ns} \chi_{ns}^2(r_M) \tag{85}$$

The integral in Eq. 85 contains only pure aromatic MOs and can therefore be evaluated following standard methods such as the LCAO-ZDO approximation [39, 44]. The coefficients $C_{x,ns}$ can be calculated using variation theory and it can be shown [96] that they are proportional to the overlap integral between the aromatic MO φ_x and the metal orbital χ_{ns}.

Calculations of such excitations of the alkali naphthalene ion pairs were performed by Corvaja [97], who found nearly vanishing spin density at the alkali cation when it is located above the center of the naphthalene skeleton. Recent calculations by Canters et al. [96] showed that this contribution is very small also for other cation locations. This is illustrated in Fig. 7 for the most probable cation position, that above the center of a benzene ring. Hence these excitations cannot account for the negative alkali HFSCs in ion pairs.

X Aromatic, Y Metallic. These excitations represent mixing of the ground state with charge-transfer states of the type Ar.M, in which the aromatic molecule is in an excited state and the metal atom is either in an excited state or in the ground state. According to Eq. 78 the contribution of these cross-excitations to the spin density at the alkali nucleus is given by

$$\rho_{Ar,M}^{(1)} = 2 \sum_{x,\mu} \frac{\langle \varphi_0' \varphi_x' \mid \chi_\mu' \varphi_0' \rangle}{E_\mu - E_x} \varphi_x'(r_M) \chi_\mu'(r_M) \tag{86}$$

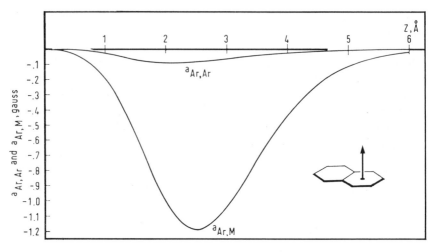

Figure 7.   Contributions to the sodium HFSC in sodium naphthalenide from local aromatic excitations, $a_{\mathrm{Ar,Ar}}$, and from aromatic to metal cross-excitations $a_{\mathrm{Ar,M}}$, versus the position of the Na$^+$ ion above the center of one of the benzene rings [96].

which is well approximated by

$$\rho_{\mathrm{Ar,M}}^{(1)} = 2 \sum_{x,\mu} \frac{\langle \varphi_0 \varphi_x \mid \varphi_x \varphi_0 \rangle}{E_\mu - E_x} C_{x,\mu} C_{\mu,x} \chi_\mu^{\,2}(r_{\mathrm{M}}) \tag{87}$$

where μ runs over all empty alkali s orbitals. It may be shown that the coefficients $C_{x,\mu}$ and $C_{\mu,x}$ have different signs, whereas all other entities of Eq. 87 are positive. Therefore $\rho_{\mathrm{Ar,M}}^{(1)}$ is negative.

Calculations by Canters et al. [96] showed that these cross-excitations cause a substantial negative spin density at the alkali nucleus (see Fig. 7). Their effect proved to be much larger than that of the local aromatic excitations. Although such calculations have been performed only on the sodium naphthalene system, it is plausible that such cross-excitations are also responsible for the observed negative alkali splittings in other systems.

A special type of cross-excitation may operate in ion pairs containing nitrogen or oxygen atoms. The excitations from the nonbonding orbitals of oxygen or nitrogen to the empty metallic orbitals may be important in these systems. However, Atherton [98] found that for alkali pyrazine ion pairs the contribution of this σ-π exchange effect to the alkali HFSC is small compared with the contribution of the overlap mechanism. On the other hand, Takeshita and Hirota [99] suggested that this σ-π exchange may be responsible for the observed negative alkali HFSCs in alkali 2,2'-dipyridyl ion pairs. The cation is likely to be located in the nodal plane of the πMO of 2,2'-dipyridyl containing the unpaired electron [100]. For this position the overlap mechanism gives no contribution to the alkali HFSC.

A similar situation was found for the lithium fluorenone ion pair studied by Nakamura and Hirota [101]. Here also the σ-π exchange could account for the observed negative alkali HFSC if the cation is located in the nodal plane, provided the bonding between oxygen and lithium has some covalent character. However, cross-excitations of the type discussed by Canters et al. [96] should be considered as well.

X Metallic, Y Aromatic. The first-order CI spin density due to these excitations is

$$\rho_{M,Ar}^{(1)} = 2 \sum_{\mu,x} \frac{\langle \varphi_0 \varphi_x \mid \varphi_x \varphi_0 \rangle}{E_x - E_\mu} C_{x,\mu} C_{\mu,x} \chi_\mu^{\ 2}(r_M) \tag{88}$$

where μ is now equal to $1s, \ldots, (n-1)s$. For the same reasons as for the aromatic to metal cross-excitations $\rho_{M,Ar}^{(1)}$ can be shown to be negative. The metal to aromatic excitations involve mixing of the ground state with charge transfer states of the type $Ar^{2-} \cdot M^{2+}$. Their effect on the metal spin density will be small because of the high energies of these charge-transfer states and because of the large number of nodes in the aromatic antibonding MOs, which greatly reduce the coefficients $C_{x,\mu}$ and $C_{\mu,x}$.

In conclusion, the aromatic-to-metal excitations are the main cause of negative alkali HFSCs. These negative coupling constants are observed if the zero-order spin density, due to the overlap mechanism, is small, as when alkali cations are placed in nodal planes of the aromatic moiety or for pairs in which the distances between the ions is large.

4.3. Temperature and Cation Dependence

The temperature dependence of the alkali HFSCs in ion pairs of radical anions was first observed by Atherton and Weissman [27] for the sodium naphthalenide ion pair. Since then similar observations were reported for many other ion pairs [36, 37]. For most systems the alkali HFSC decreases at lower temperatures.

Two models have been proposed to account for the observed temperature dependence. In the dynamic model the experimental alkali HFSC is believed to be a weighted average of the HFSCs of two or more different types of ion pairs coexisting in equilibrium. Temperature variation changes the position of the equilibrium and subsequently the measured alkali HFSC changes accordingly. Evidence for a rapid equilibrium between different ion pairs is provided by the dependence of the line-width of the alkali hyperfine lines on the magnetic quantum number M (see Section 5). However, although the behavior of some ion pairs is accounted for by this mechanism, it fails to explain change of sign of the metal HFSC observed for some systems at some temperature or the appearance of maximum or minimum in some curves representing a_M as a function of temperature, as noted for cesium biphenyl ion pairs [79].

The static model [71] relates the temperature dependence of the metal HFSC to changes in the environment of the alkali ion arising from temperature variations which affect the vibrational states of the ion pairs. For example, Atherton and Weissman [27] suggested that at equilibrium the sodium ion of the naphthalenide ion pair is located in the nodal plane passing through the central C—C bond and perpendicular to the plane of the naphthalene skeleton. The direct overlap mechanism predicts zero alkali HFSC for such a location (see Fig. 5, Section 4.2.2) and hence the positive temperature coefficient of the sodium splitting was attributed to the increased amplitude of the cation vibration through the nodal plane caused by higher temperature.

A more plausible and more general explanation of the positive temperature coefficients may be obtained in terms of the various mechanisms, each contributing to the alkali HFSC to the extent which depends on temperature. This approach explains in a natural way the sign reversal of the HFSC and eventually its negative values.

The spin density at the metal nucleus is the sum of a positive contribution determined by the overlap mechanism and a negative contribution arising from the exchange mechanisms involving the cross-excitations. The positive contribution depends on the degree of overlap of the ns metal orbital with the antibonding MO of the aromatic anion, its value rapidly decreasing with the increasing interionic distance.

On the other hand, the negative contributions of the aromatic to metal cross-excitations (see Section 4.2.3) are determined by the overlap of the bonding MOs of the radical anion and the empty metallic ns orbitals. Since the bonding MOs have a smaller number of nodes than the antibonding, this overlap is still quite substantial even at a relatively large interionic distance. Although the positive as well as the negative contributions to the alkali spin density decrease with increasing interionic distance, the negative contribution becomes relatively more important than the positive contribution because it decreases less rapidly, and this may lead to a negative spin density at a sufficiently large interionic distance. These expectations were confirmed by calculations of Canters et al. [96] on sodium naphthalenide. The overlap as well as the exchange contribution to the alkali HFSC were considered in their studies and some typical results are shown in Fig. 8 for structures in which the cation is located above the center of one of the benzene rings.

Interionic distance between the ions increases with increasing degree of solvation, and since the solvation is favored by lower temperatures, the HFSCs decrease accordingly.

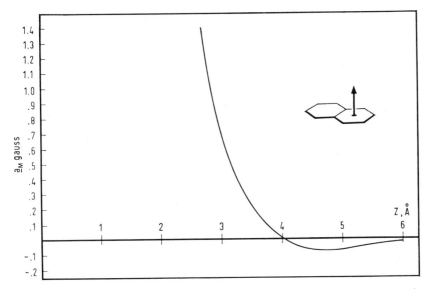

Figure 8. Sodium HFSC in sodium naphthalenide versus the position of the Na^+ ion above one of the benzene rings [96].

The preceding approach accounts well for the observed dependence of the alkali HFSCs on the nature of cation. For a series of alkali salts involving the same radical anion, the alkali spin density decreases as the atomic number of the metal increases and shows a tendency to become negative for the heaviest atoms. This trend, attributed to the increase of the interionic distance due to the increase of the cation radius, was confirmed by NMR experiments [92, 95].

In conclusion, present theories provide a reasonable explanation for the observed cation and temperature dependence of the alkali HFSCs of radical ion pairs. They permit also the calculation of the values and the sign of HFSC which agree fairly with the experimental data.

5. DYNAMIC EFFECTS IN ESR SPECTRA OF ION PAIRS

5.1. Introduction

de Boer and Mackor [33, 48, 67] observed a peculiar line-width alternation in the ESR spectra of alkali ion pairs of pyracene (see Fig. 1) and eventually concluded that this effect was caused by an intramolecular migration of the cation. Subsequently, similar effects were reported by other workers for numerous systems, and in Chapter 5 these investigations have been fully reviewed. The behavior of the alkali pyracene ion pair was comprehensively studied by Reddoch [46, 49] and by the present authors, and this system is singled out for further discussion because it illustrates all the aspects of ion pairing observed so far.

5.1.1. *Polarization Effects*

The ESR spectrum of the lithium salt of pyracene⁻ in 1,2-dimethoxyethane (DME) at −70°C is shown in Fig. 9. The splitting pattern arises from the

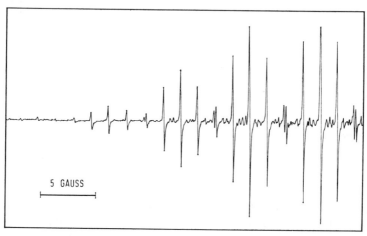

5 GAUSS

Figure 9. First derivative ESR spectrum of the lithium pyracene system in DME at −70°C. Only one-half of the spectrum is shown. Reproduced by permission from [48].

interaction of the unpaired electron with two sets of equivalent protons, the eight aliphatic protons (α-protons) and the four aromatic protons (β-protons). The magnitude of the coupling constants is given in Table 5. The small peaks seen in the spectrum are caused by ^{13}C present in its natural abundance.

Table 5   ESR Parameters of Alkali Pyracene Ion Pairs (splitting constants in gauss) [67]

Coupling Constant	Li⁺, DME −70°	Cs⁺, THF −80°	Na⁺, MTHF −80°	Li⁺, hexane/MTHF −113°
a_1	6.58	6.72	6.93	6.75
a_2	6.58	6.48	6.37	6.52
a_3	1.58	1.60	1.63	1.60
g	2.00267	2.00247	2.00265	2.00271
a_M	0	0	0.176	≤ 0.04

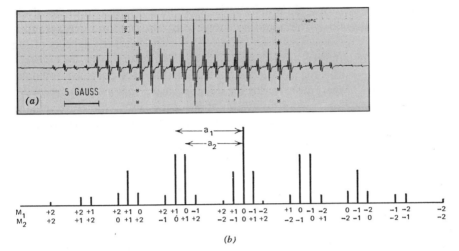

Figure 10. (a) First derivative ESR spectrum of the cesium pyracene ion pair in THF at −80°C. The cesium splitting is equal to zero. Reproduced by permission from [67], (b) Schematic stick diagram for the fine structure due to two sets of four equivalent protons with almost equal splitting factors a_1 and a_2 ($a_1 > a_2$). Note that the pattern of nine groups of lines collapses into 9 lines when $a_1 = a_2$.

In solvents of lesser solvating power, like tetrahydrofuran (THF) or 2-methyltetrahydrofuran (MTHF), the ions are tightly associated and, not surprisingly, the observed spectrum substantially differs from that shown in Fig. 9, which refers to the "free," or rather only slightly perturbed ions encountered in DME solution. In a tight ion pair the proton hyperfine lines are split due to the interaction of the odd electron with the spin of the Li nucleus, but in addition to this feature another interesting one appears. The eight aliphatic protons are no longer equivalent. Owing to the electric field of the counterion, which perturbs the charge distribution in the radical anion, two distinct aliphatic coupling constants can be discerned in the spectrum (a_1 and a_2). This is seen in Fig. 10a, which shows the ESR spectrum of the cesium pyracene ion pair in THF at −80°C. Under these conditions, the Cs splitting constant is vanishingly small and the observed spectrum is reproduced then by the set of proton-coupling constants listed in Table 5. For the sake of clarity, the respective stick diagram referring to the α protons only is depicted in Fig. 10b. The hyperfine components are characterized by the magnetic quantum numbers M_1 and M_2, each giving the sum of the z-components of the nuclear angular momenta of the two sets of four equivalent protons. When comparing the stick diagram with the experimental spectrum the additional splitting arising from the presence of the four equivalent aromatic protons should be noted. These split each line shown in the stick diagram into five lines (the coupling constant a_3) having the intensity ratio 1:4:6:4:1.

The four aromatic protons are still equivalent in spite of the ion pairing, but the g-value is affected, as shown in Table 5. The relatively large change in g-value due to variation in cation's nature arises from the increase in spin-orbit interaction (see Section 3.3).

5.1.2. Ion-Pair Structures

The observed hyperfine pattern for cesium pyracene in THF (Fig. 10) furnishes information on the structure of the ion pair. Only these locations of the counterion are acceptable which give rise to two sets of four equivalent aliphatic protons. Positions A, B, and C (see Fig. 1) meet this condition. Investigations of the ion pairs of pyracene⁻ having one or two of the CH_2 groups completely deuterated permits further differentiation between these choices [49]. The observed hyperfine pattern of the potassium pyracene-d_2 complex in THF at $-80°C$ is compatible with the structure of the ion pair in which the counterion vibrates along the X axis being on the average at position B. It was shown also [46] that the vibration along the short X axis hardly affects the spin densities on the β carbon atoms, justifying the invariance of the splitting constants of β protons. However, the vibration along the Y axis should strongly affect these splitting constants [46] (see also Section 3.11)

5.2. Intramolecular Cation Exchange

5.2.1. Experimental Evidence

An interesting spectrum was observed in THF at $-30°C$ (Fig. 11). If M denotes the magnetic quantum number for the total z-component of the

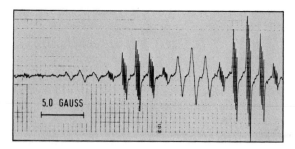

Figure 11. First derivative ESR spectrum of the sodium pyracene complex in THF at $-30°C$. Reproduced by permission from [48].

nuclear spin angular momentum of the eight aliphatic proton nuclei ($M = M_1 + M_2$), we notice that the quintuplets with even M exhibit a well-resolved alkali quartet structure, whereas in the other quintuplets the alkali quartet structure is not resolved. The alternating effect arises from a migration of the cation between two equivalent positions, in our case between B and B', where B' is situated symmetrically with respect to B. The migration from B to B' and vice versa causes time-dependent modulations in the isotropic coupling constants, particularly in the aliphatic proton couplings. The observed splitting constant for the aliphatic protons is $\bar{a} = \frac{1}{2}(a_1 + a_2)$ provided the lifetime τ of each distinct conformation is short relative

$$\{\gamma_e(a_1 - a_2)\}^{-1}$$

(γ_e is the gyromagnetic ratio of the electron), and the lines are sharp when the exchange is rapid. The spectra are complex for intermediate cases, when $\tau\gamma_e(a_1 - a_2) \approx 1$, since then some lines become broad while others remain narrow. Finally, for $\tau \gg \{\gamma_e(a_1 - a_2)\}^{-1}$, we observe again a sharp spectrum characteristic of a particular conformation of the ion pair (slow exchange). Experiments performed at various temperatures and in different solvents provided spectra corresponding to each of these situations.

The intramolecular nature of the reaction responsible for the broadening of some lines is clearly demonstrated by the spectrum depicted in Fig. 11. The width of the sharp lines of quintuplets gives the lifetime of a pair (i.e., the average time of association of a particular cation with a radical anion). This is longer than 10^{-7} sec. On the other hand, the broad lines permit to estimate the lifetime of a conformation, which is much shorter than 10^{-7} sec.

5.2.2. Line Width Alternation

The phenomenon of line-width alternation, observed in spectrum 11, is explained by the theories outlined in Section 2. The alternating effect occurs when the motion of the cation is sufficiently rapid to average the coupling constants but yet not rapid enough to eliminate the contribution of the exchange to the line-width. Since we deal with an exchange between two equivalent sites, the modified Bloch equation (16) applies with $\tau_A = \tau_B = \tau_0$.

Let us consider now two ESR lines which are interchanged by the exchange process. The corresponding frequencies are denoted by ω_B and $\omega_{B'}$ and the width of both lines is taken to be equal to T_2^{-1}. Inasmuch as τ_0 is small, the exchange terms in the numerator of Eq. 16 vanish and the total complex magnetization is, in good approximation, equal to

$$G = \frac{i\gamma_e H_1 M_0}{1/T_2 + i(\omega_0 - \omega) + (\frac{1}{8})(\omega_B - \omega_{B'})^2\tau_0} \tag{89}$$

Therefore, the line is Lorentzian, centered at $\omega_0 = \frac{1}{2}(\omega_B + \omega_{B'})$ and having an additional line-width due to the exchange process:

$$\Delta T_{2\,ex}^{-1} = \frac{1}{8}(\omega_B - \omega_{B'})^2 \tau_0 \tag{90}$$

Substituting the relevant coupling constants of aliphatic protons,

$$\omega_B = \gamma_e(a_1 M_1 + a_2 M_2) \tag{91}$$

and

$$\omega_{B'} = \gamma_e(a_2 M_1 + a_1 M_2) \tag{92}$$

into Eq. 90, we find

$$\Delta T_{2\,ex}^{-1} = \frac{1}{8}\gamma_e^2 \tau_0 (a_1 - a_2)^2 (M_1 - M_2)^2 = J(0)(M_1 - M_2)^2 \tag{93}$$

where $J(0)$ is the spectral density function at frequency zero.

The density matrix method (Section 2.2) or the relaxation matrix theory (Section 2.3) is also useful in treating the alternating line-width effect. The application of the first method is outlined in the dissertation of Neiva Correia [103], who discussed the line-width alternation effects observed in the ESR spectra of the negative ions of phthalonitrile, and terephthalonitrile [103, 104]. To familiarize the reader with the relaxation matrix method, let us derive Eq. 93 with its help. The relevant time-dependent spin Hamiltonian $\mathcal{H}_1(t)$ is (in frequency units)

$$\mathcal{H}_1(t) = F_1(t)S \cdot I_1 + F_2(t)S \cdot I_2 \tag{94}$$

where

$$F_1(t) = \gamma_e\{a_s(t) - \bar{a}\} \tag{95}$$

$$F_2(t) = \gamma_e\{a_r(t) - \bar{a}\} \tag{96}$$

and $a_s(t)$ and $a_r(t)$ are the instantaneous values of the HFSCs, their average value being $a = \frac{1}{2}(a_1 + a_2)$. The ratio of the nonsecular contributions to the line-width, arising from the terms such as $S_x I_{1x}$, $S_y I_{1y}$, to the secular contribution is

$$(1 + \omega_0^2 \tau_0^2)^{-1} : 1$$

Since $\tau_0 = 10^{-8}$ and $\omega_0 = 10^{11}$ sec^{-1}, the nonsecular contributions can be neglected and therefore in first approximation the relaxation matrix is diagonal. The exchange contribution to the line-width is then given by the diagonal matrix element, say $R_{ij,ij}$, where the pair (i,j) stands for the spin functions belonging to the energy levels between which ESR transitions occur. From Eq. 34 it follows that

$$R_{ij,ij} = 2J_{ii,jj}(0) - J_{ii,ii}(0) - J_{jj,jj}(0) \tag{97}$$

Using Eq. 35 we can calculate easily that

$$J_{ii,jj}(0) = -\frac{1}{4} \sum_{p,q=1,2} j_{pq} M_p M_q \tag{98}$$

where

$$j_{pq} = \tfrac{1}{2} \int_{-\infty}^{+\infty} F_p(t) F_q^*(t + \tau)\, dt$$

$$= \tfrac{1}{2} \int_{-\infty}^{+\infty} G_{pq}(\tau)\, dt \tag{99}$$

the function $G(\tau)$ is called the correlation function of $F(t)$, since it relates the value of $F(t)$ at one time to its value at a later time. Evaluation of $G(\tau)$ requires information on the physical process causing the modulation of the isotropic splitting constants. According to our model the instantaneous splitting constants a_s and a_r take on either of the values a_1 or a_2 and change from one value to the other randomly. As a result the functions $F_1(t)$ and $F_2(t)$ take on values of $\pm\tfrac{1}{2}\gamma_e(a_1 - a_2) = \pm a_0$. Slichter [4] showed that the correlation function for a function fluctuating randomly between two values $\pm a_0$ is given by

$$G_{11} = G_{22} = -G_{12} = -G_{21} = a_0^2 e^{-2\tau/\tau_0} \tag{100}$$

From Eqs. 99 and 100 it follows that

$$j_{11} = j_{22} = -j_{12} = -j_{21} = \tfrac{1}{2} \int_{-\infty}^{+\infty} a_0^2 e^{-2\tau/\tau_0}\, d\tau = \tfrac{1}{2} a_0^2 \tau_0 \tag{101}$$

Substitution of this equation into Eq. 98 leads to

$$J_{ii,jj}(0) = -\tfrac{1}{8} a_0^2 \tau_0 (M_1 - M_2)^2 \tag{102}$$

The other spectral density functions are found to be equal to

$$J_{ii,ii}(0) = J_{jj,jj}(0) = -J_{ii,jj}(0) \tag{103}$$

By using Eqs. 102 and 103, Eq. 97 reduces to

$$R_{ij,ij} = -\tfrac{1}{2} a_0^2 \tau_0 (M_1 - M_2)^2 \tag{104}$$

Using Eq. 36 and writing a_0 as $\tfrac{1}{2}\gamma_e(a_1 - a_2)$, we find for the exchange contribution to the line-width

$$\Delta T_{2\,\mathrm{ex}}^{-1} = \tfrac{1}{8}\gamma_e^2 \tau_0 (a_1 - a_2)^2 (M_1 - M_2)^2$$

$$= J_0(M_1 - M_2)^2 \tag{105}$$

This is the same result as obtained from the modified Bloch equations (see Eq. 93).

In Table 6 the various contributions to the line-widths of the hyperfine components are tabulated. They are expressed in the parameter $J(0)$. If $J(0)$ is large, lines with $M = \pm 3, \pm 1$ will be broadened, whereas lines with $M = \pm 4, \pm 2$ and 0 will be sharp, since for these M-values the main components correspond to $M_1 = M_2$ (see Table 6), which are not broadened by

Table 6 Exchange Contributions to the Line-Width of the Hyperfine Components Determined by the Magnetic Quantum Numbers M_1 and M_2 of the Aliphatic Protons in Pyracene[-] [48]

$M = M_1 + M_2$	M_1	M_2	n^a	Contribution to T_2^{-1}
+4	+2	+2	1	0
+3	+2	+1	4	$J(0)^b$
	+1	+2	4	$J(0)$
+2	+2	0	6	$4J(0)$
	+1	+1	16	0
	0	+2	6	$4J(0)$
+1	+2	−1	4	$9J(0)$
	+1	0	24	$J(0)$
	0	+1	24	$J(0)$
	−1	+2	4	$9J(0)$
0	+2	−2	1	$16J(0)$
	+1	−1	16	$4J(0)$
	0	0	36	0
	−1	+1	16	$4J(0)$
	−2	+2	1	$16J(0)$

^{a} n Indicates the degeneracy of the spin state level (M_1, M_2).
^{b} $J(0)$ is the secular spectral density function.

the exchange process according to Eqs. 93 and 105. The table illustrates the alternation in line-width; the agreement between the results in Table 6 and the spectrum in Fig. 11 is very satisfactory and can be considered as a confirmation for the proposed model of the ion pair.

5.2.3. Intermediate Rates of Exchange

The modified Bloch equations and the density matrix method describe the exchange process over the entire range of reaction rates. From a comparison of experimental and theoretical spectra we can determine the lifetime τ_0 and accordingly the rate constant k (for a first-order reaction $k = \tau_0^{-1}$). With the aid of the Arrhenius equation [105]:

$$k = Ae^{-\Delta E/RT} \tag{106}$$

or the Polanyi-Eyring equation [105], where k is given by

$$k = \frac{RT}{Nh} \exp\left(\frac{-\Delta G^{\ddagger}}{RT}\right) = \frac{RT}{Nh} \exp\left(\frac{\Delta S^{\ddagger}}{R}\right) \exp\left(\frac{-\Delta H^{\ddagger}}{RT}\right) \tag{107}$$

the thermodynamic quantities characterizing the process can be evaluated.

For the potassium pyracene ion pair in an equal volume mixture of MTHF and THF the spectra could only be measured within the temperature range from -60 to $-90°C$ [106], since at higher temperature rapid decomposition of the negative ion occurs and at lower temperature it dissociates into free ions. The rate constant k was found to be equal to 3.5×10^5 sec^{-1} at $-60°C$, and in the temperature range studied the activation energy $\Delta E = 0$. When Eq. 107 is used, $\Delta G^{\ddagger}$ is found to be equal to 7.0 kcal/mole, while $\Delta H^{\ddagger}$ and $\Delta S^{\ddagger}$ are -0.4 kcal/mole^{-1} and -34.7 e.u., respectively. The large negative entropy of activation indicates that the solvation of the cation increases appreciably upon migration. However, as pointed out by Sharp and Symons, (Chapter 5), these values should be considered with caution, since the structure of ion pair changes with temperature This is manifested by the dependence of the alkali HFSCs upon temperature (see Section 4.3); its value decreases at lower temperature due to the increase in the anion-cation distance. This in turn might accelerate the rate of transfer, leading to an apparent zero activation energy.

5.3. Other Cation Movements

5.3.1. Intermolecular Cation Exchange

Cation exchange may also take place via intermolecular reactions. If two equivalent sites for binding the cation are available, two types of cation exchange are possible:

$$Ar^{-}{\cdot}M_1^{+} + M_2^{+} \rightleftarrows Ar^{-}{\cdot}M_2^{+} + M_1^{+} \tag{a}$$

$$M_2^{+} + Ar^{-}{\cdot}M_1^{+} \rightleftarrows M_2^{+}Ar^{-} + M_1^{+} \tag{b}$$

In reaction a the incoming cation goes to the site already coordinated with the outgoing cation, whereas in reaction b the incoming cation becomes coordinated with the vacant site.

These types of exchange reaction, distinguished by their different effect on the line-width, have been studied for the anions of m-dinitrobenzene and 1,3-dinitro-5-t-butyl-benzene by Adams and Atherton [107] and for the anions of 9,10-anthrasemiquinone and 2,5-di-t-butyl-p-benzosemiquinone by Rutter and Warhurst [108]. On adding an excess of alkali ions it was found that the observed line-width variations were consistent with reaction b. For sodium t-butyl-m-dinitrobenzene the energy of activation of this process was found to be 6.2 kcal/mole and the preexponential factor 1.3×10^{12} mole^{-1} sec^{-1}.

At high rates of reactions a or b the alkali hyperfine structure disappears. The presence of ion pairs is then manifested by a shift of the g-value from that of the free anion.

5.3.2. *Oscillation of the Cation*

As has been argued before, it is reasonable to assume that the cation in the potential well will carry out oscillatory movements around its equilibrium position. Since this oscillating motion is rapid, its effect in the ESR spectra becomes visible only at very low temperatures and in solvents in which the intramolecular exchange is slowed down to such an extent that ESR spectroscopy portrays only a static conformation: the cation is either in position B or B'. A suitable solvent meeting these condition is MTHF. Below $-80°C$ the spectra of the sodium or potassium-pyracene ion pair in this solvent exhibit a change in hyperfine pattern, which is caused by the oscillations of the cation in the potential well.

The effect of these oscillations on the line-widths of the various hyperfine components can be illustrated most simply for the two components characterized by $M_1 = -2$, $M_2 = -1$ and $M_1 = -1$, $M_2 = -2$. In the structure diagrams shown in Fig. 12, representing the two conformations of the ion pair, we have indicated the arrangement of the nuclear spins of the aliphatic protons, which in the strongfield approximation corresponds to the two hyperfine components in question. In Fig. 12a the cation is located above the

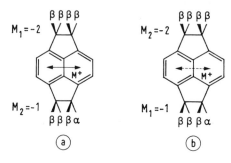

Figure 12. Spin arrangements of the aliphatic protons in pyracene$^-$ for two hyperfine components.

aromatic plane, in Fig. 12b below. Let us assume that the oscillation of the cation modulates the isotropic splitting factors of only those aliphatic protons that are at the same side of the molecular plane as the cation. Suppose the cation moves along the short axis of the molecule, so that these four protons fall into two sets of two equivalent protons with splitting factors a_1' and a_1''. If the oscillation frequency is large compared to $\gamma_e(a_1' - a_1'')$, an average splitting constant $a_1 = \frac{1}{2}(a_1' + a_1'')$ will be observed and the oscillation of the cation will not reveal itself in the spectra (rapid exchange). This situation is encountered in the spectra taken in MTHF at temperatures above $-80°C$. If the conditions for rapid exchange are not fulfilled, the spectra will undergo a change in hyperfine pattern, due to this oscillating motion of the cation. This can be illustrated for the two components under consideration as follows.

If the ion pair has the conformation of Fig. 12a, the ESR transition frequency is given by

$$\omega_a = \gamma_e H_0 - \gamma_e(a_2 + a_1' + a_2'') = \gamma_e H_0 - \gamma_e(a_2 + 2a_1) \qquad (108)$$

where a_2 is the splitting factor for the four aliphatic protons situated with respect to the cation on the opposite side of the aromatic plane. Evidently ω_a is not affected at all by the perturbing motion of the metal ion, owing to the special arrangement of the nuclear spins. On the other hand, if the ion pair has the conformation of Fig. 12b the resonance frequency ω_b will fluctuate between the values

$$\gamma_e H_0 - \gamma_e (2a_2 + a_1') \quad \text{and} \quad \gamma_e H_0 - \gamma_e (2a_2 + a_1''). \qquad (109)$$

In the limit of rapid exchange this produces a line broadening by an amount equal to

$$\tfrac{1}{8}\tau_0'\gamma_e{}^2(a_1' - a_1'')^2 \qquad (110)$$

where τ_0' is the mean time between frequency shifts.

Figure 13 shows on an extended scale the behavior of the two lines considered in relation to temperature. Hyperfine interaction with the alkali nucleus splits each line into four. Owing to the fact that

$$a_M = \frac{a_1 - a_2}{3}$$

two lines coincide and we observe groups of seven lines of which the central line has about twice the intensity of the others. The spectra show that the four alkali lines at high field remain sharp, whereas the four lines at low field are broadened with decreasing temperature. This is in accordance with the proposed mechanism, if a_1 is larger than a_2.

6. COUNTERION EFFECTS IN TRIPLET SYSTEMS

6.1. ESR of Triplet Systems

The spin Hamiltonian $\mathscr{H}(S)$ for a system of two unpaired electrons in a triplet state in the presence of a magnetic field $\mathbf{H}$ is given by [5]

$$\mathscr{H}(S) = g\beta\mathbf{H}(\mathbf{S}_1 + \mathbf{S}_2) + g^2\beta^2\left[\frac{\mathbf{S}_1 \cdot \mathbf{S}_2}{r^3} - \frac{3(\mathbf{S}_1 \cdot \mathbf{r})(\mathbf{S}_2 \cdot \mathbf{r})}{r^5}\right] \qquad (111)$$

where $\mathbf{S}_1$ and $\mathbf{S}_2$ represent the spin vectors of the two electrons and where r is the interelectronic distance. Since the two electrons spins are correlated,

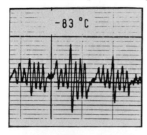

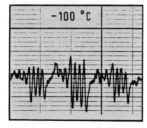

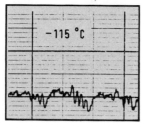

Figure 13. The group of first derivative ESR lines derived from the hyperfine components of Fig. 12, on an extended scale. Reproduced by permission from [67].

it is convenient to express $\mathscr{H}(S)$ in terms of the total spin **S** with components S_x, S_y, and S_z:

$$\mathscr{H}(S) = g\beta\mathbf{H}\cdot\mathbf{S} + g^2\beta^2\left[\sum_{i=x,y,z} S_i^2 \frac{r^2 - 3i^2}{r^5} - 3 \sum_{\substack{i\neq j \\ i,j=x,y,z}} S_i S_j \frac{ij}{r^5}\right] \quad (112)$$

By taking the average over all possible positions of the two electrons and by going from the laboratory coordinate system to the molecular principal coordinate system, $\mathscr{H}(S)$ is transformed into [5]

$$\mathscr{H}(S) = g\beta\mathbf{H}\cdot\mathbf{S} + D(S_z^2 - \tfrac{1}{3}S^2) + E(S_x^2 - S_y^2) \quad (113)$$

The parameters D and E are called zero-field splitting (ZFS) parameters and are defined as

$$D = \tfrac{3}{4}g^2\beta^2\langle\psi(1,2)|\frac{r_{12}^2 - 3z_{12}^2}{r_{12}^5}|\psi(1,2)\rangle \quad (114)$$

$$E = \tfrac{3}{4}g^2\beta^2\langle\psi(1,2)|\frac{y_{12}^2 - x_{12}^2}{r_{12}^5}|\psi(1,2)\rangle \quad (115)$$

where ψ is the orbital wave function common to the three triplet spin functions. D can be considered as a measure of the mean distance between the two triplet electrons, whereas E represents the deviation from the threefold symmetry of the electron distribution in the triplet system.

Triplet ESR studies have been performed on single crystals [109, 110], rigid glass solutions [110–112], and liquid solutions [113, 114]. For a discussion of the type of spectrum observed for the three different situations we refer to these papers. The ZFS parameters of the triplet spectra give information on the electronic properties as well as on the geometrical structure of the triplet species involved.

6.2. Biradicals

Biradicals may be formed by dimerization of monoradicals. The first experimental evidence for such dimerization was reported by Hirota and Weissman [115, 116], who studied metal ketyls in ethereal solvents and by Biloen et al. [117, 118] who investigated concentrated solutions of negative ions of aromatic hydrocarbons. The presence of paramagnetic dimers or, more generally, ion clusters was deduced from the observed rigid media triplet ESR spectra. The metal ketyl solutions showed very clearly the full-field ESR signals ("$\Delta m_s = +1$" transitions) [115, 116], whereas the concentrated solutions of aromatic hydrocarbons exhibited clearly weak half-field ESR signals ("$\Delta m_s = +2$" transitions) [117, 118]. These signals arise from the interaction between two electron spins on different radicals and were ascribed to triple ions [(Ar$^-$M$^+$Ar$^-$)] and quadrupole ions [(Ar$^-$)$_2$(M$^+$)$_2$].

The ESR spectra of rigid glass solutions have been proved to be useful in studying monomer-dimer equilibria and for the evaluation of the structure of biradicals. Comparison of the intensities of the monoradical and of the biradical ESR signals as a function of temperature and concentration may provide information on the type of equilibrium involved in the biradical formation [116–118].

Information about the structural properties of a biradical system may be obtained from the ZFS parameter measured from the triplet ESR spectrum [115–120]. According to Eq. 114 the D value is directly related to the mean distance $\bar{r}$ between the two unpaired electrons of the biradical. Since in biradical systems this distance is relatively large compared with the delocalization of the electrons, the absolute value of D may be approximated by [116]

$$|D| = \frac{\frac{3}{2}g^2\beta^2}{\bar{r}^3} \tag{116}$$

Some biradicals show a very small D value (<15 gauss), which according to

Eq. 116 corresponds to a mean distance $\bar{r}$ larger than 10 Å. This suggests the presence of solvent-shared ion clusters in accordance with the solvent dependence of the D value observed for these systems.

Other systems have D-values which are an order of magnitude larger than those observed for the solvent-shared ion clusters. This observation and the fact that the D-values for these systems are virtually independent of the solvent, led to the conclusion that these biradicals systems form contact ion clusters [116, 120]. The values for $\bar{r}$ calculated from the measured D-values by using Eq. 116 all lie in the same region, between 5 and 7 Å. As expected, $\bar{r}$ generally increases as the counterion of the biradical systems becomes smaller. The experimental D-values and the calculated $\bar{r}$ values for various biradical systems have been compiled by Hirota [36]. For a discussion of the data we refer the reader to chapter 5 by Sharp and Symons.

Considering the E-value, it has been found that in nearly all cases E is equal to zero. According to Eq. 115 a zero E-value indicates that the molecular x- and y-axes are equivalent. This suggests that the two parts of the biradical are perpendicular to each other. A nonzero E-value was found for the biradicals of the dinegative ions of dibenzoylmethane [121], dibenzamide, and benzoylacetone [122]. Using the density matrix formalism (see Section 2.2), van Willigen et al. [121, 122] were able to show that the triplet spectrum for each of these biradicals results from the presence of two different triplet species, one having axial symmetry $(E = 0)$ and the other without axial symmetry $(E \neq 0)$. The species without axial symmetry has been tentatively assigned to a planar structure with the two radical anions coordinated by two cations.

6.3. Triplet Dianions

6.3.1. *Term Schemes of Unperturbed Dianions*

Many aromatic molecules can be reduced to their dinegative ions. In most cases these ions are diamagnetic. In the MO description this corresponds to a doubly occupied lowest antibonding MO resulting in a singlet ground state. As a consequence the interaction of counterion with these dianions cannot be investigated with ESR.

A different situation may arise for the class of aromatic molecules having a threefold or sixfold symmetry axis. As can be demonstrated, with the help of group theory [123], such molecules have a set of degenerate and a set of non-degenerate π-electronic energy levels. Only in one such molecule, trinaphthylene, is the lowest antibonding level nondegenerate; its dianion has, therefore a singlet ground state in accordance with experiments [63].

Other molecules belonging to this symmetry class have a twofold degenerate lowest antibonding level, as exemplified by benzene, triphenylene (Tp),

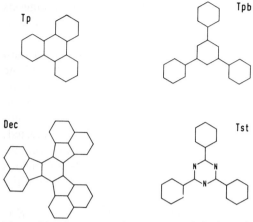

Figure 14.   Structures of some aromatic hydrocarbons with threefold symmetry.

coronene (Cor), 1,3,5-triphenylbenzene (Tpb), 2,4,6-triphenyl-sym-triazine (Tst) and decacyclene (Dec) (see Figs. 14 and 15).

Molecules possessing a twofold degenerated lowest antibonding level may form either a singlet or a triplet state dianion. In the one-electron approximation these states have the same energy, but this is no longer the case when electronic correlation is taken into account. The ground state must be a triplet state provided only the electronic correlation between the two degenerate MOs is considered. However, the relative energies of the lowest

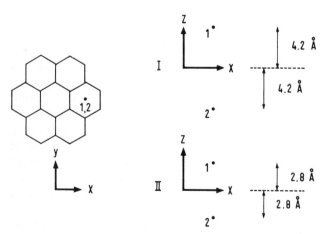

Figure 15.   Two conformations of the triple ion, $Cor^{2-} \cdot 2M^+$.

singlet and triplet states may change drastically if in the calculations the excited singlet and triplet configurations are also taken into account [113, 124, 125]. Nevertheless, the triplet state remains the lowest for most systems studies so far and, indeed, as shown by ESR, all the prepared dianions listed above have triplet ground states.

6.3.2. Perturbation by Counterion

Our discussion pertains thus far to the unperturbed dianions; however, the counterions present in the investigated solutions appreciably perturb the electronic properties of the dianions. Such a perturbation is small in solvents of high solvating power, for example, in glymes [CH_3—$(O—CH_2—CH_2)_n$—OCH_3], which form tight solvation shells around the cations [127]. Indeed, glyme solutions of the dianions of Tp, Tpb, Dec, Cor, and Tst have a triplet ground state irrespective of the cation's nature [54, 114] in agreement with these predictions [113, 125].

In poorer solvents like 2-methyltetrahydrofuran (MTHF) the cations may strongly perturb the electronic distribution in the dianions and consequently the relative energies of the lowest singlet and triplet states may change drastically. This phenomenon was demonstrated unambiguously by Glasbeek et al. [128] for Cor^{2-}, which has a singlet ground state and a thermally accessible triplet state when dissolved in MTHF, indicating a reversal, arising from the cation perturbation, of the lowest states of the term scheme. Glasbeek et al. [114, 129] thus calculated the term scheme of Cor^{2-} taking into account the interaction with one or two alkali cations treated as point charges. The matrix elements of the perturbation operator $-e^2/r$ were evaluated as indicated in Section 3.1.2.

It was shown that the order of energies of the lowest singlet and triplet states was reversed for several noncentric conformations. For example, Fig. 15 depicts the results for two conformations of $Cor^{2-} \cdot 2M^+$. In Fig. 16 the term scheme of the four lowest states of the free dianion is given in a. The influence of the two cations on the energy levels is shown in b and c.

The most important feature of the cation perturbation results from the lowering of the symmetry from D_{6h} to C_{2v}. Consequently, the two $^1E_{2g}$ levels become split into a $^1B_1$ level and a $^1A_1$ level, where A_1 and B_1 are representations of the C_{2v} group. Strong mixing may now occur with the nearby $^1A_1$ level resulting in an appreciable lowering of the lowest singlet state with respect to the lowest triplet state (see the schemes b and c in Fig. 16). Thus for a sufficiently strong cation perturbation the ground state changes from a triplet to a singlet.

The singlet-triplet separation (ΔE_{ST}) calculated for the scheme c agrees fairly well with the experimental value of about 0.05 eV. This value has been

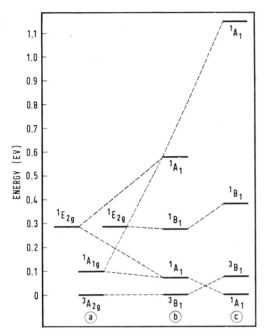

Figure 16. Term scheme of Cor^{2-} (*a*) corresponds to the unperturbed ion [113]. (*b*) and (*c*) correspond to the conformations I and II, respectively of Fig. 15 [114, 129].

determined from the variation of the triplet ESR intensity (I) with temperature, using the conventional equation [114]:

$$\ln(IT) = -\frac{\Delta E_{ST}}{kT} + \text{constant} \qquad (117)$$

Counterion effects on the dianions of Tp, Tpb, Dec, and Tst have been studied experimentally by van Broekhoven et al. [54, 130–132] and theoretically by Sommerdijk et al. [125, 132–134]. Modified Hückel MOs, constructed as outlined in Section 3.1.1, were used in the calculation of terms according to the combined SCF-CI procedure (see Section 3.1.3).

The occurrence of a thermally excited triplet state for Tst^{2-} in MTHF has been tentatively attributed to a conformation in which the cations are located above the nitrogen atoms. The placement of the cation over the center of Tst^{2-} is not acceptable, since this conformation predicts a triplet ground state in contradiction with experiment. The placement of the cations above the outer rings is also excluded because these configurations lead to a thermally nonaccessible triplet state contradicting the experimental results.

The dianions of Tp^{2-}, Tpb^{2-}, and Dec^{2-} have triplet state ground state even in MTHF solution [54, 130, 131]. Indeed, the calculations demonstrate that for all possible conformations of $Tp^{2-} \cdot 2M^+$ the lowest triplet state has still an appreciably lower energy than the lowest singlet state [134]. On the other hand, for Tpb^{2-} and Dec^{2-} a triplet ground state is predicted only for the conformations having the cations located over the center of the dianion, whereas a singlet ground state and a thermally nonaccessible triplet state is predicted for several noncentric conformations [134]. Hence the experimentally observed triplet ground state requires the cations to be located over the centers of Tpb^{2-} and Dec^{2-}. More arguments favoring such a conformation will be given in the next section.

6.3.3. Counterion Perturbations of ZFS Parameters

An interesting counterion effect is revealed by the ESR spectra shown in Fig. 17. Strikingly, the triplet spectrum of Tp^{2-} in MTHF (Fig. 17a) does not reflect the trigonal symmetry of the parent molecule ($E \neq 0$). On adding a small amount of diglyme to the system, a new spectrum (II) appears, which still shows a nonzero E value (Fig. 17b), but eventually a spectrum (III) showing trigonal symmetry is obtained on further addition of glyme (see Figs. 17c and 17d).

In contrast to spectra I and II, spectrum III was independent of the kind of counterion. On this basis spectrum III was attributed to free Tp^{2-}, whereas the spectra I and II were ascribed to Tp^{2-} perturbed by two cations and one cation, respectively. These ideas have been tested theoretically by carrying out ZFS calculations on the various species. For the free dianion the triplet SCF MOs were derived according to the Pariser-Parr-Pople (PPP) approximation [52, 53]. These were used for calculation of the ZFS parameters by the semiempirical method proposed by van der Waals and ter Maten [135]. In this method, only the doubly excited configurations with respect to the ground configuration are considered and the σ-π interaction is accounted for in a semiempirical way.

The perturbing cations were represented by one or two point charges. The modified Hückel MOs were calculated, as indicated in Section 3.1.1, and used for construction of the triplet SCF MOs. The ZFS calculations proceed then in a similar way like for the unperturbed Tp^{2-} ion. Since the cation positions are unknown, several possible conformations have to be considered. On the basis of the total energy calculations and from the comparison of the calculated values with the experimental ZFS parameters it could be concluded that in $Tp^{2-} \cdot M^+$ the cation is located above an outer ring, whereas in $Tp^{2-} \cdot 2M^+$ the two cations are located above and below the same noncentral ring. The cation distance to the aromatic plane is estimated between 2 and 3 Å.

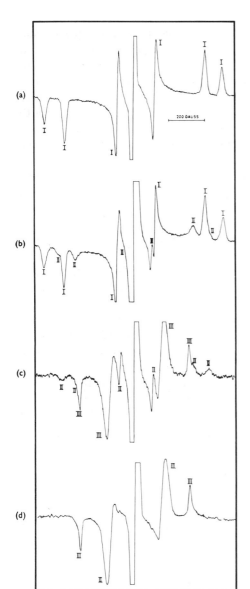

Figure 17. First derivative ESR spectrum of Tp^{2-} with K^+ as counterion at $-160°C$ [130]. (*a*) In pure MTHF, (*b*) After introduction of a trace of diglyme. (*c*) After addition of more diglyme (MTHF:diglyme $\approx$ 20:1). (*d*) Final spectrum (MTHF:diglyme $\approx$ 5:1). The central peak in the spectra is due to Tp^-. Reproduced by permission from [130].

371

The most important results of the ZFS calculations are presented in Table 7 and show a fairly good agreement with the experimental observations. The results support therefore the proposed models.

Three types of spectra have been also observed for Tpb^{2-} [131]. In contrast to Tp^{2-}, all the spectra of Tpb^{2-} reveal a zero E-value corresponding to a trigonal spin distribution in Tpb^{2-}. This is not so surprising for Tpb^{-2} dissolved in glymes, since in these solvents Tpb^{2-} is likely to exist as free ion. However, in pure MTHF or in MTHF-glyme mixtures Tpb^{2-} must be considerably perturbed by the cations. The increasing D-value found in lower-polarity as solvents compared with Tpb^{2-} in glymes, and the dependence of the D-value on the nature of cation provide evidence for the formation of ion pairs.

The zero E-values imply that the perturbation by the cations does not lead to a loss of trigonal symmetry in the spin distribution of Tpb^{2-}. van Broekhoven et al. [131] suggested that the cation either is located on the trigonal axis of Tpb^{2-} or jumps rapidly around Tpb^{2-}, preserving in this way the average trigonal symmetry. However, in the previous section we have seen that the noncentric conformations of $Tpb^{2-} \cdot M^+$ or $Tpb^{2-} \cdot 2M^+$ require a singlet ground state, and hence it was concluded that the cations have to be located on the trigonal axis of Tpb^{2-}, confirming the total energy calculations. A further support for the proposed structures was obtained from the ZFS calculations (see Table 8).

In conclusion, the present theories provide satisfactory explanations for the observed counterion effects on the triplet ESR spectra of Tp^{2-}, Tpb^{2-}, and Dec^{2-}. Although no detailed calculations have been performed on Dec^{2-}, we suggest, by analogy with the Tpb^{2-}, that the cations are located over the center of Dec^{2-}.

The preference of the cations for the noncentral positions in Tp^{2-} and

Table 7   ZFS Parameters of Tp^{2-} (in cm^{-1}) [125, 133]

Species	Theoretical			Experimental	
	D	E		D	E
$Tp^{2-} \cdot 2M^+$	0.042	0.003	(3 Å)[a]	0.046	0.009
	0.047	0.005	(2 Å)[a]		
$Tp^{2-} \cdot M^+$	0.036	0.002	(3 Å)[b]	0.037	0.007
	0.039	0.004	(2 Å)[b]		
Tp^{2-}	0.033	0		0.029	0

[a] Cations above and below the center of the same outer ring.
[b] Cation above the center of an outer ring.

Table 8 ZFS Parameters of Tpb^{2-} (in cm^{-1}) [125, 134]

Species	Theoretical		Experimental	
	D	E	D	E
Tpb^{2-} · 2M^+	0.040	0 (3 Å)^a	0.046	0 (K$^+$)
	0.046	0 (2 Å)^a	0.049	0 (Na$^+$)
Tpb^{2-} · M^+	0.034	0 (3 Å)^a	0.038	0 (K$^+$)
	0.037	0 (2 Å)^a	0.041	0 (Na$^+$)
Tpb^{2-}	0.029	0	0.032	0

^{a} Cation(s) located over the center of the dianion.

for the central positions in Tpb^{2-} and Dec^{2-} can be justified by calculating the charge densities in the free dianions [125, 134]. In Tp^{2-} the outer ring atoms have a much larger charge density than the central ring atoms. On the other hand, the central ring atoms have the largest charge density in Tpb^{2-} and Dec^{2-}. Since cations prefer positions close to the highest electron density, they seek the noncentral positions in Tp^{2-} and the central positions in Tpb^{2-} and Dec^{2-}.

In contrast to the latter dianions the triplet spectra of Cor^{2-} and Tst^{2-} in glyme still depend on the nature of the counterion, suggesting that these dianions are associated with the cations even in glyme solutions. This conclusion was confirmed by ZFS calculations on Cor^{2-} [114] and on Tst^{2-} [132]. In both cases the D-value calculated for the free dianion is 0.020–0.030 cm^{-1}, much too low when compared with the experimental value in glymes. However, the calculations on the perturbed systems did not give good results either, although the experimental increase of the D-value in less solvating solvents is theoretically justified. It seems that the discrepancy is closely connected with the cation effects on term schemes because a reversal in the order of the lowest triplet and singlet state as a result of the cation perturbation has been observed only for these two systems (see Section 6.3.2). Although a satisfactory explanation is still lacking, we suggest that the observed anomalies in the ZFS parameters of Cor^{2-} and Tst^{2-} may be attributed to enhanced cation-anion interactions. Further experimental and theoretical studies will be necessary to settle this point.

ACKNOWLEDGMENT

The authors are thankful to Dr. A. H. Reddoch and to Dr. G. W. Canters for communicating their results to us before they were published.

REFERENCES

1. J. A. Pople, W. G. Schneider, and H. J. Bernstein, *High Resolution Nuclear Magnetic Resonance*, McGraw-Hill, New York, 1959.
2. A. Abragam, *The Principles of Nuclear Magnetism*, Clarendon Press, Oxford, 1961.
3. G. E. Pake, *Paramagnetic Resonance*, Benjamin, New York, 1962.
4. C. P. Slichter, *Principles of Magnetic Resonance*, Harper and Row, New York, 1963.
5. A. Carrington and A. D. McLachlan, *Introduction to Magnetic Resonance*, Harper and Row, New York, 1967.
6. F. Bloch, *Phys. Rev.*, **70**, 460 (1946).
7. M. S. Gutowsky, D. W. McCall, and C. P. Slichter, *J. Chem. Phys.*, **21**, 279 (1953).
8. H. M. McConnell, *J. Chem. Phys.*, **28**, 430 (1958).
9. R. M. Lynden-Bell, *Progress in Nuclear Magnetic Resonance Spectroscopy*, Vol. 2, J. W. Emsley, J. Feeney, and L. H. Sutcliffe, Eds., Pergamon Press, Oxford, 1967, Chapter 4.
10. J. I. Kaplan, *J. Chem. Phys.*, **28**, 278 (1958); **29**, 462 (1958).
11. S. Alexander, *J. Chem. Phys.*, **37**, 967 (1962); **37**, 974 (1962).
12. E. de Boer and C. MacLean, *J. Chem. Phys.*, **44**, 1334 (1966).
13. Y. Ayant, *J. Phys. Radium*, **16**, 411 (1955).
14. F. Bloch, *Phys. Rev.*, **102**, 104 (1956); **105**, 1206 (1957).
15. A. G. Redfield, *I.B.M. J. Res. Develop.*, **1**, 19 (1957).
16. A. G. Redfield, *Advances in Magnetic Resonance*, Vol. 1, J. S. Waugh, Ed., Academic Press, New York, 1965, Chapter 1.
17. J. H. Freed and G. K. Fraenkel, *J. Chem. Phys.*, **39**, 326 (1963).
18. J. H. Freed, *J. Chem. Phys.*, **41**, 2077 (1964).
19. G. K. Fraenkel, *J. Phys. Chem.*, **71**, 139 (1967).
20. H. M. McConnell, *J. Chem. Phys.*, **24**, 632 (1956).
21. E. de Boer, *J. Chem. Phys.*, **25**, 190 (1956).
22. T. R. Tuttle, R. L. Ward, and S. I. Weissman, *J. Chem. Phys.*, **25**, 189 (1956).
23. S. I. Weissman, T. R. Tuttle, and E. de Boer, *J. Phys. Chem.*, **61**, 28 (1957)
24. E. de Boer and S. I. Weissman, *J. Am. Chem. Soc.*, **80**, 4549 (1958).
25. A. C. Aten, J. Dieleman, and G. J. Hoytink, *Disc. Faraday Soc.*, **29**, 182 (1960); R. V. Slates and M. Szwarc, *J. Phys. Chem.*, **69**, 4124 (1965).
26. H. V. Carter, B. J. McClelland, and E. Warhurst, *Trans. Faraday Soc.*, **56**, 455 (1960).
27. N. M. Atherton and S. I. Weissman, *J. Am. Chem. Soc.*, **83**, 1330 (1961).
28. E. W. Stone and A. H. Maki, *J. Chem. Phys.*, **36**, 1944 (1962).
29. N. Hirota, *J. Chem. Phys.*, **37**, 1884 (1962).
30. P. B. Ayscough and R. Wilson, *J. Chem. Soc.*, **1963**, 5412.
31. J. Q. Chambers, T. Layloff, and R. N. Adams, *J. Phys. Chem.*, **68**, 661 (1964).
32. T. Kitagawa, T. Layloff, and R. N. Adams, *Anal. Chem.*, **36**, 925 (1964).
33. E. de Boer and E. L. Mackor, *Proc. Chem. Soc.*, **1963**, 23.
34. J. R. Bolton and G. K. Fraenkel, *J. Chem. Phys.*, **40**, 3307 (1964).

35. A. H. Reddoch, *J. Chem. Phys.*, **41**, 444 (1964).

36. N. Hirota, *Metal Ketyls and Related Radical Ions—Electronic Structures and Ion Pair Equilibria, Radical Ions*, E. T. Kaiser and L. Kevan, Eds., Interscience, New York, 1968, Chapter 2.

37. J. H. Sharp and M. C. R. Symons, Chapter 5 of this book.

38. A. Streitwieser, *Molecular Orbital Theory for Organic Chemists*, Wiley, New York, 1961.

39. L. Salem, *The Molecular Orbital Theory of Conjugated Systems*, Benjamin, New York, 1966.

40. P. B. Ayscough, *Electron Spin Resonance in Chemistry*, Methuen, London, 1967.

41. H. M. McConnell, *J. Chem. Phys.*, **24**, 764 (1956).

42. S. I. Weissman, *J. Chem. Phys.*, **25**, 890 (1956).

43. R. Daudel, R. Lefebvre, and C. Moser, *Quantum Chemistry: Methods and Applications*, Interscience, New York, 1959.

44. R. G. Parr, *The Quantum Theory of Molecular Electronic Structure*, Benjamin, New York, 1963.

45. B. J. McClelland, *Trans. Faraday Soc.*, **57**, 1458 (1961).

46. A. H. Reddoch, *Actes Coll. Intern.* 164, 1966, Centre National de la Recherche Scientifique, Paris, 1967, p. 419.

47. M. Iwaizumi, M. Suzuki, I. Isobe, and H. Azumi, *Bull. Chem. Soc. Japan*, **40**, 1325 (1967); **40**, 2754 (1967).

48. E. de Boer and E. L. Mackor, *J. Am. Chem. Soc.*, **86**, 1513 (1964).

49. A. H. Reddoch, *Chem. Phys. Lett.*, **10**, 108 (1971).

50. P. H. Rieger and G. K. Fraenkel, *J. Chem. Phys.*, **37**, 2811 (1962).

51. A. H. Reddoch, *J. Chem. Phys.*, **43**, 225 (1965).

52. R. Pariser and R. G. Parr, *J. Chem. Phys.*, **21**, 466 (1953); **21**, 767 (1953).

53. J. A. Pople, *Trans. Faraday Soc.*, **49**, 1375 (1953).

54. J. A. M. van Broekhoven, Thesis, Nijmegen (1970).

55. G. J. Hoytink, *Mol. Phys.*, **1**, 157 (1958).

56. J. P. Colpa and E. de Boer, *Mol. Phys.*, **7**, 333 (1964).

57. E. de Boer and J. P. Colpa, *J. Phys. Chem.*, **71**, 21 (1967).

58. G. K. Fraenkel, *Pure Appl. Chem.*, **4**, 143 (1962).

59. D. T. Wilkinson and H. R. Crane, *Phys. Rev.*, **130**, 852 (1963).

60. A. J. Stone, *Mol. Phys.*, **6**, 509 (1963).

61. M. S. Blois, H. W. Brown, and J. E. Maling, *Arch. Sci.*, **13**, (9e Colloque Ampère) 243 (1960).

62. B. G. Segal, M. Kaplan, and G. K. Fraenkel, *J. Chem. Phys.*, **43**, 4191 (1965).

63. J. L. Sommerdijk, E. de Boer, F. W. Pijpers, and H. van Willigen, *Z. Phys. Chem. (Frankfurt)*, **63**, 183 (1969).

64. W. D. Hobey and A. D. McLachlan, *J. Chem. Phys.*, **33**, 1695 (1960).

65. H. M. McConnell and A. D. McLachlan, *J. Chem. Phys.*, **34**, 1 (1961).

66. H. M. McConnell, *J. Chem. Phys.*, **34**, 13 (1961).

67. E. de Boer, *Rec. Trav. Chem.*, **84**, 609 (1965).

68. C. L. Dodson and A. H. Reddoch, *J. Chem. Phys.*, **48**, 3226 (1968).

69. W. G. Williams, R. J. Pritchett, and G. K. Fraenkel, *J. Chem. Phys.*, **52**, 5584 (1970).

70. M. Szwarc, *Makromol. Chem.*, **89**, 44 (1965).

71. P. Chang, R. V. Slates, and M. Szwarc, *J. Phys. Chem.*, **70**. 3180 (1966); M. Szwarc; Account of Chem. Res. **2**, 87 (1969).

72. E. Grunwald, *Anal. Chem.*, **26**, 1696 (1954).

73. S. Winstein and G. C. Robinson, *J. Am. Chem. Soc.*, **80**, 169 (1958).

74. N. Hirota and R. Kreilick, *J. Am. Chem. Soc.*, **88**, 614 (1966).

75. T. E. Hogen-Esch and J. Smid, *J. Am. Chem. Soc.*, **87**, 669 (1965); **88**, 307 (1966); **88**, 318 (1966); **89**, 2764 (1967).

76. A. H. Reddoch, C. L. Dodson, and D. H. Paskovich, *J. Chem. Phys.*, **52**, 2318 (1970).

77. N. Hirota, *J. Phys. Chem.*, **71**, 127 (1967).

78. C. Carvajal, K. J. Tolle, J. Smid, and M. Szwarc, *J. Am. Chem. Soc.*, **87**, 5548 (1965).

79. G. W. Canters, E. de Boer, B. M. P. Hendriks, and A. A. K. Klaassen, *Proc. Coll. Ampère XV, 1968*, P. Averbuch, Ed., North-Holland, Amsterdam, 1969, p. 242.

80. M. C. R. Symons, *J. Phys. Chem.*, **71**, 172 (1967).

81. G. W. Canters, E. de Boer, B. M. P. Hendriks, and H. van Willigen, *Chem. Phys. Lett.*, **1**, 627 (1968).

82. S. Aono and K. Oohashi, *Prog. Theor. Phys.*, **30**, 162 (1963).

83. I. R. Goldberg and J. R. Bolton, *J. Phys. Chem.*, **74**, 1965 (1970).

84. M. Wolfsberg and L. Helmholz, *J. Chem. Phys.*, **20**, 837 (1952).

85. N. M. Atherton and A. E. Goggins, *Trans. Faraday Soc.*, **62**, 1701 (1966).

86. L. Pedersen and R. G. Griffin, *Chem. Phys. Lett.*, **5**, 373 (1970).

87. J. A. Pople, D. L. Beveridge, and P. A. Dobosh, *J. Chem. Phys.*, **47**, 2026 (1967).

88. K. H. J. Buschow, Thesis, Amsterdam (1963).

89. K. H. J. Buschow, J. Dieleman, and G. J. Hoytink, *J. Chem. Phys.*, **42**, 1993 (1965).

90. H. M. McConnell and D. B. Chesnut, *J. Chem. Phys.*, **28**, 107 (1958).

91. A. D. McLachlan, *Mol. Phys.*, **3**, 233 (1960).

92. G. W. Canters, Thesis, Nijmegen (1969).

93. D. A. Goodings, *Phys. Rev.*, **123**, 1706 (1961).

94. P. Kusch and H. Taub, *Phys. Rev.*, **75**, 1477 (1949).

95. B. M. P. Hendriks, G. W. Canters, C. Corvaja, J. W. M. de Boer, and E. de Boer, *Mol. Phys.*, **20**, 193 (1971).

96. G. W. Canters, C. Corvaja, and E. de Boer, *J. Chem. Phys.*, **54**, 3026 (1971).

97. C. Corvaja, private communication.

98. N. M. Atherton, *Trans. Faraday Soc.*, **62**, 1707 (1966).

99. T. Takeshita and N. Hirota, *Chem. Phys. Lett.*, **4**, 369 (1969).

100. A. Zahlen, F. W. Heineken, M. Bruin, and F. Bruin, *J. Chem. Phys.*, **37**, 683 (1962).

101. K. Nakamura and N. Hirota, *Chem. Phys. Lett.*, **3**, 137 (1969).

102. S. Aono and K. Oohashi, *Prog. Theor. Phys.*, **32**, 1 (1964).

103. A. F. Neiva Correia, Thesis, Amsterdam (1967).

104. K. Nakamura and Y. Deguchi, *Bull. Chem. Soc. Japan*, **40**, 705 (1967); K. Nakamura, *Bull. Chem. Soc. Japan*, **40**, 1019 (1967).

105. S. Glasstone and D. Lewis, *Elements of Physical Chemistry*, Macmillan, London, 1963.

106. A. W. J. Raaymakers and F. W. Pijpers, forthcoming, to be published.

107. R. F. Adams and N. M. Atherton, *Trans. Faraday Soc.*, **64**, 7 (1968).

108. A. W. Rutter and E. Warhurst, *Trans. Faraday Soc.*, **64**, 2338 (1968).

109. C. A. Hutchison and B. W. Mangum, *J. Chem. Phys.*, **29**, 952 (1958); **34**, 908 (1961).

110. J. H. van der Waals and M. S. de Groot, *Mol. Phys.*, **2**, 333 (1959); M. S. de Groot and J. H. van der Waals, *Mol. Phys.*, **3**, 190 (1960).

111. E. Wasserman, L. C. Snyder, and W. A. Yager, *J. Chem. Phys.*, **41**, 1763 (1964).

112. P. Kottis and R. Lefebvre, *J. Chem. Phys.*, **39**, 393 (1963); **41**, 379 (1964).

113. R. E. Jesse, Thesis, Amsterdam (1966).

114. M. Glasbeek, Thesis, Amsterdam (1969).

115. N. Hirota and S. I. Weissman, *J. Am. Chem. Soc.*, **83**, 3533 (1961); **86**, 2537 (1964).

116. N. Hirota, Thesis, Washington (1963); *J. Am. Chem Soc.*, **89**, 32 (1967).

117. P. Biloen, Thesis, Amsterdam (1968).

118. P. Biloen, R. Prins, J. D. W. van Voorst, and G. J. Hoytink, *J. Chem. Phys.*, **46**, 4149 (1967).

119. I. M. Brown, S. I. Weissman, and L. C. Snyder, *J. Chem. Phys.*, **42**, 1105 (1965).

120. J. D. W. van Voorst, W. G. Zijlstra, and R. Sitters, *Chem. Phys. Lett.*, **1**, 321 (1967).

121. H. van Willigen and S. I. Weissman, *Mol. Phys.*, **11**, 175 (1966).

122. F. W. Pijpers, H. van Willigen, and J. J. Th. Gerding, *Rec. Trav. Chim.*, **86**, 511 (1967).

123. F. A. Cotton, *Chemical Applications of Group Theory*, Interscience, New York, 1963.

124. J. N. Murrell and A. Hinchliffe, *Mol. Phys.*, **11**, 101 (1966).

125. J. L. Sommerdijk, Thesis, Nijmegen (1970).

126. R. E. Jesse, P. Biloen, R. Prins, J. D. W. van Voorst, and G. J. Hoijtink, *Mol. Phys.*, **6**, 633 (1963).

127. F. Cafasso and B. R. Sundheim, *J. Chem. Phys.*, **31**, 809 (1959).

128. M. Glasbeek, J. D. W. van Voorst, and G. J. Hoytink, *J. Chem. Phys.*, **45**, 1852 (1966).

129. M. Glasbeek, A. J. W. Visser, G. A. Maas, J. D. W. van Voorst, and G. J. Hoytink, *Chem. Phys. Lett.*, **2**, 312 (1968).

130. H. van Willigen, J. A. M. van Broekhoven, and E. de Boer, *Mol. Phys.*, **12**, 533 (1967).

131. J. A. M. van Broekhoven, H. van Willigen, and E. de Boer, *Mol. Phys.*, **15**, 101 (1968).

132. J. A. M. van Broekhoven, J. L. Sommerdijk, and E. de Boer, *Mol. Phys.*, **20**, 993 (1971).

133. J. L. Sommerdijk and E. de Boer, *J. Chem. Phys.*, **50**, 4771 (1969).

134. J. L. Sommerdijk, J. A. M. van Broekhoven, H. van Willigen, and E. de Boer, *J. Chem. Phys.*, **51**, 2006 (1969).

135. J. H. van der Waals and G. ter Maten, *Mol. Phys.*, **8**, 301 (1964).

Author Index

Abragam, A., 290, 292, 294 [1]; 323 [2]
Accascina, F., 5 [9]; 10 [20]; 169 [10]
Adams, D. G., 269 [24]; 272, 273 [29]
Adams, R. F., 18 [39]; 102 [35]; 196, 227 [136]; 233 [93, 95, 96]; 234 [95]; 243 [151]; 361 [107]
Adams, R. N., 217, 218 [66]; 243[108]; 329 [31, 32]
Adrian, F. J., 3[3]; 189[26]
Al-Baldawi, S. A., 204 [145, 146]
Al-Joboury, M. I., 38 [17]
Alei, M., 313 [18]
Alexander, S., 325, 327 [11]
Alger, T. D., 277 [44]
Amis, E. S., 314 [22]
Aono, S., 186 [20]; 343, 344 [82, 102]
Arnett, E. M., 103 [37]
Anshadi, M., 57 [58]; 61, 62, 63 [62]; 61, 62, 63 [63]; 76 [75]
Aten, A. C., 329 [25]
Atherton, N. M., 157 [3]; 180, 186, 195, 232, 233 [7]; 195, 196, 222, 241 [37]; 195, 227 [88]; 222, 241 [73]; 233 [93]; 233, 234 [95], 233 [96]; 305 [34]; 329, 339, 340, 342, 344, 351, 352 [27]; 344 [85]; 350 [98]; 361 [107]
Atkins, P. W., 182, 215, 224 [13]
Ayant, Y., 327 [13]
Ayscough, P. B., 182 [14]; 195, 227 [87]; 194, 213 [60]; 329 [30, 40]
Azumi, H., 180, 213, 231 [10]; 195, 212; 222, 238 [76]; 330, 334 [47]

Bafus, D. A., 267 [14]; 269 [14, 20]

Baker, E. B., 267 [15, 16]; 268 [16]; 270 [16, 25]; 271, 275, 276, 277 [25]
Barbetta, A., 153, 172 [1]
Bar-Eli, K., 257 [132]
Barker, R. S., 45 [26]
Barnes, J. D., 195 [29, 30]; 196 [29, 30, 31]; 212 [30]; 217 [29, 30]
Barrow, R. F., 77 [77]
Bates, R. B., 284 [62]
Bauld, N. L., 197 [43]
Bayliss, N. S., 87 [5]
Beauchamp, J. L., 44; 47 [29]
Beck, D., 41 [22]
Beeker, R. S., 50 [42a]
Beckey, H. D., 52 [51]
Belousova, M. I., 105, 137 [40]
Benderson, B., 32 [8]
Bennett, J. E., 198, 199 [49]; 253 [122]
Berge, A., 126 [74]
Bergman, G. A., 76, 77 [76]
Bergmann, E., 137 [90]
Berkowitz, J., 77 [85, 86]
Bernal, J. P., 4[6]; 169[14]
Bernheim, R. A., 292 [20]
Bernstein, H. J., 291 [14]; 323 [1]
Berry, R. S., 32, 36, 38 [7]; 48, 49 [36]
Beveridge, D. L., 344, 346 [87]
Bhattacharyya, D. N., 5 [10]; 23 [53]; 92, 133 [18]
Biloen, P., 143 [98]; 246 [113]; 365 [117, 118]
Birrell, R., 275 [37]
Bjerrum, N., 6, 7 [14]; 169 [9]
Blackledge, J., 253 [119]
Blair, L. K., 75 [74]
Blandamer, M. J., 90, 91 [14, 15]; 92 [14, 15, 17]; 93 [14, 15, 17, 21, 22]; 189 [26]; 257 [131]
Blander, M., 78 [88, 90]

Subject Index